Natural Numbers = $\{1, 2, 3, 4, 5, 6, 7, 8, 9, 10, 11, 12, 13,...\}$
The natural numbers are also known as the counting numbers or the positive integers.

Whole numbers = $\{0, 1, 2, 3, 4, 5, 6, 7, 8, 9, 10, 11, 12, 13,...\}$

Integers = $\{...,-7, -6, -5. -4, -3, -2, -1, 0, 1, 2, 3, 4, 5, 6, 7,...\}$

Rational number : A number which can be written as the ratio of two integers.

Signed number operations: **1.** -7 - 9 = -16 (add); **2.** -7 - (-9) = -7 + 9 = 2 (subtract); **3.** -7(-9) = +63 (multiply).

Slope, m, of the line passing through the points (x_1, y_1) and (x_2, y_2) is given by $m = \dfrac{y_2 - y_1}{x_2 - x_1}$

Slope-intercept form of the **equation** of a line with slope m and y-intercept b, is $y = mx + b$.

An **equation** of the line with y-intercept b, and x-intercept a, is given by $y = -\dfrac{b}{a}x + b$.

Factoring: $a^2 - b^2 = (a + b)(a - b)$;   **2.** $3a^2 - 12 = 3(a^2 - 4) = 3(a + 2)(a - 2)$.

Quadratic equations: **1.** If $(x - 2)(x + 3) = 0$, then $x = 2$ or $x = -3$.

Radicals: **1.** $\sqrt{12} = \sqrt{4}\sqrt{3} = 2\sqrt{3}$;   **2.** $\sqrt{18} = \sqrt{9}\sqrt{2} = 3\sqrt{2}$.

Scientific notation: **1.** $436.7 = 4.367 \times 10^2$;   **2.** $.00721 = 7.21 \times 10^{-3}$

Ratio: The ratio $a{:}b$ is the fraction $\dfrac{a}{b}$

Proportion: If a is to b as c is to d, then $\dfrac{a}{b} = \dfrac{c}{d}$

Exponents: **1.** $x^2 \cdot x^3 = x^5$;   **2.** $(x^2)^3 = x^6$;   **3.** $\dfrac{x^6}{x^2} = x^4$;   **4.** $(xy)^3 = x^3 y^3$.

Factoring: 3. $a^3 + b^3 = (a + b)(a^2 - ab + b^2)$.   **4.** $a^3 - b^3 = (a - b)(a^2 + ab + b^2)$.

Quadratic equations: **1.** If $(x - 2)(x + 3) = 0$, then $x = 2$ or $x = -3$; **2.** If $x^2 = 4$, then $x = \pm 2$.

Quadratic formula: $x = \dfrac{-b \pm \sqrt{b^2 - 4ac}}{2a}$

Complex numbers: **1.** $i = \sqrt{-1}$; $i^2 = -1$;   **2.** If $x^2 = -4$, then $x = \pm 2i$

Logarithms: $\log_2 16 = 4$, because $2^4 = 16$.

$\log_b x = y$ if and only if $b^y = x$

Absolute value: $|x| = \begin{cases} x \text{ if } x \geq 0 \\ -x \text{ if } x < 0 \end{cases}$

Absolute value equations: If $|x - 2| = 6$, then $x - 2 = 6$ or $-(x - 2) = 6$; and from which $x = 8$ or $x = -4$.

Absolute value inequalities

Case **1**: $|ax + b| < c$. where c is positive. Example: $|x - 2| < 1$. (There is no solution if c is negative)

Case **2**: $|ax + b| > c$, where c may be positive or negative. Example: $|x - 2| > 1$.

Equation of the **circle** with center at (h,k) and radius r is given by $(x - h)^2 + (y - k)^2 = r^2$

Area & Perimeter of a Circle: Area, A , is given by $A = \pi r^2$; Perimeter, P, is given by $P = 2\pi r$

Heron's Formula : For any triangle of sides a , b , c , the area A is given by

$$A = \sqrt{s(s - a)(s - b)(s - c)} \text{ where } s = \tfrac{1}{2}(a + b + c)$$

Distance Formula: $d = \sqrt{(x_2 - x_1)^2 + (y_2 - y_1)^2}$

The **mid-point** of the line, $P_1 P_2$, connecting the points $P_1(x_1, y_1)$ and $P_2(x_2, y_2)$ has the coordinates given by the following formulas:

x-coordinate, x_m, of the midpoint is given by $x_m = \dfrac{x_1 + x_2}{2}$

y-coordinate, y_m of the midpoint is given by $y_m = \dfrac{y_1 + y_2}{2}$

Trigonometric Identities

Reciprocal Identities

1. $\sec\theta = \dfrac{1}{\cos\theta}$

2. $\csc\theta = \dfrac{1}{\sin\theta}$

3. $\cot\theta = \dfrac{1}{\tan\theta}$

Ratio Identities

1. $\tan\theta = \dfrac{\sin\theta}{\cos\theta}$

2. $\cot\theta = \dfrac{\cos\theta}{\sin\theta}$

Pythagorean Identities

1. $\sin^2\theta + \cos^2\theta = 1$

2. $1 + \tan^2\theta = \sec^2\theta$

3. $1 + \cot^2\theta = \csc^2\theta$

Addition formulas

1. $\cos(A - B) = \cos A \cos B + \sin A \sin B$

2. $\cos(A + B) = \cos A \cos B - \sin A \sin B$

3. $\sin(A + B) = \sin A \cos B + \cos A \sin B$

4. $\sin(A - B) = \sin A \cos B - \cos A \sin B$

5. $\tan(A - B) = \dfrac{\tan A - \tan B}{1 + \tan A \tan B}$

6. $\tan(A + B) = \dfrac{\tan A + \tan B}{1 - \tan A \tan B}$

Double angle identities

1. $\sin 2\theta = 2\sin\theta \cos\theta$;

2. $\cos 2\theta = \cos^2\theta - \sin^2\theta$;

3. $\cos 2\theta = 2\cos^2\theta - 1$

4. $\cos 2\theta = 1 - 2\sin^2\theta$;

5. $\tan 2\theta = \dfrac{2\tan\theta}{1 - \tan^2\theta}$

If you can remember a needed information, you make decisions faster, you learn faster, you work faster, and you are more productive.

Yes, you can memorize them.

Squares of Natural Numbers

$1 \times 1 = 1$	$26 \times 26 = 676$
$2 \times 2 = 4$	$27 \times 27 = 729$
$3 \times 3 = 9$	$28 \times 28 = 784$
$4 \times 4 = 16$	$29 \times 29 = 841$
$5 \times 5 = 25$	$30 \times 30 = 900$
$6 \times 6 = 36$	$31 \times 31 = 961$
$7 \times 7 = 49$	$32 \times 32 = 1024$
$8 \times 8 = 64$	$33 \times 33 = 1089$
$9 \times 9 = 81$	$34 \times 34 = 1156$
$10 \times 10 = 100$	$35 \times 35 = 1225$
$11 \times 11 = 121$	$36 \times 36 = 1296$
$12 \times 12 = 144$	$37 \times 37 = 1369$
$13 \times 13 = 169$	$38 \times 38 = 1444$
$14 \times 14 = 196$	$39 \times 39 = 1521$
$15 \times 15 = 225$	$40 \times 40 = 1600$
$16 \times 16 = 256$	$41 \times 41 = 1681$
$17 \times 17 = 289$	$42 \times 42 = 1764$
$18 \times 18 = 324$	$43 \times 43 = 1849$
$19 \times 19 = 361$	$44 \times 44 = 1936$
$20 \times 20 = 400$	$45 \times 45 = 2025$
$21 \times 21 = 441$	$46 \times 46 = 2116$
$22 \times 22 = 484$	$47 \times 47 = 2209$
$23 \times 23 = 529$	$48 \times 48 = 2304$
$24 \times 24 = 576$	$49 \times 49 = 2401$
$25 \times 25 = 625$	$50 \times 50 = 2500$

**FREMPONG'S STEP-BY-STEP SERIES IN
MATHEMATICS**

Intermediate Mathematics (US)
(Algebra, Geometry & Trigonometry)

Excellent for Exam Preparation

SIXTH EDITION

A.A. FREMPONG

Intermediate Mathematics (Algebra, Geometry & Trigonometry)

ISBN 978-1-946485-14-4

Sixth Edition

Printed in the United States of America

In Memory of My Parents

Mom:
She was a devoted mother, sharing, kind, kinder to strangers and generous
to a fault. She never cursed, she never hated; she never cheated, and she never
envied. She never lied, and she never got angry. Once, she nursed an almost
dying stranger renting a room in her house back to good health to the extent
that the relatives of this renter later travelled one hundred miles just to thank
mom. She was always peaceloving and forever forgiving.
An angel once lived on this earth to serve others.

Dad:
A great dad, kind, generous and forgiving. He emphasized and was an example
of both formal education and self-education. A veterinarian, a bacteriologist,
an Associate of the Institute of Medical Laboratory Technology (UK), a Fellow
of the Royal Society of Health (UK); an incorruptible civil servant; his book on
ticks has always inspired me to write whenever the need arises.

NOTE TO THE STUDENT

This book was written with you in mind at all times. You may use this book as the course textbook or as a review book since the book gets to the point quickly on all relevant topics and yet covers these topics in detail.

Begin to master the definitions and the solutions of the sample problems thoroughly. (You have mastered a sample problem if you can solve the sample problem and similar problems without any reference to this book or any other source). For some problems, two or more methods are presented. Read the
various methods and decide which methods you would like to remember; but always be aware of the existence of the other methods, in case the need arises. After having mastered the sample problems, try the exercise problems. The answers to these problems are presented immediately after the problems. You may cover the answers with paper before you attempt these problems, if the answers are too obvious. You may refer back and forth to the solved problems when you do not remember how to proceed.

You may also attempt some of the sample problems first, before reading the solution methods, and in this approach, the sample problems become more practice problems for you.

As a reminder, in any book, do not dwell on the few inadvertent errors you may find, but rather concentrate on what is useful to you.

For this book to be useful both as the course textbook, as well as review for exams, it is **important** to **Understand, Remember, Apply**, and **Remember** the material covered.

Wishing you Good Luck on all the exams
A.A.Frempong

Books in the series by the author: Integrated Arithmetic; Elementary Algebra; Intermediate Algebra, Elementary Mathematics; **Intermediate Mathematics;** Elementary & Intermediate Mathematics (combined); College Algebra; College Trigonometry; College Algebra & Trigonometry and Calculus 1 & 2

PREFACE TO THE SIXTH EDITION

This edition retains the spirit of the previous editions. The topics have been expanded and covered quite extensively, especially those topics students find difficult to understand or sort out. For such topics, I have endeavored to organize the coverage in such a way as to minimize student confusion. One such topic whose coverage is well-organized is "finding an equation of a straight line" (page 64-76).

   In writing this edition, I was encouraged by comments from students who had used the previous editions. One student remarked: "Your book gets to the point. It does not beat about the bush."
In addition to the regular programs, this book can also be used for short term programs such as mini-sessions and immersion programs, and distance learning programs, with the instructor, perhaps, supplementing with hand-out homework problems, when necessary.

I gratefully acknowledge the help and encouragement of students, colleagues, and friends.

A. A. Frempong
New York.

CONTENTS

CHAPTER 15 195
Complex Numbers

CHAPTER 16 204
Logarithms

CHAPTER 17 211

CHAPTER 18 216
Basic Review For Geometry

CHAPTER 19 232
Areas and Perimeters

CHAPTER 20 245
CONGRUENT TRIANGLES

CHAPTER 1
Review of Basic Elementary Algebra 1

Lesson 1: **Signed Numbers and Real Number Operations**

Lesson 2**: Rules of Exponents and Applications**

Lesson 3: **Simplification of Algebraic Expressions**
Like terms; Grouping Symbols; Addition, Subtraction, Multiplication and
Division of Polynomials

Signed Numbers

A signed number is a number with either a plus sign " +" or a minus sign "-" preceding it
(in front of it). If there is no sign in front of a number , we will assume that the number has a plus sign. We call a
number with a plus sign a positive number, and we call a number with a minus sign a negative number. For a positive
number, we sometimes do not write the plus sign, but for a negative number we **must** always write the minus sign.

The Real Number Line and Real numbers

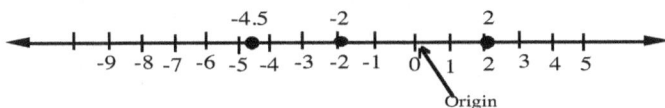

Figure 1

The real number line is a horizontal straight line with equally spaced intervals as in Figure 1 above. We label a point
called the origin, 0 (zero). Points to the right of the origin are labeled positive and points to the left of the origin are
labeled negative. The numbers increase as one moves from the left to the right on the real number line. Roughly
speaking, a real number is a number that can be represented by a point on the real number line. The real numbers
consists of the integers, fractions, mixed numbers, decimals, and radicals. In Figure 1, if the real numbers, -4.5, -2,
and 2 are of interest, we can represent them by the dots shown. Every point on this line is associated with a real
number; and every real number is associated with a point on this line. We can also say that the set of real numbers
consists of the signed numbers and zero.

Absolute Value

The absolute value of a signed number may be defined as the number without its sign.
Examples: .

1. The absolute value of -4 , symbolized $|-4| = 4$

2. The absolute value of 6, symbolized $|6| = 6$.

3. The absolute value of 0 is 0.

Note: The absolute value of a signed number is also its distance from zero on the number line.

Addition of Two Signed Numbers

To add any two signed numbers, first determine which of the following two cases is involved.

Case 1: The two numbers have the same sign
Step 1: Add the absolute values of the numbers (the unsigned parts)
Step 2: Write the common sign in front of the sum from Step 1.

Examples 1. $(+2) + (+3) = +5$ or 5;   **2.** $(-3) + (-4) = -7$;   **3.** $(-.5) + (-.25) = (-.75)$

Case 2: The two numbers have different signs

Step 1 : Subtract the smaller absolute value from the larger absolute value
Step 2. Write the sign of the number with the larger absolute value in front of the difference
(result) from Step 1.   Examples: **1.** $(+5) + (-7) = -2$;   **2.** $(-6) + (+9) = +3$ or 3.

Absolute value defined more formally
The absolute value of a real number x is x if x is a positive number or zero, but it is $-x$ if x is a
negative number . (i.e. the negative of a negative number).

Subtraction of Signed Numbers

Procedure: **Change the sign** of the number being subtracted, **and add** the changed number to the number from which you are subtracting (i.e., add the negative of the subtrahend to the minuend).

1.

(+5) - (+7)   (subtract)

Step 1: We change +7 to -7
Step 2: We add +5 and -7

(+5) + (-7)   (add)
= - 2

2.

(-6) - (-9)   (subtract)

Step 1: We change -9 to +9
Step 2: We add -6 and +9

(-6) + (+9)   (add)
= +3

Multiplication of Two Signed Numbers

Implication of Multiplication   Example: $2 \times 3 = 2(3) = (2)(3) = (2) \cdot (3) = (2)3 = 2 \cdot 3 = 6$

Step 1: Multiply the absolute values.

Step 2: The product is positive (that is, use a plus sign or no sign) if the two numbers have the same sign; but the product is negative (that is, use a minus sign) if the two numbers have different signs.

Examples.   **1.** $(+2)(+3) = +6$ or 6;   **2.** $(-2)(-3) = +6$ or 6;   **3.** $-4(-2) = +8$

   4. $(-4)(+2) = -8$;   **5** $(+5)(-3) = -15$;   **6.** $-6(4) = -24$;   **7.** $(3)(-7) = -21$

Division of Signed Numbers

The rules of signs for division are the same as those for multiplication.

Step 1: Divide the absolute values of the numbers.

Step 2: The quotient is positive (that is, use a plus sign or no sign) if the two numbers have the same sign; but the quotient is negative (that is, use a minus sign) if the two numbers have different signs.

Examples: **1.** $\frac{+8}{+4} = 2$;   **2.** $\frac{-8}{-4} = 2$;   **3.** $\frac{-12}{-9} = \frac{4}{3}$;   **4.** $\frac{-6}{+2} = -3$;   **5.** $\frac{8}{-2} = -4$;   **6.** $\frac{-13}{3} = -\frac{13}{3}$

Operations Involving Zero

Addition

Examples 1. $6 + 0 = 6$;   **2.** $0 + 4 = 4$;   **3.** $-7 + 0 = -7$;   **4.** $0 - 5 = -5$;   **5.** $b + 0 = b$

Multiplication Examples 1. $5(0) = 0$;   **2.** $(7)(9)(0)(14) = 0$;.   **Note:**   **3.** $0 \times 0 = 0$

Division

Examples 1. $\frac{0}{5} = 0$ (because $5 \times 0 = 0$)   **2.** $\frac{0}{-6} = 0$ (because $-6 \times 0 = 0$

   But 3. $\frac{5}{0}$ is undefined.(There is **no** number such that $0 \times$ "that number" = 5. Do not divide by 0)

   4. $\frac{0}{0}$ is indeterminate. (Any number will do since $0 \times$ "any number" = 0)

   Square root: $\sqrt{0} = 0$

Powers of Signed Numbers

Examples 1. $2^5 = (2)(2)(2)(2)(2) = 32$

2^5 is read as 2 to the fifth power. (or 2 raised to the fifth power)
The exponent "5" indicates how many times the base "2" is being used as a factor.

2. $(-2)^3 = (-2)(-2)(-2)$
$= -8$

3. $(-3)^2 = (-3)(-3)$
$= 9$

Zero as an Exponent

Any (nonzero) number raised the to power zero is 1.

Examples 1. $b^0 = 1$;   **2.** $4^0 = 1$;   **3.** $(7x^2dz)^0 = 1$

Note that 0^0 is indeterminate. Example: $0^0 = \dfrac{0^2}{0^2} = \dfrac{0}{0}$ which is indeterminate (see previous page)

Roots of Signed Numbers

Root finding and Power finding are inverse operations (in much the same way as multiplication and division are inverse operations).

Square Root: Symbol " $\sqrt{}$ " or " $\sqrt[3]{}$ "

Finding the square roots of perfect squares by inspection or by factoring

1. We may cheaply define the square root (principal square root) of a nonzero number as one of the
   two equal **positive** factors of that number. We exclude negative roots. The square root of a
   negative number is not real.

2. The square root of 0 is 0.

Examples

1. The square root of 9 is symbolized $\sqrt{9}$, and

$$\sqrt{9} = 3$$
because $3^2 = 9$

Note that , according to the above definition,
$$\sqrt{9} = \sqrt{(3)(3)} = 3$$

2. The square root of 64 is written $\sqrt{64}$, and

$\sqrt{64} = 8$, because $8^2 = 64$.   Also, $\sqrt{64} = \sqrt{(8)(8)} = 8$

3.   **Note:** $\sqrt{0} = 0$, because $0^2 = 0$

If the given number is not a perfect square, we will use tables or a calculator to find the square root.

The **Cube Root** : Symbol" $\sqrt[3]{}$ "

From the above square root definition, we can similarly define the cube root of a number as
one of the three equal factors of that number. (In this definition, we do **not** specify **positive root,**
since the cube root may be positive or negative, depending on whether the given number is positive or negative.)

Examples

1. $\sqrt[3]{8} = 2$; because $2^3 = 8$.   Note that $\sqrt[3]{8} = \sqrt[3]{(2)(2)(2)} = 2$

2. $\sqrt[3]{-8} = -2$; because $(-2)^3 = -8$.   Note that $\sqrt[3]{(-2)(-2)(-2)} = -2$

More formal definition of square root

1. $\sqrt{A} = r$ if $r^2 = A$, where A, and r are both real and positive.

Order of Operations and Evaluation of Algebraic Expressions

Perform the operations according to the following order
1. Grouping Symbols: If there are grouping symbols, evaluate within the grouping symbols first;
2. Then, evaluate the powers and the roots in any order ; followed by
3. Division and Multiplication from left to right (or invert the divisors and multiply in any order); and then
4. Addition and Subtraction from left to right.

Note: If grouping symbols occur within other grouping symbols, evaluate within the innermost grouping symbols first, followed by the next innermost symbols and so on.

Example 1
$5 + 2(4 - 7) - 9$
$= 5 + 2(-3) - 9$ Evaluating within the parentheses first
$= 5 - 6 - 9$ Multiplying next
$= -1 - 9$ Adding from left to right
$= -10$

Example 2
$2^3 - 9(4 + 3) + 8$ Evaluating within the parentheses first
$= 2^3 - 9(7) + 8$
$= 8 - 9(7) + 8$ **Scrapwork:** $2^3 = (2)(2)(2) = 8$
$= 8 - 63 + 8$
$= -47$

Example 3
$18 \div 6 \times 3$ From left to right, we encounter division first and
$= 3 \times 3$ therefore, we divide first, according to
$= 9$ the order of operations.

Note in Example 3 above that, if we had multiplied first, we would **not** obtain the correct answer of 9.
In multiplying first, $18 \div 6 \times 3 = 18 \div 18 = 1$ <-----**This is the wrong solution.**

Example 4
$6 \times 3 \div 18$
$= 18 \div 18$ (From left to right, we encounter multiplication
$= 1$ first, therefore, we multiply first and then divide)
Note however that, if we divide first, we will still get the correct answer.

$$\text{Thus}, 6 \times 3 \div 18 = \frac{6}{1} \times \frac{3}{18} = \frac{6}{1} \times \frac{1}{6} = 1$$

We conclude, from Examples 3 and 4, that if we always divide first (whether or not division comes first) , we will obtain the correct answer; but if multiplication does not come first in going from left to right, then we should **not** multiply first. Thus if we always perform division first, we will obtain the correct answer. Note that Example 3 and Example 4 are different problems.

5 $10 - [7 - 3(5 - 1)]$
$= 10 - [7 - 3(4)]$
$= 10 - [7 - 12]$
$= 10 - [-5]$
$= 10 + [+5]$
$= 15$

6. Simplify: $20 \div \frac{1}{2}$ of 10
(Hint: Perform the "of" operation first before dividing.)
Solution $20 \div \frac{1}{2}$ of 10
$= 20 \div (\frac{1}{2} \times 10)$
$= 20 \div 5 = 4.$

Review: **Evaluation of Algebraic Expressions**

Procedure:

Step 1: Replace each letter by its given numerical value in the expression, (enclosing each value in parentheses during the replacement).

Step 2: Evaluate the resulting expressions using the order of operations learned previously (see p.4).

Example 1   Given that $x = -3$, evaluate $x^2 + 7x + 2$

$$\textbf{Solution:}\quad (-3)^2 + 7(-3) + 2$$
$$= 9 - 21 + 2 \qquad \text{Scrapwork:}\ (-3)^2 = (-3)(-3)$$
$$= -12 + 2 \qquad\qquad\qquad\qquad\quad = 9$$
$$= -10$$

Example 2   Find the value of $5 - 4[3x - 1]$, if $x = 2$.

Solution   Substituting 2 for x, we obtain
$$5 - 4[3(2) - 1]$$
$$= 5 - 4[6 - 1]$$
$$= 5 - 4[5]$$
$$= 5 - 20$$
$$= -15$$

Example 3   (a) Evaluate: $200\ (2^n)$ when $n = 3$

Procedure: Substituting n = 3 in $200(2^n)$ and applying the order of operations, then we obtain:

$$200\ (2^3)$$
$$= 200\ (8)$$
$$= 1600$$

Scrapwork
$$2^3 = (2)(2)(2) = 8$$

(b) Evaluate: $1000\ (4)^n$ if $n = 2$

Procedure: Substitute $n = 2$ in $1000\ (4)^n$

$$1000\ (4)^2$$
$$= 1000\ (16)$$
$$= 16000$$

Scrapwork
$$(4)^2 = (4)(4) = 16$$

Lesson 1 Exercises

Attempt all questions on clean sheets of paper, **Do not write in the book** Show all necessary work.

---

1. (a) $(-2)^4 =$

 (b) $(-3)^3 =$

2. (a) $\sqrt{81}$

 (b) $\sqrt[3]{-8}$

---

3. Evaluate (a) $7 - 2(3 + 4)$

 (b) $16 \times 4 \div 4$

4. (a) $5\sqrt{49} - 3(4)(-2)$

 (b) $-3[8 - (-4 - 3) + 8]$

---

5. $\dfrac{5 - 3}{-10 + 6}$

6. (a) $5 + 3(4 + 1)^2$

 (b) $2(-3)^3 - 7(-5)$

---

7. (a) $(-5)^2 =$

 (b) $-5^2 =$

8. (a) $6 \times 12 \div 2$

---

9. (a) $2 - 4\{ - (4)(-2) + 2[4 - 7]$

10. $\dfrac{-6 - 3}{-2 - 3}$

---

11. (a) $0 - 4^2$
 (b) $0 - (-4)^2$
 (c) $0 + (-4)^2$

12. (a) $\sqrt{4^2 + 3^2}$

 (b) $\left(\dfrac{2}{3}\right)^3 + \dfrac{1}{9}$

---

13. (a) $(-4)2(0)(9) =$

 (b) $-7^2 + 2(0) =$

14. (a) $-3(2 - 7) - (-3)$

 (b) $(-5)^2 - (-3)^2$

---

15. (a) $6 - 2[4 + 3\}^2 - 8$

16. (a) Find the value of $b^2 - 6b + 7$ when $b = -2$

17. Given that $A = \frac{1}{2}bh$ find the value of A if $b = 10, h = 3$

18. Find the value of F if $C = -20$, given that $F = \frac{9}{5}C + 32$

19. Given that $A = L \times W$, find the value of A when $L = 7, W = 4$.

20. $\dfrac{(-4) - (-1)}{-1}$

Questions 21-24
Evaluate if $a = -2, b = 1, c = 3, x = -5, y = -1$
21. $6x + 5y$

22. $abcxy$

23. $\dfrac{4a - 2}{ax + 2}$

24. $a^2 - b^2$

25 Find the value of F if $C = -40$, given that $F = \frac{9}{5}C + 32$

.

Bonus:

Answers: **1.** (a) 16; (b) -27 ; **2.** (a) 9; (b) -2 ; **3.** (a) -7; (b) 16 ; **4.** (a) 59; (b) -69 ; **5.** $-\frac{1}{2}$;
6. (a) 80; (b) -19 ; **7.** (a) 25; (b) -25 ; **8.** 36 ; **9.** -6 ; **10.** $\frac{9}{5}$; **11.** (a) -16; (b) -16; (c) 16 ;
12. (a) 5 ; (b) $\frac{11}{27}$; **13.** (a) 0; (b) -49 ; **14.** (a) 18; (b) 16 ; **15.** -100 ; **16.** 23 ; **17.** 15 ;
18. -4 ; **19.** 28 ; **20.** 3; **21.** -35 ; **22.** -30 ; **23.** $-\frac{5}{6}$; **24.** 3 ; **25.** -40.

Lesson 2
Review: **Rules of Exponents and Applications**

Example 1. $2^5 = 2 \cdot 2 \cdot 2 \cdot 2 \cdot 2 = 32$

2^5 is read as 2 raised to the fifth power.(or 2 to the fifth power)
The exponent "5" indicates how many times the base "2" is being used as a factor.

Example 2. $x^6 = x \cdot x \cdot x \cdot x \cdot x \cdot x$

Mnemonic Examples for the Rules of Exponents
(Match the following examples with the rules of exponents which are presented later, below.)

1. $x^2 x^3 = x^5$
2. $\dfrac{x^7}{x^4} = x^{7-4} = x^3$

3. $(x^4)^2 = x^8$
4. $(xy)^5 = x^5 y^5$

5. $(\frac{x}{y})^6 = \dfrac{x^6}{y^6}$
6. $x^{-2} = \dfrac{1}{x^2}$

7. $x^0 = 1$
8. $\dfrac{x^4}{x^7} = x^{4-7} = x^{-3} = \dfrac{1}{x^3}$

Rules of Exponents

1. $x^a x^b = x^{a+b}$ (Product of powers of the same base rule)   Example $x^2 x^3 = x^5$

2. $(x^a)^b = x^{ab}$ (Power to a power rule)   Example $(x^4)^2 = x^8$

3. $x^{-a} = \dfrac{1}{x^a}$; also $\dfrac{1}{x^{-a}} = x^a$ (where a is positive) Negative exponent rule .Ex.1: $x^{-2} = \dfrac{1}{x^2}$; Ex.2: $\dfrac{1}{x^{-2}} = x^2$

4. $\dfrac{x^a}{x^b} = x^{a-b}$ (Quotient of powers of the same base rule) Example $\dfrac{x^7}{x^4} = x^{7-4} = x^3$

5. $(xy)^n = x^n y^n$ (Product to a power rule)   Example $(xy)^3 = x^3 y^3$

6.. $\left(\frac{x}{y}\right)^n = \dfrac{x^n}{y^n}$ ($y \neq 0$)   (Quotient to a power rule) Example $(\frac{x}{y})^5 = \dfrac{x^5}{y^5}$

Zero as an exponent: $x^0 = 1$ ($x \neq 0$) Example: $5^0 = 1$

Applications of the rules of exponents

Example 1   Simplify: $(3x^2 y)^2$

$= 3^2 x^4 y^2$ (Using the rules of exponents)

$= 9x^4 y^2$

(**Note**: also that $(3x^2 y)^2 = (3x^2 y)(3x^2 y) = 9x^4 y^2$)

Example 2   Simplify: $(2x^4 y^6)^4$

$= 2^4 x^{16} y^{24}$

$= 16x^{16} y^{24}$

Scrapwork: $2^4 = (2)(2)(2)(2) = 16$

Lesson 2 Exercises

A. Simplify the following: **1.** $x^6 x^3$; **2.** $(y^7)^3$; **3.** $(bc^4)^6$; **4.** $\dfrac{x^9}{x^5}$;

5. $\dfrac{(y^3)^5 x^9}{(y^7)^2 x^4}$; **6.** $(x^3 y^6)^4$; **7.** $(-\dfrac{2}{3})^3$

Answers: **1.** x^9 ; **2.** y^{21}; **3.** $b^6 c^{24}$; **4.** x^4 ; **5.** $x^5 y$; **6.** $x^{12} y^{24}$; **7.** $-\dfrac{8}{27}$

B. Simplify the following: **1.** $(3x^5 y)^3$; **2.** $(-2x^2 y^7)^5$; **3.** Multiply: $\left(\dfrac{-8b^2 c}{c}\right)\left(\dfrac{9d}{3bcd}\right)\left(\dfrac{-15}{27c}\right)$;

4. $(-2x^2 y^7)^4$

Answers: **1.** $27x^{15} y^3$; **2.** $-32x^{10} y^{35}$; **3.** $\dfrac{40b}{3c^2}$; **4.** $16x^8 y^{28}$

C. Simplify, leaving answers with positive exponents.

1. $x^8 \cdot x^2$

2. $x^{-4} \cdot x^6$

3. $x(x^2)$;

4. $\dfrac{x^9}{x^6}$

5. $\dfrac{x^6}{x^9}$

6. $(a^2 b^3)^5$

7. $(xy)^0$;

8. $(x^{-2})^4$

9. $10^{-2} \cdot 10^6$

10. $\dfrac{m^2}{n^{-4}}$

11. $\dfrac{10^{-3} \cdot 10^8}{10^3}$;

12. $(x^2 y^{-3} z)^4$

13. $x^5 x^7$

14. $(x^8)^4$

15. $(-3x)^3$;

16. $(2x^2 y^3)^4$

17. $(-3ab^2 c^3)^3$

18. $\dfrac{x^2 y^4}{xy}$

19. $\dfrac{xy}{x^2 y^5}$;

20. $2x(x)$

21. $\dfrac{(x^3 y^4)^2}{x^2 y}$

22. $(a^3)^{12}(a^2)^5$

23. $\dfrac{b^5 (b^2)^3}{b^8 (b^3)^4}$;

24. $\dfrac{(x^3)^5 (x^4)^2}{(x^5)^2 x^9}$

Answers: **1.** x^{10} ; **2.** x^2 ; **3.** x^3 ; **4** x^3 ; **5.** $\dfrac{1}{x^3}$; **6.** $a^{10} b^{15}$; **7.** 1 ; **8.** $\dfrac{1}{x^8}$; **9.** 10^4 ; **10.** $m^2 n^4$;

11. 10^2 or 100; **12.** $\dfrac{x^8 z^4}{y^{12}}$; **13.** x^{12} ; **14.** x^{32} ; **15.** $-27x^3$; **16.** $16x^8 y^{12}$; **17.** $-27a^3 b^6 c^9$; **18.** xy^3;

19. $\dfrac{1}{xy^4}$; **20.** $2x^2$; **21.** $x^4 y^7$; **22.** a^{46} ; **23.** $\dfrac{1}{b^9}$; **24.** x^4

Lesson 3

Review: **Simplification of Algebraic Expressions**

Definitions

Constant: a symbol representing a quantity that does not change in value in a certain discussion or operation. We usually represent a constant by a number.

Example: In $4x^2$ -$3xy$, the constants are 4, 2 and -3.

We may also represent a constant by a letter, such as one of the first few letters of the alphabet. In a discussion in which letters are used to represent constants, it is good practice to state which letters represent constants.

Example: A quadratic trinomial is of the form $ax^2 + bx + c$, where **a**, **b**, and **c** are **constants**.

Variable: a variable is a symbol representing a quantity that changes in value. We usually represent a variable by a letter.

Example: In $4x^2$ -$3xy$, the variables are x and y.

Algebraic Expression

An algebraic expression is any variable (letter) or constant (number) or any combination of variables and constants related by signs of operation (such as $+, -, \times$, etc.) and grouping symbols.

Examples 1. $5x^2 + 8x$ - 9.

          2. $-2x(x - y) + 8$.

The plus and the minus signs separate an algebraic expression into parts. Each part is called a **term**. $5x^2 + 8x$ - 9 has three terms : The first term is $5x^2$; the second term is $+ 8x$, and the third term is - 9. $-2x(x - y) + 8$ has two terms : The first term is $-2x(x - y)$ and the second term is $+8$.

Monomial: A monomial in x is a one-term algebraic expression of the form ax^n, where a is a real number and n is a whole number. Examples: **1**. $4x^3$ **2**. $5x$;   **3**. $6x^3$;   **4**. 8

A monomial in x and y is a one-term algebraic expression of the form $ax^m y^n$, where a is a real number, and m and n are whole numbers.

Polynomial: A polynomial is an algebraic expression consisting of two or (a finite number of) more unlike monomials.

Examples: **1**. $7x^4$ -$5x^3$ - $4x + 7$ **2**. $4x + 8$;   **4**. $\frac{2}{3}x^5 + 7x + 4$;   **5**. $3x^5 y^2 + 7x^3 y$.

An example of an expression which is **not** a polynomial: $3x^2 + 6x^4 + \frac{2}{x}$ - 9 is **not** a polynomial because of the

x in the denominator of the third term. Note also that $\frac{2}{x} = 2x^{-1}$ (has a negative exponent).

Another definition

A polynomial in x is an algebraic expression consisting of a finite number of terms with each term being of the form ax^n, where a is a real number and n is a whole number. (In a polynomial, there are **no** variables in the denominator or **no** negative exponents; no variables with fractional exponents or the radical equivalents)

Types of polynomials

Binomial : a polynomial of exactly two unlike monomials. Examples: **1.** $5x^2$ - $3x^4$;   **2.** $a + b$

Trinomial: a polynomial of exactly three unlike monomials. Examples: **1.** $2x^3$ - $5xy$ - 8; **2.**$x^2 + 7x + 12$.

Note that a monomial is a special type of term. All monomials are terms but not all terms are monomials.

A polynomial may also be in terms of two or more variables. It may also be in terms of any letters. 1 1

Definition: A polynomial in x and y is an algebraic expression in which all the terms are of the form $ax^m y^n$, where a is any real number and m and n are whole numbers.

The **degree** of a **term** is the sum of the exponents of the variables of that term.

Example: Find the degree of the following: **1.** $3x^5 y^2$, **2.** $7x^4$, **3.** x, **4.** 8
Solution:

1. degree = 5 + 2 =7, **2.** degree = 4; **3.** degree = 1 ; **4.** degree = 0 (since $8 = 8x^0$ and $x^0 = 1$)

The **degree** of a **polynomial** is the degree of the term with the highest degree.

Example: Find the degree of the following polynomials

1. $7x^4 - 5x^3 - 4x + 7$, **2.** $4x + 8$, **3.** $\frac{2}{3}x^5 + 7x + 4$, **4.** $3x^5 y^2 + 7x^3 y$

Solution   1. Degree = 4 (degree of $7x^4$);   2. degree = 1 (degree of $4x$)

 3. degree =5 (degree of $\frac{2}{3}x^5$) ; **4.** degree = 7 (degree of $3x^5 y^2$).

Like Terms

Like terms have exactly the same literal (letter) factors, with each letter raised to same power.
Examples $3x$ and $-7x$ are like terms.
          $5x^2 y$ and $6x^2 y$ are like terms.
          $2bc$ and $-5cb$ are like terms.
 However, $3x^2 y^2$ and $-8x^2 y$ are **unlike** terms.

Note: Like terms **can** be added or subtracted,   Unlike terms **cannot** be added or subtracted.

Numerical Coefficient of a Term

Examples   The numerical coefficient of $8x$ is 8
          The numerical coefficient of $-8x$ is -8
          The numerical coefficient of $-7xy$ is -7
          The coefficient of $6x^2$ is 6.

The Distributive Property $\boxed{a(b + c) = ab + ac}$

Also, $a(b + c + d) = ab + ac + ad$
     $a(b - c - d) = ab - ac - ad$
     $a(b + c) = (b + c)a = ba + ca = ab + ac$

Examples   **1.** $5(2x + y)$ **2.** $3(a - b)$
             $= 10x + 5y$ $= 3a - 3b$

Addition of Like Terms

To add like terms, add the numerical coefficients and keep the literal parts.
                                You may skip writing this step
Example 1 $4x + 2y + 6x + 3y$ Scrapwork
              $= (4 + 6)x + (2 + 3)y$ <--------   For the x-terms: 4 + 6 = 10
              $= 10x + 5y$ For the y-terms: 2 + 3 =5

Example 2 $-8x + 5x + 2y$
              $= (-8 + 5)x + 2y$<----------------------------You may skip this step
              $= -3x + 2y$ For the x-terms: -8 + 5 = -3

How to Remove Grouping Symbols

We will consider four cases.

Case 1: If there is a plus sign or no sign preceding the grouping symbols, delete the plus sign, delete the grouping symbols, and copy the expression within the grouping symbols.

Case 2: If a minus sign precedes the grouping symbols, delete the minus sign, delete the grouping symbols, and change the sign of every term within the grouping symbols.

Case 3: If a factor precedes the grouping symbols, multiply each term within the grouping symbols by this factor. This case is the application of the distributive property:
$$a(b + c) = ab + ac$$

Case 4: If there are grouping symbols within other grouping symbols, remove the innermost grouping symbols first, followed by the next innermost, and so on.

Addition of Polynomials

Procedure: To add polynomials, add the like terms.

Example 1   Find the sum of $-2b + 3c$ and $5b - 9c$

Procedure: Add the like terms.
$$-2b + 3c + 5b - 9c$$
$$= 3b - 6c$$

Scrapwork: For the b- terms:
$$-2 + 5 = 3$$
For the c-terms:
$$3 - 9 = -6$$

Example 2   Add $b^2 + 4c$ and $-3b^2 - 9c$

Procedure: Add the like terms: $b^2 + 4c - 3b^2 - 9c$

$$= -2b^2 - 5c$$

Subtraction of Polynomials

Example   Subtract $5x^2 - 3$ from $2x - 4$

Procedure: Write the "from" expression first, followed by the subtracted expression in parentheses with a minus sign in front of the parentheses.
$$2x - 4 - (5x^2 - 3)$$
Now simplify by removing the parentheses and adding any like terms.

or

Change the signs of all the terms of the subtracted expression and add to the expression from which you are subtracting.

$$2x - 4 - (5x^2 - 3)$$
$$= 2x - 4 - 5x^2 + 3$$
$$= -5x^2 + 2x - 1$$

Note: The subtracted expression (the subtrahend) follows the word subtract and the "from" expression (the minuend) follows the word "from".

Multiplication of Polynomials

Example 1   Find the product of $4xy^2$ and $-3y^2$
Multiply: $(4xy^2)(-3y^2) = -12xy^4$

Example 2 $(-2x^3y)(3xy^5)$
$$= (-2 \cdot 3)x^3 \cdot x \cdot y \cdot y^5$$
$$= -6x^4 y^6$$

Example 3 $2x(x + 4)$
$$= 2x^2 + 8x$$

Example 4 $(x + 2)(x + 5)$ <------------ Multiplication of two binomials   1 3

$x(x + 5) + 2(x + 5)$ (Distribute x over $(x + 5)$, followed by distribution of $+2$ over $(x + 5)$.
You may skip writing this step, and write the next step.

$= x^2 + 5x + 2x + 10$

$= x^2 + 7x + 10$

Example 5 $(3x + 2)(4x - 3)$

$= 12x^2 - 9x + 8x - 6$

$= 12x^2 - x - 6$

Special Products: **1.** $(x + y)^2 = (x + y)(x + y)$

$= x^2 + xy + xy + y^2$ (Always read the letters alphabetically)

$= x^2 + 2xy + y^2$

2. $(x - y)^2 = (x - y)(x - y)$

$= x^2 - xy - xy + y^2$

$= x^2 - 2xy + y^2$

3. $(a + b)(a - b) = a^2 - ab + ab - b^2$

$= a^2 - b^2$ $(- ab + ab = 0)$

The expression $a^2 - b^2$ is known as the difference between two squares. We will meet this expression again when we factor polynomials in the future (Lesson 47).

Division of a monomial by a monomial

Example 1 $\dfrac{8x^3y}{2xy}$

$= \dfrac{8}{2} \cdot \dfrac{x^3}{x} \cdot \dfrac{y}{y}$

$= \dfrac{4}{1} \cdot \dfrac{x^2}{1} \cdot 1$

$= 4x^2$

Example 2 $\dfrac{-6x^3y^4}{-2x^8y}$

$= \dfrac{-6}{-2} \cdot \dfrac{x^3}{x^8} \cdot \dfrac{y^4}{y}$

$= \dfrac{3}{1} \cdot \dfrac{1}{x^5} \cdot \dfrac{y^3}{1}$

$= \dfrac{3y^3}{x^5}$

(Leaving answer with positive exponents only)

Division of a polynomial by a monomial

Example 1   Simplify: $\dfrac{12x^2 - 8x}{4x}$

Step 1: $\dfrac{12x^2}{4x} + \dfrac{-8x}{4x}$

Step 2: $3x + (-2)$

Step 3: $3x - 2$

Example 2 Simplify: $\dfrac{15y^3 + 20y^2}{-5y}$

Step 1: $\dfrac{15y^3}{-5y} + \dfrac{20y^2}{-5y}$

Step 2: $-3y^2 + (-4y)$

Step 3: $-3y^2 - 4y$

Note that in Examples 1 and 2 (above), you may skip writing Steps 1 and 2 if you **make sure** that you divide **every term** (taking into account its sign) in the **numerator** by the denominator. The author suggests skipping Steps 1 and 2, provided attention is paid to the above **note**.

Lesson 3 Exercises A

A. **1.** Find the product of $-5ab^2$ and $(6b^4)$;   **2.** Simplify: $-4x(x-5)$;   **3.** $(x-3)(x+7)$.

4. Find the product: $(4x+2)(3x-6)$;   **5.** Simplify: $(x-y)(x+y)$

Answers: **1.** $-30ab^6$; **2.** $-4x^2+20x$; **3.** $x^2+4x-21$; **4.** $12x^2-18x-12$; **5.** x^2-y^2

B. Carry out the indicated operations:

1. $2x(bx)$ **2.** $-3x(-7x)$ **3.** $x(2x+6y)$

4. $2(3a-b)$ **5.** $(x+2)(x+3)$ **6.** $(x-2)(x+3)$

7. $(2a+3b)(a-b)$ **8.** $(a+b+c)(a-b+c)$ **9.** $(3x-2)(2x+5)$

10. $(2x-3)^2$ **11.** $(a+b)^2$ **12.** $2a^2(5a-5b-c)$

13. Simplify: $(x+4)(x-5)+(x-6)(x+3)$

Answers: **1.** $2bx^2$; **2.** $21x^2$; **3.** $2x^2+6xy$; **4.** $6a-2b$; **5.** x^2+5x+6; **6.** x^2+x-6;

7. $2a^2+ab-3b^2$; **8.** $a^2+2ac-b^2+c^2$; **9.** $6x^2+11x-10$; **10.** $4x^2-12x+9$;

11. $a^2+2ab+b^2$; **12.** $10a^3-10a^2b-2a^2c$; **13.** $2x^2-4x-38$.

C. Simplify the following **1.** $-(-4x+6y+6)$; **2.** $2+4(3x-5y-2)$; **3.** $-4(3a+c)-3(5a-4)$

4. $+(6x+4y)+4(7x+8)$;   **5.** $4xy(5x^2y-3y)-7xy^2$

Answers: **1.** $4x-6y-6$; **2.** $12x-20y-6$; **3.** $-27a-4c+12$;

4. $34x+4y+32$; **5.** $20x^3y^2-19xy^2$

D. Simplify and leave answers with positive exponents:

1. $\dfrac{10a^7}{2a^2}$ **2.** $\dfrac{12x^4y^6}{-2x^3y^2}$ **3.** $\dfrac{-18a^5b^7}{-3a^2b^4}$; **4.** $\dfrac{16a^2bc}{14a^3bc^2}$

5. $\dfrac{-10x^5y^7}{20x^6y^6}$ **6.** $\dfrac{16c^3d}{-12cd^2}$ **7.** $\dfrac{.75a^4bc}{2.5abc}$

Answers: **1.** $5a^5$; **2.** $-6xy^4$; **3.** $6a^3b^3$; **4.** $\dfrac{8}{7ac}$; **5.** $-\dfrac{y}{2x}$; **6.** $-\dfrac{4c^2}{3d}$; **7.** $.3a^3$

E. Simplify the following:

1. $\dfrac{-8x^5y^3}{2x^2y}$; **2.** $\dfrac{15a^4b^2-20a^3b^2-5ab}{5ab}$; **3.** $\dfrac{14x^3y^2-21x^2y}{7xy}$

Answers: **1.** $-4x^3y^2$; **2.** $3a^3b-4a^2b-1$; **3.** $2x^2y-3x$

Division of a polynomial by a polynomial (The long division method)   1 5

We will consider four cases.

Case 1: The divisor divides the dividend exactly (i.e. without a remainder).

Example: Divide $7x^2 + x^3 + 6 + 13x$ by $x + 2$

Note : $7x^2 + x^3 + 6 + 13x$ is the dividend; $x + 2$ is the divisor.

We will use a long division method which is similar to the long division in arithmetic.

Step 1: Check the dividend and the divisor to see if the terms of each expression are arranged in descending powers of the variable, x. Descending powers of x means the term with the highest power of x is written first, followed in decreasing orders by terms with lower powers of x.
The divisor $x + 2$ is already arranged in descending powers of x.
The dividend $7x^2 + x^3 + 6 + 13x$ is not arranged in descending powers of x. In descending powers of x, the dividend becomes $x^3 + 7x^2 + 13x + 6$.

Step 2: Arrange the divisor and the dividend as done in arithmetic long division, as shown below.

$$x + 2 \, \overline{)\, x^3 + 7x^2 + 13x + 6.}$$

Step 3: Divide the first term (leading term) of the dividend by the first term (leading term) of the divisor, and write quotient x^2 above the x^3 (as we do in arithmetic).

$$\begin{array}{r} x^2 \\ x + 2 \, \overline{)\, x^3 + 7\, x^2 + 13 \; x + 6} \end{array}$$ Scrapwork: $\dfrac{x^3}{x} = x^2$

Step 4: As done in arithmetic, multiply the divisor, $x + 2$ by the x^2, and write the product under the terms of the dividend, lining up the like terms vertically.

$$\begin{array}{r} x^2 \\ x + 2 \, \overline{)\, x^3 + 7\, x^2 + 13 \; x + 6} \\ +x^3 + 2x^2 \end{array}$$ Scrapwork: $x^2(x + 2) = x^3 + 2x^2$

Step 5: Subtract the $x^3 + 2x^2$ from the like terms above them; but recall that to subtract we change the signs of the terms being subtracted and add the changed terms to the terms from which we are subtracting.

$$\begin{array}{r} x^2 \\ x + 2 \, \overline{)\, x^3 + 7\, x^2 + 13 \; x + 6} \end{array}$$

Subtract: $\dfrac{x^3 + 2x^2}{5x^2}$

Scrapwork:

$$\begin{array}{cc} x^3 & 7x^2 \\ \underline{-x^3} & \underline{-2x^2} \\ 0 & 5x^2 \end{array}$$

Step 6: Bring down the other terms; divide, multiply and subtract as done in Steps 3, 4, and
   5 above. Remember, always the first term of divisor divides the first term of dividend.

$$x + 2 \enclose{longdiv}{x^3 + 7x^2 + 13x + 6}$$

quotient: $x^2 + 5x$

$$+ x^3 + 2x^2$$

$$+ 5x^2 + 13x + 6$$

Scrapwork: $\dfrac{5x^2}{x} = 5x$

Subtract: $+ 5x^2 + 10x$

Scrapwork: $5x(x + 2) = 5x^2 + 10x$

$$+ 3x$$

$(5x^2 + 13x - 5x^2 - 10x = 3x)$

Step 7: Bring down the 6, divide, multiply, and subtract.

$$x + 2 \enclose{longdiv}{x^3 + 7x^2 + 13x + 6}$$

quotient: $x^2 + 5x + 3$

$$+x^3 + 2x^2$$

$$+ 5x^2 + 13x + 6$$

Subtract: $+ 5x^2 + 10x$

$$+ 3x + 6$$

$$+ 3x + 6$$

$$0 + 0$$

Scrapwork: $\dfrac{3x}{x} = 3$

$(3x + 6 - 3x - 6 = 0 + 0)$

$\therefore \qquad \dfrac{x^3 + 7x^2 + 13x + 6}{x + 2} = x^2 + 5x + 3$

Normally, when we perform the division process, we show only Step 7 above.

Note: In the above division process, we divide, multiply, subtract, bring down the remaining terms of the dividend and repeat the process. Note also that always, the **first term** of divisor divides the **first term** of the dividend.

Case 2: The division process leaves a remainder.

Solution: The steps are exactly like those of Example 1; however we will show only the compact form in Step 7 of Example 1.

Divide $6x^4 + x^2 - x^3 + 30x + 7$ by $3 + 2x$

Step 1 : Arrange the divisor and dividend in descending powers of x; and set up the problem as in arithmetic long division.

$$
\begin{array}{r}
3x^3 \; - 5x^2 + 8x + 3 \text{ R } {-2} \\
\text{Step 2:} \quad 2x + 3 \overline{)\; 6x^4 \; - \; x^3 + x^2 + 30x + 7}
\end{array}
$$

Scrapwork: $\dfrac{6x^4}{2x} = 3x^3$

$$
+6x^4 + 9x^3
$$
$$
\overline{-10x^3 + x^2 + 30x + 7}
$$
$$
-10x^3 - 15x^2
$$
$$
\overline{16x^2 + 30x + 7}
$$
$$
+16x^2 + 24x
$$
$$
\overline{+6x + 7}
$$
$$
+6x + 9 \qquad 6x + 7 - 6x - 9
$$
$$
\overline{-2} \; \text{<---------Remainder}
$$

There is a remainder of -2. The division process ends when the degree of the divisor is larger than the degree of the dividend.

We may write the answer as $3x^3 \; - 5x^2 + 8x + 3$ R -2 or $3x^3 \; - 5x^2 + 8x + 3 \; + \dfrac{-2}{2x + 3}$

$$
= 3x^3 \; - 5x^2 + 8x + 3 \; - \dfrac{2}{2x + 3}
$$

To check the answer: Multiply $3x^3 - 5x^2 + 8x + 3$ by $2x + 3$ and add the remainder -2. If you obtain the original dividend the answer is correct.

Case 3: The dividend has missing powers

Example Divide $x^5 - 2x^3 - 10x^2 - 30$ by $x - 3$.

In setting up the problem vertically, we may either leave a column space for the missing powers or we may write zero-coefficient terms for the powers. We choose to write zero coefficient powers.

Then the dividend becomes $x^5 + 0x^4 - 2x^3 - 10x^2 + 0x - 30$

Solution Arrange the divisor and dividend in descending powers of x and set up the problem as in arithmetic long division, and divide.

$$
\begin{array}{r}
x^4 + 3x^3 + 7x^2 + 11x + 33 \\
x-3\overline{)\ x^5 + 0x^4 - 2x^3 - 10x^2 + 0x - 30} \\
\end{array}
$$

$+x^5 - 3x^4$ Scrapwork: $\dfrac{x^5}{x} = x^4$

$+ 3x^4 - 2x^3 - 10x^2 + 0x - 30$ Scrapwork: $\dfrac{3x^4}{x} = 3x^3$

$+ 3x^4 - 9x^3$

$+ 7x^3 - 10x^2 + 0x - 30$

$+ 7x^3 - 21x^2$

$+ 11x^2 - 0x - 30$ $(7x^3 - 10x^2 - 7x^3 + 21x^2)$

$+ 11x^2 - 33x$

$+33x - 30$

$+ 33x - 99$

69 <---------Remainder $(33x - 30 - 33x + 99)$

In this problem, the division process ends when the degree of the first term of the divisor is larger than the degree of the first term of the dividend (if the terms are in descending powers of x).

$\therefore \dfrac{x^5 - 2x^3 - 10x^2 - 30}{x - 3} = x^4 + 3x^3 + 7x^2 + 11x + 33 \text{ R } 69$

or

$= x^4 + 3x^3 + 7x^2 + 11x + 33 + \dfrac{69}{x - 3}$

Case 4: Division of polynomials involving two or more variables.

In this case, we may base the descending order arrangement on one of the variables.

Example   Divide $- 3y^3 + 2x^2y - 5xy^2$ by $2x + y$

We base the order of arrangement of the terms on x.

$$
\begin{array}{r}
xy - 3y^2 \\
2x + y \;|\; \overline{2x^2y - 5xy^2 - 3y^3} \\
+ 2x^2y + xy^2 \\
\hline
-6xy^2 - 3y^3 \\
-6xy^2 - 3y^3 \\
\hline
0 + 0
\end{array}
$$

Note above that if we base the order of arrangement of the terms on y, we will obtain the same quotient, (except for the order of the terms) $-3y^2 + xy$.

Lesson 3 Exercises B

A. 1. Find the product of $-5ab^2$ and $(6b^4)$;   **2.** Simplify: $-4x(x - 5)$;   **3.** $(x - 3)(x + 7)$.

4. Find the product: $(4x + 2)(3x - 6)$;   **5.** Simplify: $(x - y)(x + y)$

Answers: **1.** $-30ab^6$;   **2.** $- 4x^2 + 20x$;   **3.** $x^2 + 4x - 21$;   **4.** $12x^2 - 18x - 12$;   **5.** $x^2 - y^2$

B. Carry out the indicated operations:

1. $2x(bx)$ **2.** $-3x(-7x)$ **3.** $x(2x + 6y)$

4. $2(3a - b)$ **5.** $(x + 2)(x + 3)$ **6.** $(x - 2)(x + 3)$

7. $(2a + 3b)(a - b)$ **8.** $(a + b + c)(a - b + c)$ **9.** $(3x - 2)(2x + 5)$

10. $(2x - 3)^2$ **11.** $(a + b)^2$ **12.** $2a^2(5a - 5b - c)$

13. Simplify: $(x + 4)(x - 5) + (x - 6)(x + 3)$

Answers: **1.** $2bx^2$;   **2.** $21x^2$;   **3.** $2x^2 + 6xy$;   **4.** $6a - 2b$;   **5.** $x^2 + 5x + 6$;   **6.** $x^2 + x - 6$;

7. $2a^2 + ab - 3b^2$;   **8.** $a^2 + 2ac - b^2 + c^2$;   **9.** $6x^2 + 11x - 10$;   **10.** $4x^2 - 12x + 9$;

11. $a^2 + 2ab + b^2$;   **12.** $10a^3 - 10a^2b - 2a^2c$;   **13.** $2x^2 - 4x - 38$.

C. Simplify the following **1.** $- (-4x + 6y + 6)$;   **2.** $2 + 4(3x - 5y - 2)$;   **3.** $-4(3a + c) - 3(5a - 4)$

4. $+(6x + 4y) + 4(7x + 8)$;   **5.** $4xy(5x^2y - 3y) - 7xy^2$

Answers: **1.** $4x - 6y - 6$;   **2.** $12x - 20y - 6$;   **3.** $- 27a - 4c + 12$;

4. $34x + 4y + 32$;   **5.** $20x^3y^2 - 19xy^2$

D. Simplify and leave answers with positive exponents:

1. $\dfrac{10a^7}{2a^2}$ 2. $\dfrac{12x^4y^6}{-2x^3y^2}$ 3. $\dfrac{-18a^5b^7}{-3a^2b^4}$;   4. $\dfrac{16a^2bc}{14a^3bc^2}$

5. $\dfrac{-10x^5y^7}{20x^6y^6}$ 6. $\dfrac{16c^3d}{-12cd^2}$ 7. $\dfrac{.75a^4bc}{2.5abc}$

Answers: **1.** $5a^5$; **2.** $-6xy^4$; **3.** $6a^3b^3$; **4.** $\dfrac{8}{7ac}$; **5.** $-\dfrac{y}{2x}$; **6.** $-\dfrac{4c^2}{3d}$; **7.** $.3a^3$

E. Simplify the following:

1. $\dfrac{-8x^5y^3}{2x^2y}$; 2. $\dfrac{15a^4b^2 - 20a^3b^2 - 5ab}{5ab}$;   3. $\dfrac{14x^3y^2 - 21x^2y}{7xy}$

Answers: **1.** $-4x^3y^2$;   **2.** $3a^3b - 4a^2b - 1$;   **3.** $2x^2y - 3x$

F. Divide:

1. $\dfrac{16x^2y - 8x}{2x}$ 2. $\dfrac{16x^2y - 8x}{-2x}$ 3. $\dfrac{14c^5d^2 - 21cd}{-7cd}$

4. $\dfrac{20x^4yz - 5xy}{5xy}$ 5. $\dfrac{12a^2bc - 14ab^3c}{6}$

Answers: **1.** $8xy - 4$;   **2.** $-8xy + 4$;   **3.** $-2c^4d + 3$;   **4.** $4x^3z - 1$;   **5.** $2a^2bc - \dfrac{7ab^3c}{3}$

G. Carry out the indicated operation.
1. Divide $6x^2 + x^3 + 6 + 11x$ by $x + 2$ 2. Divide $8x^4 + 9x^2 + 14x^3 + 19x + 12$ by $3 + 2x$

3. $(x^4 + 4x^2 + 5x + 2) \div (x + 3)$ 4. $(4x^3y - 6x^2y^2 + 5y^4 + 6xy^3) \div (2x + y)$

Answers   **1.** $x^2 + 4x + 3$;   **2.** $4x^3 + x^2 + 3x + 5$ R $- 3$;
3. $x^3 - 3x^2 + 13x - 34$ R 104; **4.** $2x^2y - 4xy^2 + 5y^3$

H. 1. Divide $x^3 + 2x^2 - 5x - 6$ by $x - 2$ **2.** Divide $6x^2 - 8x + 5$ by $x + 3$

3. Divide $x^3 + 10x^2 + 33x + 36$ by $x + 4$ **4.** Divide $4x^4 - 2x^2 - 3$ by $x - 3$

5. Find the remainder when $x^3 + 6x^2 + 6x - 12$ is divided by $x + 2$

Answers: **1.** $x^2 + 4x + 3$;   **2.** $6x - 26$ R 83 ;   **3.** $x^2 + 6x + 9$;
4. $4x^3 + 12x^2 + 34x + 102$ R 303;   **5.** Remainder $= -8$

Lesson 3: Review: Simplification of Algebraic Expressions

Review Test #1 - Student's Self-Test (Always, Test yourself before you are tested)

Attempt all questions on clean sheets of paper, **Do not write in the book** Show all necessary work.

Add the like terms:

1. (a) $4x - 6y + 7x - 4y$

   (b) $2a - 4b - 2a - 5b$

Simplify:

2. (a) $-7x + 6y - 3x + 4$

   (b) $5a + 3b - 5a + 4b - 9$

---

3. (a) Subtract $-8x^3 + 2x - 5$ from $6x - 2$.

   (b) From $9x - 2$, subtract $-6x^3 + 4x - 4$

4. (a) Subtract 5 from 3.

   (b) Subtract 3 from 7.

---

5. (a) Subtract $5y - 4x + 2$ from $3x - 6y - 5$

   (b) Subtract $2a - 7b - 8$ from $a - 36$

6. (a) Simplify: $3a^2 - 4ab + b^2 - (5a^2 + 2ab - 3b^2)$

   (b) Subtract $a - 32$ from $2a - 7b - 8$

---

7. (a) Find the product of $-4ab^2$ and $(5b^4)$.

   (b) Find the product: $(3x + 2)(5x - 6)$.

8. Simplify :

   (a) $6x^2y^3 - (x^2y^3 + x^2y) + xy^2 - 5x^2y$;

   (b) $5a^4b^5 + 3ab - (5a^4b^5 + 3ab)$

---

9. Simplify:

   (a) $2(3a - b)$

   (b) $(x + 2)(x + 3)$

10. Simplify:

   (a) $(6x - 8y + 7) - (-3x + 5y - 1)$

   (b) $-4x(x - 5)$;

---

11. Simplify:

   (a) $(x - 4)(x + 6)$.

   (b) $(x - y)(x + y)$

12. Simplify: (a) $(a + b + c)(a - b + c)$

   (b) $(3x - 2)(2x + 5)$

---

13. Simplify: (a) $(4x - 3)^2$

   (b) $(a + b)^2$

14. Simplify:

   (a) $2a^2(5a - 5b - c)$

   (b) $(x + 2)(x - 5) + (x - 6)(x + 3)$

15. Carry out the indicated operations:

    (a) $(x - 2)(x + 5)$

    (b) $(4a + 3b)(a - b)$

16. Simplify: (a) $2x(bx)$

    (b) $-4x(-7x)$

---

17. Simplify the following
    (a) $- (-4x + 6y + 6)$

    (b) $2 + 5(3x - 5y - 2)$.

18. Simplify: (a) $-4(3a + c) - 3(5a - 4)$

    (b) $+(2x + 4y) + 3(7x + 8)$;

---

19. Simplify: (a) $3xy(5x^2y - 3y) - 7xy^2$

           (b) $7ab - 8ba + 2ab + 6b$

20 Simplify:
    (a) $6x - 7y + 3y - 4x$

    (b) $\frac{1}{2}a^2b + 8 + \frac{1}{4}a^2b$

---

21. Add the following:

    (a) $2x - 6y - 5$ and $2y - 4x + 3$

    (b) $3x^2y - 9xy^2$ and $5x^2y + 3xy^2$

22. Simplify: (a) $-3a^2b + 5ab^2 + 3a^2b + ab^2$

    (b) $- 4x^2y - 5xy + 2x^2y + y^2$

Simplify:

23. $\dfrac{.75a^4bc}{2.5abc}$

Simplify:

24.   (a) $\dfrac{10a^7}{2a^2}$; (b) $\dfrac{12x^4y^6}{-2x^3y^2}$

Simplify:

25. (a) $\dfrac{-18a^5b^7}{-3a^2b^4}$;   (b) $\dfrac{16a^2bc}{14a^3bc^2}$

Simplify:

Bonus: (a) $\dfrac{-10x^5y^7}{20x^6y^6}$; (b) $\dfrac{16c^3d}{-12cd^2}$

Answers: **1**. (a) $11x - 10y$; (b) $-9b$; **2**. (a) $-10x + 6y + 4$; (b) $7b - 9$; **3**. (a) $8x^3 + 4x + 3$;
(b) $6x^3 + 5x + 2$; **4**. (a) -2; (b) 4; **5**. (a) $7x - 11y - 7$; (b) $-a + 7b - 28$; **6**. (a) $-2a^2 - 6ab + 4b^2$;
(b) $a - 7b + 24$; **7**.(a) $-20ab^6$; (b) $15x^2 - 8x - 12$; **8** (a) $5x^2y^3 - 6x^2y + xy^2$; (b) 0; **9**. (a) $6a - 2b$;
(b) $x^2 + 5x + 6$; **10**. (a) $9x - 13y + 8$; b) $-4x^2 + 20x$; **11**. (a) $x^2 + 2x - 24$; (b) $x^2 - y^2$;
12. (a) $a^2 + 2ac - b^2 + c^2$; (b) $6x^2 + 11x - 10$; **13**. (a) $16x^2 - 24x + 9$; (b) $a^2 + 2ab + b^2$;
14. (a) $10a^3 - 10a^2b - 2a^2c$; (b) $2x^2 - 6x - 28$; **15**. (a) $x^2 + 3x - 10$; (b) $4a^2 - ab - 3b^2$;
16. (a) $2bx^2$; (b) $28x^2$; **17** (a) $4x - 6y - 6$; (b) $15x - 25y - 8$; **18**. (a) $-27a - 4c + 12$; (b)
$23x + 4y + 24$; **19**. (a) $15x^3y^2 - 16xy^2$; (b) $ab + 6b$; **20**. (a) $2x - 4y$; (b) $\frac{3}{4}a^2b + 8$;
21. (a) $-2x - 4y - 2$; (b) $8x^2y - 6xy^2$; **22**. (a) $6ab^2$; (b) $-2x^2y - 5xy + y^2$;
23. $.3a^3$ or $\frac{3}{10}a^3$;**24**. (a) $5a^5$; (b) $-6xy^4$;**25**. (a) $6a^3b^3$; (b) $\dfrac{8}{7ac}$

; **Bonus**: (a) $-\dfrac{y}{2x}$; (b) $-\dfrac{4c^2}{3d}$.

Lesson 3: Review: Simplification of Algebraic Expressions

Review Test #2- Student's Self-Test (Always, Test yourself before you are tested)   2 3
Attempt all questions on clean sheets of paper, **Do not write in the book** Show all necessary work.

1. Simplify: $\dfrac{42x^3y^2 - 21x^2y}{7xy}$

2. Simplify: $\dfrac{30a^4b^2 - 20a^3b^2 - 5ab}{5ab}$

3. Divide:

(a) $\dfrac{24x^2y - 8x}{2x}$;   (b) $\dfrac{14x^2y - 8x}{-2x}$

Divide:

4. (a) $\dfrac{14c^5d^2 - 21cd}{-7cd}$;   (b) $\dfrac{20x^4yz - 5xy}{5xy}$

5. Divide $6x^2 + x^3 + 6 + 11x$ by $x + 3$

6. Simplify: $(x^4 + 4x^2 + 5x + 2) \div (x + 3)$

7. Divide $6x^2 - 8x + 5$ by $x - 3$

8. Divide $x^3 + 10x^2 + 33x + 36$ by $x + 4$

9. Divide $4x^4 - 2x^2 - 3$ by $x - 2$

10. $(4x^3y - 6x^2y^2 + 5y^4 + 6xy^3) \div (2x + y)$

11.
Divide $8x^4 + 9x^2 + 14x^3 + 19x + 12$ by $3 + 2x$

12.   Divide $x^3 + 2x^2 - 5x - 6$ by $x - 2$

Answers: 1. $6x^2y - 3x$; 2. $6a^3b - 4a^2b - 1$; 3. (a) $12xy - 4$; (b) $-7xy + 4$; 4. (a) $-2c^4d + 3$;

(b) $4x^3z - 1$; 5. $x^2 + 3x + 2$; 6. $x^3 - 3x^2 + 13x - 34 + \dfrac{104}{x + 3}$; 7. $6x + 10 + \dfrac{35}{x - 3}$; 8. $x^2 + 6x + 9$;

9. $4x^3 + 8x^2 + 14x + 28 + \dfrac{53}{x - 2}$; 10. $2x^2y - 4xy^2 + 5y^3$; 11. $4x^3 + x^2 + 3x + 5 - \dfrac{3}{2x + 3}$;

12. $x^2 + 4x + 3$.

CHAPTER 2

Lesson 4
REVIEW OF BASIC ELEMENTARY ALGEBRA 2
First Degree Equations Containing One Variable
(Simple Linear Equations)

Axioms for Solving Equations

Axioms are general mathematical statements that we accept as true, without any proof, in order to deduce other less obvious statements. The following are very useful in constructing proofs and solving equations.

1. A quantity is equal to itself. (reflexive property of equality, also identity principle)
   $$a = a$$
2. An equality may be reversed. (symmetric property of equality)
   If $a = b$, then
   $$b = a$$
3. Quantities equal to the same quantity are equal to each other.(transitive property of equality)
   If $a = b$, and $b = c$, then
   $$a = c.$$
4. A quantity may be substituted for its equal in any expression or equation. (substitution axiom)
5. A whole equals the sum of all its parts. (partition axiom)
6. If equal quantities are added to equal quantities, the sums are equal (addition axiom)

   If $a = b$,   (If $a = b$, and
   then $a + c = b + c$ **ALSO** $c = d$, then
                                                  $a + c = b + d$)

7. If equal quantities are subtracted from equal quantities, the differences are equal.(subtraction axiom)

   If $a = b$,   (If $a = b$, and
   then $a - c = b - c$ **ALSO** $c = d$, then
                                                  $a - c = b - d$)

8. If equal quantities are multiplied by equal quantities, the products are equal. (multiplication axiom)

   If $a = b$,   (If $a = b$, and
   then $ac = bc$ **ALSO** $c = d$, then
                                                  $ac = bd$.)

9. If equal quantities are divided by equal quantities (not zero), the quotients are equal. (division axiom)

   If $a = b$,   (If $a = b$, and
   then $\dfrac{a}{c} = \dfrac{b}{c}$ **ALSO** $c = d$, then

                                                  $\dfrac{a}{c} = \dfrac{b}{d}$)

10. Like powers of equals are equal . (powers axiom)
    Example: If $a = 3$, then
    $$a^2 = 3^2 \text{ or } a^2 = 9.$$
11. Like roots of equals are equal. (roots axiom)
    Example: if $a^3 = 8$, then
    $$\sqrt[3]{a^3} = \sqrt[3]{8}$$
    $$a = 2.$$

Solving First Degree Equations

An equation is a statement of the equality between two expressions.

Example: $2x + 3 = x - 10$

The equality symbol "=" breaks up an equation into two sides or members , namely the left-hand side and the right-hand side of the equation. To solve an equation involving a single variable, say x, means we are to find values of x which satisfy the equation. **A value of x is said to satisfy an equation if this value when substituted in the equation makes both the left-hand side and the right-hand side of the equation equal to each other.**

A value of x which satisfies a given equation is said to be a solution or a root of the given equation. **To obtain a value for the variable, x, we will get x by itself on one side of the equation .** We agree that a value of x has been obtained if we have x by itself on one side of the equation and all the other quantities on the other side of the equation do **not** involve x. To get x by itself, we will use inverse operations. Addition and subtraction are inverse operations. Multiplication and division are inverse operations. Inverse operations "undo" each other. For example, to undo multiplication by a number, we will use division by the same the number. We will keep the above discussion in mind when we solve linear equations.

Example 1   Solve for x

$$3(x + 2) = 5(x - 6)$$

First, remove the parentheses on both sides of the equation

$$3(x + 2) = 5(x - 6)$$
$$3x + 6 = 5x - 30$$
$$\underline{-5x \qquad -5x}$$
$$-2x + 6 = -30$$
$$\underline{-6 \quad -6}$$
$$-2x = -36$$
$$\frac{-2x}{-2} = \frac{-36}{-2}$$
$$x = 18$$

The solution is 18.

Example 2   Solve for x: $\dfrac{x}{6} = \dfrac{11}{12}$

Method 1	**Method 2**
To get x by itself on the left-hand side of the equation, we undo the "6" (which is a divisor) by multiplying each side of the equation by 6.	Multiply each term of the equation by the LCD of the fractions. The LCD is 12.

$$\text{Method 1: } (6)\frac{x}{6} = (6)\frac{11}{12}_2 \qquad\qquad \text{Method 2: } (12)\frac{x}{6} = (12)\frac{11}{12}$$

$$(\cancel{6})\frac{x}{\cancel{6}}\,^1 = (\cancel{6})\frac{11}{\cancel{12}}_2 \qquad\qquad 2\,(\cancel{12})\frac{x}{\cancel{6}} = (\cancel{12})\frac{11}{\cancel{12}}$$

$$x = \frac{11}{2} \qquad\qquad 2x = 11$$

$$\frac{2x}{2} = \frac{11}{2}$$

$$x = \frac{11}{2}$$

Method 3 Cross-multiplication

$$\frac{x}{6} = \frac{11}{12}$$
$$12x = 6(11)$$
$$\frac{\cancel{12}x}{\cancel{12}} = \frac{\cancel{6}(11)}{\cancel{12}\,2} \quad \text{and} \quad x = \frac{11}{2}$$

Note: We can cross-multiply if there is only a **single** fraction on the left-hand side of the equation, 26 and only a **single** fraction on the right-hand side of the equation. For instance, in the following example, we cannot cross-multiply immediately, but rather, we will have to combine the right-hand side into a single fraction before cross-multiplying.

Example 3   Solve for x: $\dfrac{x}{4} = \dfrac{x}{20} + \dfrac{2}{5}$

Method 1   The LCD method. Multiply each term of the equation by the LCD of the fractions. The LCD is 20.

$$(20)\frac{x}{4} = (20)\frac{x}{20} + (20)\frac{2}{5}$$

$$\overset{5}{\cancel{(20)}}\frac{x}{\cancel{4}} = \cancel{(20)}\frac{x}{\cancel{20}} + \overset{4}{\cancel{(20)}}\frac{2}{\cancel{5}}$$

$$5x = x + 8$$

$$\underline{-x \quad -x}$$

$$4x = 8$$

$$\frac{4x}{4} = \frac{8}{4}$$

$$x = 2$$

The solution is 2.

Method 2 By cross-multiplication

To apply cross-multiplication to the above problem, first, combine the two terms on the right-hand side into a single fraction. (We have a single fraction if we have a single denominator.)

Step 1   Combine the right sides of the equation into a single fraction.

$$\frac{x}{4} = \frac{x}{20} + \frac{2}{5}$$

$$\frac{x}{4} = \frac{x}{20} + \frac{2(4)}{5(4)}$$

$$\frac{x}{4} = \frac{x}{20} + \frac{8}{20}$$

$$\frac{x}{4} = \frac{(x+8)}{20}$$

Step 2: Now, we cross-multiply:

$20x = 4(x + 8)$ <---The parentheses are important

$20x = 4x + 32$

$$\underline{-4x \ - 4x}$$

$$16x = 0 + 32$$

$$\frac{16x}{16} = \frac{32}{16}$$

$$x = 2$$

Method 3   You may undo one denominator at a time.

$$\frac{x}{4} = \frac{x}{20} + \frac{2}{5}$$

Step 1: Undo the "4" : (Multiply each term by 4)

$$\overset{1}{\frac{\cancel{(4)}x}{\cancel{4}_1}} = \overset{1}{\frac{\cancel{(4)}x}{\underset{5}{\cancel{20}}}} + \frac{(4)2}{5}$$

$$x = \frac{x}{5} + \frac{8}{5}$$

Step 2:   To undo the "5" multiply by 5:

$$(5)x = \overset{1}{\frac{\cancel{(5)}x}{\cancel{5}_1}} + \frac{\cancel{(5)}8}{\cancel{5}_1}$$

$$5x = x + 8$$

$$4x = 8$$

$$x = 2$$

Note: In method 3, multiplying the equation first by 4 and then by 5 is equivalent to multiplying the equation by 20 (as was done in Method 1).

Checking Solutions of Equations (see also Lesson 13)

Example (a) Determine if - 5 is a solution of the equation
$$4x + 18 - x = 8x + 32 + 2x.$$
- 5 will be a solution of the given equation if this value when substituted in the given equation makes the left-hand side of the equation equal to the right-hand of the equation.

Checking: Replace x by -5 in the equation.

Then, $4(-5) + 18 - (-5) \overset{?}{=} 8(-5) + 32 + 2(-5)$

$-20 + 18 + 5 \overset{?}{=} - 40 + 32 - 10$

$-2 + 5 \overset{?}{=} - 8 - 10$

$3 = -18 \quad$ False

The left-hand side of the equation and the right-hand side of the equation are **not** equal. Therefore, -5 is **not** a solution of the given equation.

(b) Determine if -2 is a solution of $4x + 18 - x = 8x + 32 + 2x.$

Checking: Replace x by -2 in the equation.

Then, $4(-2) + 18 -(-2) \overset{?}{=} 8(-2) + 32 + 2(-2)$

$-8 + 18 + 2 \overset{?}{=} -16 + 32 - 4$

$10 + 2 \overset{?}{=} 16 - 4$

$12 = 12 \quad$ True

Since the left-hand side of the equation equals the right-hand side of the equation, -2 is a solution of the given equation.

Solving Decimal Equations

Discussion: There are a number of methods for solving decimal equations.

Method **1:** We can solve a decimal equation by strictly using decimals.

Method **2:** We can also change a decimal equation to an equivalent equation in which all the constants are integers, by multiplying every term of the equation by a power of 10. The power of 10 to use is based on the constant with the most number of decimal places or decimal digits.

Method **3:** We can make a partial conversion by multiplying the equation by a power of 10 so that only the numerical coefficients of the variables are necessarily changed to integers; some of the constant terms may consequently be changed partially or wholly, since we have to multiply every term of the equation by a power of 10. In multiplying by a power of 10, just move the decimal point accordingly..

Example 1 Solve for x:
$$.5x = 2$$
Solution: **Using Method 2** above, multiply both sides of the equation by 10.

$5x = 20$
$x = 4$

Example 2 Solve for x:
$$.02x - 6 = .1$$
Solution; Multiply every term by 100. (**Using Method 2**)

$2x - 600 = 10$
$2x = 610$
$x = 305$

Example 3 Solve for x:

$$.7x + .002 + .04 = .01x$$

Solution: **Method 3** , multiply every term 100.

$$70x + .2 + 4 = x$$
$$70x + 4.2 = x$$
$$69x = -4.2$$
$$x = \frac{-4.2}{69}$$
$$x = -.06 \text{ (to the nearest hundredth)}$$

Note that in Example 3, if we used **Method 2**. we would have

$$700x + 2 + 40 = 10x$$
$$700x + 42 = 10x$$
$$690x = -42$$
$$x = -\frac{42}{690} = x = -.06$$

Or if we used **Method 1** (strictly decimals) we would have

$$.7x - .01x = -.042$$

$$.69x = -.042 ----> x = \frac{-.042}{.69} = -.06$$

Solving Literal Equations

Here, note that the letters represent numbers and therefore the principles and techniques we have used, so far, in solving linear equations are applicable.

Example 1   Solve for x: $4x + 3y = 7$

We want to get x by itself on one side of the equation. Let us use some common sense here. We will undo (remove) the $3y$ and the 4 on the left-hand side of the equation.

$$4x + 3y = 7$$
$$\underline{\quad -3y \qquad -3y}$$
$$4x + 0 = 7 - 3y$$

$$\frac{4}{4}x = \frac{7 - 3y}{4}$$
$$x = \frac{7 - 3y}{4} = \frac{-3y + 7}{4} = \frac{-3y}{4} + \frac{7}{4} = -\frac{3}{4}y + \frac{7}{4}$$

If the question is of the multiple-choice type, then, be on the look out for any of these equivalent forms of answers.

Example 2   Solve Example 1 for y.
i.e., Solve $4x + 3y = 7$ for y.

Note in this case that, we want y by itself one side of the equation.

$$4x + 3y = 7$$
$$\underline{-4x \qquad\qquad -4x}$$
$$0 + 3y = 7 - 4x$$
$$\frac{3y}{3} = \frac{7 - 4x}{3}$$
$$y = \frac{-4x + 7}{3} = \frac{-4}{3}x + \frac{7}{3} = -\frac{4}{3}x + \frac{7}{3}$$

Example 3 Solve for A: $B = AC + CE$

We will obtain A by itself on one side of the equation.

$$B = AC + CE$$
$$B - CE = AC + CE - CE \quad \text{(subtract } CE\text{)}$$
$$B - CE = AC$$
$$\frac{B - CE}{C} = \frac{AC}{C} \qquad \text{(dividing both sides by } C\text{)}$$
$$\frac{B - CE}{C} = A$$
$$A = \frac{B - CE}{C} \, or \, \frac{B}{C} - \frac{CE}{C} = \frac{B}{C} - E$$

Review Test # 3- Student's Self-Test **(Always, Test yourself before you are tested)**   29
Attempt all questions on clean sheets of paper, **Do not write in the book** Show all necessary work.

Problems 1- 23: Solve the equations unless otherwise instructed:

1. (a) $x - 5 = 12$

   (b) $x + 4 = -9$

2. (a) $4x + 5 = 35$

   (b) $2x + 3 = 9$

3. (a) $5x = 20$

   (b) $-2x = 20$

4. (a) $8x = -32$

   (b) $-8x = 32$

5. (a) $7x - 8 = 3x + 24$

6. (a) $\frac{x}{6} + 3 = 7$

   (b) $8x = 22$

7. (a) $\frac{x}{6} = 8$ (b) $3x = 18$

8. (a) $5x - 7 + 3x + 9 = 6x + 5 + 8x$;

   (b) $\frac{x}{4} - 6 = 14$

9. (a) $2(x - 4) + 3 = 15$

   (b) $3(x + 5) + 1 = 5(x + 4) - 6$

10. (a) $3(2x - 5) + 4 = -4(x + 7)$

    (b) $5x - 4 + x = 4x + 6 - 3x$

11. $2x - 5 + 3x + 2 = 7x - 5 - 3x$

12. $\frac{-x}{3} + 5 = 7$

13.   Is 2 a solution of $5x + 2 = x - 4$?

14.   Is -3 a solution of $4x - 1 = 15$?

15.   Is 2 a solution of $3x + 1 = 7$?

16. (a) $x - \frac{1}{3} = \frac{1}{4}$

    (b) $\frac{x}{3} = \frac{x}{6} + \frac{3}{5}$

17. (a) $\frac{x}{-2} = 7$ (b) $\frac{-x}{4} = 9$

18. (a) $\frac{x}{6} + \frac{x}{3} = \frac{3}{4}$

    (b) $\frac{x}{8} = \frac{x}{2} - 6$

19.   (a) $\frac{x}{2}+1=\frac{x}{3}$

   (b) $\frac{x-2}{4}+\frac{2}{3}=\frac{1}{2}$

20. (a) $\frac{x+3}{4}+\frac{x-2}{8}=2$

   (b) $\frac{x}{4}+\frac{1}{2}=\frac{x}{5}+1$

21.   (a) $\frac{x-2}{3}+1=\frac{x}{5}$

   (b) $\frac{x}{2}+\frac{2}{3}=\frac{x}{4}$

22. (a) $2x+.6=1$

   (b) $5x+.04=.2$

23. (a) Solve for x $4x+.1=2$.
     (b) Solve for x: $2x-3y=12$

24.   (a) Solve for C: $F=\frac{9}{5}C+32$
     (b) Solve for y: $5x+4y=20$

25. (a) Solve for y: $3x+8y=32$

   (b) Solve for x: $2y-6x=21$

Bonus: (a) Solve for P $nRT=PV$

   (b) Solve for R: $nRT=PV$

   (c) Solve for b: $abc=x$

   (d) Solve for x: $5x+4y=20$

Answers: **1.** (a) $x=17$; (b) $x=-13$; **2.** (a) $x=\frac{15}{2}$; (b) $x=3$; **3.** (a) $x=4$; (b) $x=-10$;

4. (a) $x=-4$; (b) $x=-4$; **5.** $x=8$; **6.** (a) $x=24$; (b) $x=\frac{11}{4}$; **7.** (a) $x=48$; (b) $x=6$;

8. (a) $x=-\frac{1}{2}$; (b) $x=80$; **9.** (a) $x=10$; (b) $x=1$; **10.** (a) $x=-\frac{17}{10}$; (b; $x=2$; **11.** $x=-2$;

12. $x=-6$; **13.** No ; **14.** No; **15.** Yes ; **16.** (a) $x=\frac{7}{12}$; (b) $x=\frac{18}{5}$; **17.** (a) $x=-14$;

(b) $x=-36$; **18.** (a) $x=\frac{3}{2}$; (b) $x=16$; **19.** (a) $x=-6$; (b) $x=\frac{4}{3}$; **20.** (a) $x=4$; (b) $x=10$;

CHAPTER 3A

REVIEW OF BASIC ELEMENTARY ALGEBRA 3
Factoring Polynomials

Lesson 5: Greatest Common Factor and Common Monomial Factoring
Lesson 6: Difference between two Squares; Factoring by Grouping
Lesson 7: Factoring Monic Quadratic Trinomials (Coefficient of the x^2-term is 1)
Lesson 8 Non-monic Quadratic Trinomials (Coefficient of the x^2-term is **not** 1)
Lesson 9: Factoring trinomials quadratic in form (Substitution method)
Lesson 10: Solving Quadratic Equations by Factoring and Applications

You are responsible for three main types of factoring:

1. Common monomial factoring. Example: $2x + 8 = \mathbf{2(x+4)}$
2. Factoring quadratic trinomials. Example: $x^2 + 7x + 12 = \mathbf{(x + 3)(x + 4)}$
3. Factoring the difference between two squares. Example: $x^2 - y^2 = \mathbf{(x + y)(x - y)}.$

Factoring Completely:

Note: Step 1. In any factoring, perform common monomial factoring first, if this is possible. (That is, factor out the greatest common divisor.)

Step 2. Look out for the difference between squares or quadratic trinomials and factor accordingly.

Lesson 5

Greatest Common Factor and Common Monomial Factoring

The Greatest Common Factor (GCF)

Definition: The **greatest common factor** of two or more given numbers is the largest **common** divisor of the given numbers.

We will cover two methods which are similar to the methods used in finding the LCM of numbers.
Note that the greatest common factor is also the greatest common divisor (GCD) or the highest common factor (HCF).

Example   Find the greatest common factor (GCF) of 36 and 48.

Method 1
We will use a direct method which is similar to a prime factorization technique.. We will use **prime numbers** successively as divisors. At each step of the division process, each prime number used **must divide all** the dividends. The division process ends when there are no more common divisors to be used. The GCF is the product of the divisors used.

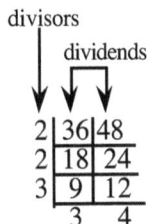

```
divisors
   │   dividends
   │   ↓   ↓
   ↓   ↓   ↓
  2 │ 36│48
  2 │ 18│ 24
  3 │  9 │ 12
      3    4
```

Answer: The GCF of 36 and 48 = $2 \times 2 \times 3$
                                      =12.

Method 2

Solution   Step 1 : Factor each number into primes, and write repeated factors in power form.

$$36 = 2 \times 2 \times 3 \times 3 = 2^2 \times 3^2$$
$$48 = 2 \times 2 \times 2 \times 2 \times 3 = 2^4 \times 3$$

Step 2 : Consider each prime factor which is common to both 36 and 48 (i.e., consider 2 and 3) and choose the lowest power of each prime factor.

Step 3: The GCF = the product of the **lowest powers** of the prime factors considered.

$$= 2^2 \times 3$$
$$= 4 \times 3$$

The GCF of 36 and 48 = 12

Note : 2×2 or 2^2 is common to both 36 and 48; and also 3 is common to both 36 and 48.

Also, if a prime factor is **not common** to all the given numbers, we do not "consider" it.

Common Monomial Factoring

Example   Factor completely: $9x^3 - 6x$

$$9x^3 - 6x$$

Step 1: $= 3x\left(\dfrac{9x^3}{3x} - \dfrac{6x}{3x}\right)$ <---You may skip Step 1. ($3x$ is the greatest common factor (divisor) of $9x^3$ and $-6x$)

Scrapwork:

Step 2: $= 3x(3x^2 - 2)$ $\left(\textbf{1.}\ \dfrac{9x^3}{3x} = 3x^2;\ \textbf{2.}\ \dfrac{-6x}{3x} = -2\right)$

Step 3:   We perform two checks (The checking may be done by mere inspection)

Check # 1:   Multiply the factors:
$$3x(3x^2 - 2)$$
$$= 9x^3 - 6x \text{ <----------- This must agree with the original expression.}$$

Check # 2: Inspect $3x^2 - 2$ (the expression within the parentheses to make sure the terms do not have anymore common factors (divisors).
This checking is usually done mentally (by inspection).

$\therefore\ 9x^3 - 6x = \boxed{3x(3x^2 - 2)}$ <----**answer**

Incomplete Factorization:

Example: $9x^3 - 6x$
$$= 3(3x^3 - 2x) \text{ <-----------This is incomplete factorization because}$$
$3x^3$ and $-2x$ still have common divisors (factors).

To complete the factorization:
$$= 3(3x^3 - 2x)$$
$$= 3x(3x^2 - 2)$$
and then , $9x^3 - 6x = 3x(3x^2 - 2)$ as obtained previously.

Lesson 5 Exercises

A.

Find the GCF of the following: **1.** 24 and 30 **2.** 56, 70 and 105 **3.** 12, 64 and 72.

Answers: **1.** 6 ; **2.** 7 ; **3.** 4

B.

Factor completely: **1.** $4x^4y^6 - 20xy^2$ **2.** $2a^5b^3c^7 + a^4b^2c^6 - a^2b^2c^5$

3. $4x^4y^2 - 2x^2y$; **4.** $6a^3bc - 2x^2y$; **5.** $7x^5y^2z - 21x^4y$; **6.** $8a^2b^2 - ab$; **7.** $4x^2 - 3xy$

Answers **1.** $4xy^2(x^3y^4 - 5)$; **2.** $a^2b^2c^5(2a^3bc^2 + a^2c - 1)$; **3.** $2x^2y(2x^2y - 1)$;
4. $2(3a^3bc - x^2y)$; **5.** $7x^4y(xyz - 3)$; **6.** $ab(8ab - 1)$; **7.** $x(4x - 3y)$

Lesson 6
Difference between two Squares; Factoring by Grouping

Factoring the Difference Between Two Squares

Examples **1.** $a^2 - b^2 = (a + b)(a - b)$;

   2. $c^2 - d^2 = (c + d)(c - d)$

   3. $16 - x^2 = (4 + x)(4 - x)$

   4. $9x^2 - 1 = (3x + 1)(3x - 1)$

   5. $6x^4 - 6 = 6(x^4 - 1) = 6(x^2 + 1)(x^2 - 1) = 6(x^2 + 1)(x + 1)(x - 1)$

   6. $a^4 - b^4 = (a^2 + b^2)(a^2 - b^2) = (a^2 + b^2)(a + b)(a - b)$

   7. $3x^2 - 75 = 3(x^2 - 25) = 3(x + 5)(x - 5)$

   8. $9y^2 - 16 = (3y + 4)(3y - 4)$

Factoring by Grouping

Sometimes, a seemingly non-factorable expression can be factored after a proper grouping of the terms of the expression. Look out for this type of factoring when there are four, six, or an even number of terms.

Example 1:   Factor by grouping: $ab + cd + ac + bd$

Step 1:   Pick the first term and then pick another term (of the expression)
            which has a common factor with the first term picked.
            $\underline{ab + ac}$ $+ \underline{cd + bd}$

Step 2: Perform common monomial factoring on the first two terms, and then on the
           other terms.
            $a(b + c) + d(c + b)$
            $= a(b + c) + d(b + c)$ (Note that $c + b = b + c$)

Step 3:   Perform "common binomial" factoring. (This is similar to common monomial factoring)
               Since $(b + c)$ is common , we pick it first.
               Then we obtain $(b + c)(a + d)$

   Note : Step 3 is a common monomial factoring if we consider $(b + c)$ as a single quantity.

Example 2: Factor by grouping (without adding any like terms).

$$6x^2 + 15x - 4x - 10$$

Step 1: Perform common monomial factoring on the first two terms and then on the
other terms.

$3x(2x + 5) - 2(2x + 5)$ <------- Adjust the signs here so that we obtain
$- 4x - 10x$ when the parentheses are removed.

Step 2: Pick the common factor $2x + 5$ and factor as done in common monomial factoring.

$(2x + 5)(3x - 2)$

You can check this factoring by multiplying the factors.
Note: We will obtain the same factors if we group the first and the third terms.
i.e. $6x^2 - 4x + 15x - 10$
$2x(3x - 2) + 5(3x - 2)$

Remove or factor out the common factor $(3x - 2)$, and then we obtain

$(3x - 2)(2x + 5)$
Again, we obtain the same binomial factors.

Lesson 6 Exercises

A
Factor completely: **1.** $y^2 - 16$;   **2.** $4x^2 - 36$;   **3.** $4x^2 - 9$;   **4.** $25 - 4y^2$

Answers: **1.** $(y + 4)(y - 4)$; **2.** $4(x + 3)(x - 3)$; **3.** $(2x + 3)(2x - 3)$; **4.** $(5 + 2y)(5 - 2y)$ or $- (2y + 5)(2y - 5)$

B.
Factor completely :   **1.** $x^2 + 2y + xy + 2x$;   **2.** $15x^2 - 9x + 20x - 12$
3. $ac - bc + ad - bd$ **4.** $ax + by + bx + ay$ **5.** $8x^2 + 12x + 2x + 3$

Answers: **1.** $(x + y)(x + 2)$;   **2.** $(3x + 4)(5x - 3)$; **3.** $(a - b)(c + d)$; **4.** $(a + b)(x + y)$; **5.** $(2x + 3)(4x + 1)$

Lesson 7
*Factoring Monic Quadratic Trinomials
(Coefficient of the x^2-term is 1)

Case 1: All the terms are positive

Factor completely:

Example 1: $x^2 + 7x + 12$

Step 1: $(x \quad)(x \quad)$

Step 2: $(x + 3)(x + 4)$

Scrapwork:
Find two numbers whose product is 12 and whose sum is 7
$12 = 6 \times 2$
$12 = \boxed{4 \times 3}$

Always check:

$(x + 3)(x + 4) = x^2 + 4x + 3x + 12$ (multiplying the factors)

$= x^2 + 7x + 12$ <------------------- This agrees with the original expression.

$\therefore x^2 + 7x + 12 = (x + 3) \ (x + 4)$

Case 2: The last term is positive and the middle term is negative

Example 2:   Factor completely: $x^2 - 5x + 6$

Step 1: $(x \quad)(x \quad)$

These two terms must have the same sign because this is positive.

Step 2: $(x - 2)(x - 3)$

Scrapwork : $6 \times 1 = 6$
$\boxed{2 \times 3} = 6$

Always check by multiplying.

$(x - 2)(x - 3)$
$= x^2 - 3x - 2x + 6$
$= x^2 - 5x + 6$

(We want two numbers whose
product is +6, but whose sum is -5)

$\therefore \ \ x^2 - 5x + 6 = (x - 2)(x - 3)$

*The coefficient of the x^2-term of the quadratic trinomial is also called the leading coefficient of the quadratic trinomial. Also, a quadratic trinomial in which the coefficient of the x^2-term is 1 is called a **monic** quadratic trinomial.

Case 3: The last term is negative and the middle term is negative or positive

Example 3 Factor completely:

$$x^2 - 7x - 18$$

Step 1: $(x \quad)(x \quad)$ ⟶ These two terms must have different signs because this term is negative

Step 2: $(x + 2)(x - 9)$

Scrapwork:

$9 \times 2 = 18$

$3 \times 6 = 18$

We want two numbers whose product is -18 and whose sum is -7
(Give the sign of the middle term to the larger of 2 and 9 (absolute values))

After checking, $x^2 - 7x - 18 = (x + 2)(x - 9)$.

Note that $x^2 + 7x - 18$ would be factored as $(x - 2)(x + 9)$

Example 4   Factor completely: $2x^2 - 12x + 16$

$$2x^2 - 12x + 16$$

Step 1: $2(x^2 - 6x + 8)$ (Factoring out the greatest common divisor)

Step 2: $2(x - 2)(x - 4)$ (Factoring the quadratic trinomial)

After checking, $2x^2 - 12x + 16 = 2(x - 2)(x - 4)$

Lesson 7 Exercises

A. Factor completely: **1.** $x^2 + 9x + 20$;   **2.** $x^2 + 16x + 28$;   **3.** $x^2 + 11x + 30$

Answers **1.** $(x + 4)(x + 5)$; **3.** $(x + 2)(x + 14)$;   **4.** $(x + 5)(x + 6)$

B. Factor completely: **1.** $x^2 - 9x + 18$;   **2.** $x^2 - 15x + 36$;   **3.** $x^2 - 13x + 40$.

Answers **1.** $(x - 3)(x - 6)$ **2.** $(x - 3)(x - 12)$ **3.** $(x - 5)(x - 8)$

C. Factor completely: **1.** $x^2 + 5x - 24$;   **2.** $x^2 - 5x - 24$;   **3.** $x^2 - 13x - 48$

Answers: **1.** $(x - 3)(x + 8)$;   **2.** $(x + 3)(x - 8)$;   **3.** $(x + 3)(x - 16)$

D. Factor completely: **1.** $3x^2 - 24x + 36$;   **2.** $2x^2 + 2x - 84$

Answers **1.** $3(x - 2)(x - 6)$;   **2.** $2(x - 6)(x + 7)$

E. Factor completely

1. $x^2 + 9x + 18$ 2. $x^2 + 8x + 15$ 3. $x^2 + 17x + 72$

4. $x^2 + 3x + 2$ 5. $x^2 + 14x + 33$ 6. $x^2 + 14x + 45$

7. $x^2 + 15x + 36$ 8. $x^2 - 4x - 21$ 9. $x^2 + 4x - 21$

10. $x^2 - 11x + 18$ 11. $x^2 - 10x + 21$ 12. $x^2 - 9x + 18$

13. $x^2 + 8x - 33$ 14. $4x^2 + 44x + 72$ 15. $2x^2 + 28x + 90$

Answers: 1. $(x + 3)(x + 6)$; **2.** $(x + 3)(x + 5)$; **3.** $(x + 8)(x + 9)$; **4.** $(x + 1)(x + 2)$; **5.** $(x + 3)(x + 11)$;
6. $(x + 5)(x + 9)$; **7.** $(x + 3)(x + 12)$; **8.** $(x + 3)(x - 7)$; **9.** $(x - 3)(x + 7)$; **10.** $(x - 2)(x - 9)$;
11. $(x - 3)(x - 7)$; **12.** $(x - 3)(x - 6)$; **13.** $(x - 3)(x + 11)$; **14.** $4(x + 2)(x + 9)$; **15.** $2(x + 5)(x + 9)$.

F. Factor completely:

1. $x^2 - 9x + 20$;   2. $x^2 + 3x - 18$ 3. $6x^3y - 14x^2y$;   4. $x^2 - 11x + 24$;

5. $2x^2 - 12x - 54$;   6. $a^2 + 3a - 28$;   7. $b^2 - 2b - 48$;   8. $x^2 - 49$;   9. $2y^2 - 32$;

10. $4x^2y^2 - 9b^2$;   11. $b^4 - a^4$;   12. $9y^2 - 36$;   13. $9 - x^2$ 14. $-x^2 - 7x - 12$.

Answers: 1. $(x - 5)(x - 4)$;   **2.** $(x + 6)(x - 3)$;   **3.** $2x^2y(3x - 7)$ **4.** $(x - 8)(x - 3)$;   **5.** $2(x + 3)(x - 9)$; **6.** $(a + 7)(a - 4)$;

7. $(b + 6)(b - 8)$;   **8.** $(x + 7)(x - 7)$;   **9.** $2(y + 4)(y - 4)$;   **10.** $(2xy + 3b)(2xy - 3b)$; **11.** $(b^2 + a^2)(b + a)(b - a)$

12. $9(y + 2)(y - 2)$; **13.** $(3 + x)(3 - x)$ or $-(x + 3)(x - 3)$;   **14.** $-(x + 3)(x + 4)$

Test # 8 – Student's Self-Test (**Always, Test yourself before you are tested**)

Attempt all questions on clean sheets of paper, **Do not write in the book** Show all necessary work.

---

1. Find the GCF of the following:

   (a). 32 and 36;   (b) 14, 70 and 182 ;

   (c). 12, 56 and 72.

2. Factor completely: (a) $x^2 + 11x + 18$;

   (b) $x^2 - 11x + 18$.

---

3. **Factor** completely: (a) $x^2 + 9x + 20$;

   (b) $x^2 + 13x + 42$.

4. Factor completely: (a) $x^2 + 12x + 20$;

   (b) $x^2 - 17x + 72$.

---

5. **Factor** completely: (a) $x^2 - 10x + 24$;

   (b) $2x^2 + 22x + 60$.

6. Factor completely: (a) $x^2 - 15x + 36$;

   (b) $4x^2 + 44x + 72$.

---

7. Factor completely: (a) $x^2 - 12x + 32$;
   (b) $x^2 + 4x - 77$.

8. Factor completely: (a) $x^2 + 5x - 24$;
   (b) $x^2 - 9x + 18$.

---

9. Factor completely: (a) $x^2 - 5x - 24$;

   (b) $x^2 - 10x + 9$.

10. **Factor** completely: (a) $x^2 - 13x - 48$;

    (b) $x^2 + 4x - 21$.

---

11. Factor completely: (a) $10x^2 + 4x - 6$;

    (b) $x^2 + 14x + 33$

12. Factor completely: (a) $2x^2 - 4x - 96$;

    (b) $x^2 + 14x + 45$.

---

13. Factor completely: (a) $x^2 - 5x - 36$;

    (b) $x^2 + 3x + 2$.

14. Factor completely: (a) $x^2 + 9x + 18$;

    (b) $x^2 + 8x + 15$.

---

15. Factor completely: (a) $4x^2 - 16$;

    (b) $4 - 9x^2$.

16. **Factor** completely: (a) $y^2 - 16$;

    (b) $25 - 4y^2$.

17. Factor completely: (a) $9y^2 - 36$;

(b) $-x^2 - 7x - 12$.

18. Factor completely: (a) $4x^2y^2 - 9b^2$;

(b) $b^4 - a^4$.

19. Factor completely: (a) $x^2 - 49$;

(b) $2y^2 - 32$.

20. Factor completely: (a) $a^2 + 3a - 28$;

(b) $b^2 - 2b - 48$.

21. Factor completely: (a) $x^2 - 11x + 24$;

(b) $2x^2 - 12x - 54$.

22. Factor completely: (a) $-x^2 - 3x + 18$

(b) $x^2 + 3xy + 2y^2$

23. Factor completely: (a) $x^2 - 17x + 72$;

(b) $2x^2 + x - 15$

Factor completely:

24. **(a)** $ad - bc - ac + bd$

(b) $ay + bx + by + ax$

25. Factor completely: $a^2 + 2x + ax + 2a$

Bonus: Write down your first and last names, and place a plus sign between these names. Assuming that each letter in your name represents a real number, factor completely, if possible, the expression obtained.

Answers: 1 (a) 4; (b) 14; (c) 4; **2.** (a) $(x+2)(x+9)$; (b) $(x-2)(x-9)$; **3.** (a) $(x+4)(x+5)$;
(b) $(x+6)(x+7)$; **4.** (a) $(x+2)(x+10)$; (b) $(x-8)(x-9)$; **5.** (a) $(x-6)(x-4)$;
(b) $2(x+5)(x+6)$; **6.** (a) $(x-3)(x-12)$; (b) $4(x+2)(x+9)$; **7.** (a) $(x-4)(x-8)$;
(b) $(x-7)(x+11)$; **8.** (a) $(x-3)(x+8)$; (b) $(x-3)(x-6)$ **9.** (a) $(x-8)(x+3)$; (b) $(x-1)(x-9)$;
10. (a) $(x+3)(x-16)$; (b) $(x+7)(x-3)$; **11.** (a) $2(5x-3)(x+1)$; (b) $(x+3)(x+11)$;
12. (a) $2(x-8)(x+6)$; (b) $(x+9)(x+5)$; **13.** (a) $(x-9)(x+4)$; (b) $(x+1)(x+2)$;
14. (a) $(x+3)(x+6)$; (b) $(x+3)(x+5)$; **15.** (a) $4(x+2)(x-2)$; (b) $(2+3x)(2-3x)$;
16. (a) $(y+4)(y-4)$; (b) $(5+2y)(5-2y)$; **17.** (a) $9(y+2)(y-2)$; (b) $-(x+3)(x+4)$;
18. (a) $(2xy+3b)(2xy-3b)$; (b) $(b^2+a^2)(b+a)(b-a)$; **19.** (a) $(x+7)(x-7)$;
(b) $2(y+4)(y-4)$; **20.** (a) $(a-4)(a+7)$; (b) $(b+6)(b-8)$; **21.** (a) $(x-3)(x-8)$;
(b) $2(x+3)(x-9)$; **22.** (a) $-(x-3)(x+6)$; (b) $(x+y)(x+2y)$; **23.** (a) $(x-8)(x-9)$;
(b) $(2x-5)(x+3)$; **24** (a) $(d-c)(a+b)$; (b) $(a+b)(x+y)$; **25.** $(a+x)(a+2)$.
Bonus: Answer varies.

Lesson 8
Factoring Non-monic Quadratic Trinomials
(Coefficient of the x^2-term is **not** 1)

Example   Factor completely: $6x^2 + 11x - 10$

Solution

There are a number of methods for factoring non-monic (the coefficient of the x^2-term is not 1) quadratic trinomials, namely

1. A general method (Method 1)
2. A Substitution Method (Method 2) and
3. A third method (Method 3, the **ac**-method), which involves factoring by grouping.

Method 1   The General Method

This method involves considering the products of the various possible binomial factors to determine which factors yield the original quadratic trinomial.

We will now factor $6x^2 + 11x - 10$
We consider the possible binomials
(a) $(3x + 2)(2x - 5)$
(b) $(3x - 2)(2x + 5)$
(c) $(3x + 5)(2x - 2)$
(d) $(3x - 5)(2x + 2)$
(e) $(6x + 5)(x - 2)$
(f) $(6x - 5)(x + 2)$
(g) $(6x + 2)(x - 5)$
(h) $(6x - 2)(x + 5)$

Multiply the above factors. The pair of binomial factors which yield the original quadratic expression are the correct factors. **Note** that since the original quadratic trinomial did not have any common factors, each possible binomial factor will not have any common factors; and therefore, we could have eliminated the factors in (c),(d) and (h) without having to multiply them out, and then check only the products for cases (a), (b), (e) and (f).

In the above example, the correct factors are $(3x - 2)$ and $(2x + 5)$.

$$\therefore \quad 6x^2 + 11x - 10 = (3x - 2)(2x + 5)$$

Factorability of a quadratic trinomial: $ax^2 + bx + c$

A quadratic trinomial is factorable if the discriminant $b^2 - 4ac$ is a perfect square.

Example Is $6x^2 + 11x - 10$ factorable?

      Check: $a = 6, b = 11, c = -10$

$$b^2 - 4ac = 11^2 - 4(6)(-10)$$
$$= 121 + 240$$
$$= 361$$

      361 is a perfect square because $\sqrt{361} = 19$, a rational number.

Answer: Yes. $6x^2 + 11x - 10$ is factorable.

Method 2: The Substitution Method

In this method, we change the quadratic trinomial to a monic trinomial which is then easily factored.

Example   Factor $6x^2 + 11x - 10$

Step 1: Multiply the expression by the coefficient of the x^2-term.

$6(6x^2) + 6(11x) - 6(10)$

Step 2: Write the first term as a square and interchange the 6 and the 11 in the second term.

$36x^2 + 11(6x) - 60$

$(6x)^2 + 11(6x) - 60$ ....(A)

Step 3: Let $6x = s$ (That is, replace $6x$ by s in expression (A))

Then, we obtain $s^2 + 11s - 60$

Now, since this is a monic trinomial (coefficient of s^2 is 1), factor easily.

$(s - 4)(s + 15)$ ....(B)

Scrapwork
$60 \times 1 = 60$
$30 \times 2 = 60$
$\boxed{15 \times 4 = 60}$

Step 4:   Replace s by $6x$,and then, expression (B) becomes

$(6x - 4)(6x + 15)$..............................(C)

Since we multiplied the original trinomial by 6, we must divide expression (C) by 6 (that is we must undo the "6" we introduced in Step 1.)

In order to divide (C) by 6, we perform common monomial factoring on the two binomial factors (in some cases, this factoring is performed only on one of the binomial factors).

$(6x - 4)(6x + 15)$
$2(3x - 2)\ \ 3(2x + 5)$
$2(3)(3x - 2)(2x + 5)$
$6(3x - 2)(2x + 5)$

Now, we divide by 6: $\dfrac{6(3x - 2)(2x + 5)}{6}$

and then the complete factorization of

$6x^2 + 11x - 10$ is $(3x - 2)(2x + 5)$

Method 3 Factoring by the ac-Method

The "a" is the coefficient of the x^2-term, and the "c" is the constant term of the general quadratic trinomial $ax^2 + bx + c$.

Example   Factor completely. $6x^2 + 11x - 10$
   $a = 6, \ c = -10.$

Step 1: Multiply the coefficient of the x^2-term by the last term (the constant term).

$$6x^2 + 11x - 10 \ \dots\dots\dots\dots\dots\dots\dots\dots\dots(A)$$

$$6(-10) = -60$$

Step 2: Replace the $+11x$ (the middle-term) of expression (A) by two terms whose coefficients add up to 11 and whose product is -60.

Then, we obtain

$$6x^2 + 15x - 4x - 10$$

Scrapwork

$60 \times 1 = 60$
$30 \times 2 = 60$
$\boxed{15 \times (-4) = -60}$
$20 \times 3 = 60$

Step 3: Factor the expression by grouping. (Break up the expression into two pairs and factor)
   $6x^2 + 15x - 4x - 10$
   $3x(2x + 5) - 2(2x + 5)$

   $(2x + 5)(3x - 2)$
Therefore, $6x^2 + 11x - 10 = (2x + 5)(3x - 2)$

Factoring by any method

Example   Factor by any method

   $3x^2 - 7x - 10$
We will use Method 2 and Method 3.

Method 2
Step 1: Multiply every term by 3 (the coefficient of the x^2-term)

   $3(3x^2) - 7x(3) - 30$
   $9x^2 - 7(3x) - 30$
   $(3x)^2 - 7(3x) - 30 \ \dots\dots\dots\dots\dots\dots\dots\dots(A)$

Step 2: Let $3x = s$ in expression (A) and factor the resulting monic trinomial.
   Then $s^2 - 7s - 30$
   $(s + 3)(s - 10)$

Step 3:   Replace s by $3x$.
   $(3x + 3)(3x - 10)$

Step 4:   Factor the first binomial and divide by 3

   $$\frac{3(x + 1)(3x - 10)}{3}$$ This "3" was used as a multiplier in Step 1

$\therefore 3x^2 - 7x - 10 = (x + 1)(3x - 10)$

Method 3   Factor $3x^2 - 7x - 10$.

Step 1: Multiply the coefficient of the x^2-term by the last term.

$$3x^2 - 7x - 10$$

$$3(-10) = -30$$

Step 2: Replace $-7x$ by two terms whose coefficients add up to -7 and whose product is -30.
$$3x^2 - 10x + 3x - 10$$

Step 3:   Factor by grouping
$$3x^2 - 10x + 3x - 10$$
$$= x(3x - 10) + 1(3x - 10)$$
$$= (3x - 10)(x + 1)$$
$$\therefore\ 3x^2 - 7x - 10 = (x + 1)(3x - 10)$$

Lesson 8 Exercises

A. Factor completely: **1.** $12x^2 - x - 6$;   **2.** $18x^2 - 3x - 10$;   **3.** $10x^2 - 29x + 21$

Answers: **1.** $(3x + 2)(4x - 3)$;   **2.** $(6x - 5)(3x + 2)$;   **3.** $(2x - 3)(5x - 7)$

B. Factor completely: **1.** $12x^2 - x - 6$;   **2.** $18x^2 - 3x - 10$;   **3.** $10x^2 - 29x + 21$

Answers: **1.** $(3x + 2)(4x - 3)$;   **2.** $(6x - 5)(3x + 2)$;   **3.** $(2x - 3)(5x - 7)$

C Factor completely:

1. $8x^2 - 2x - 15$ **2.** $4x^2 - 12x + 9$ **3.** $6x^2 - 19x + 10$

4. $15x^2 - 14x - 8$ **5.** $12x^2 + x - 6$

Answers: **1** . $(2x - 3)(4x + 5)$ **2.** $(2x - 3)(2x - 3)$ or $(2x - 3)^2$ **3.** $(2x - 5)(3x - 2)$
4. $(3x - 4)(5x + 2)$ **5** . $(3x - 2)(4x + 3)$

Lesson 9

Factoring trinomials quadratic in form (Substitution method)

Example (a) Factor completely: $x^4 + 7x^2 + 12$ ....(1)

Solution Step 1: Let $x^2=u$. Then, expression (1) becomes $u^2 + 7u + 12$................(2)

Step 2 : Factoring (2), we obtain $(u + 3)(u + 4)$................................(3)

Step 3: Replacing u by x^2 in expression (3), we obtain $(x^2 + 3)(x^2 + 4)$
$\therefore x^4 + 7x^2 + 12 = (x^2 + 3)(x^2 + 4)$.

Example (b) Factor completely: $2a^4 - 5a^2 - 12$ ....(1)

Solution Let $a^2 = u$. Then, we obtain $2u^2 - 5u - 12$ ....(2)

Factoring (2), we obtain $(2u + 3)(u - 4)$ .... (3)

Replacing u by a^2 in expression (3), we obtain $(2a^2 + 3)(a^2 - 4) = (2a^2 + 3)(a + 2)(a - 2)$

$\therefore 2a^4 - 5a^2 - 12 = (2a^2 + 3)(a + 2)(a - 2)$.

Lesson 9 Exercises

Factor completely: **1.** $x^4 - 7x^2 + 12$;   **2.** $2a^4 + 5a^2 - 12$

Answers. **1.** $(x^2 - 3)(x + 2)(x - 2)$;   **2.** $(2a^2 - 3)(a^2 + 4)$

EXTRA
Factoring Perfect Square Trinomials. See Appendix A. p.303

Lesson 10
Solving Quadratic Equations by Factoring and Applications

Standard form of the quadratic equation: $ax^2 + bx + c = 0$, where a, b, and c are constants and $a \neq 0$

Examples 1. $x^2 - 3x - 28 = 0$ <---------in standard form
         2. $x^2 - 2x = 0$ <--------- in standard form
         3. $x^2 = 5x + 14$ <---------**not** in standard form
         4. $5x^2 = 8x$ <---------**not** in standard form

Principle of zero products: If $ab = 0$ then either $a = 0$, or $b = 0$ (or both $= 0$).

To solve by factoring, the quadratic equation must be in standard form before factoring

Example 1   Solve by factoring
$$x^2 - 3x - 28 = 0$$

Step 1:   Factor the quadratic trinomial.
$$(x + 4)(x - 7) = 0$$
Step 2:   Set each factor equal to zero and solve for x.

      $x + 4 = 0$ $x - 7 = 0$
        $\underline{-4 \ \ -4}$ $\underline{+ 7 \ \ +7}$
         $x = -4$ $x = 7$
                $\therefore \ \ x = -4, x = 7$

The solutions are -4 and 7.

Example 2   Solve for x by factoring.
$$3x^2 - 4x = 0$$

Step 1: Factor. $3x^2 - 4x = 0$

           $x(3x - 4) = 0$ (performing common monomial factoring)

Step 2:   Set each factor equal to zero and solve for x.   (We set only factors having a **variable** to zero)

       $x = 0$ or $3x - 4 = 0$
                         $\underline{+ 4 \ \ +4}$
                        $\frac{3}{3}x = \frac{4}{3}$
                         $x = \frac{4}{3}$
            $\therefore x = 0, \ \ x = \frac{4}{3}$ **Note** that 0 is also a solution .

The solutions are 0 and $\frac{4}{3}$.

Note: A common **wrong** approach in the above problem:
     $3x^2 - 4x = 0$
        $3x^2 = 4x$
   $3x = 4$ (Canceling, that is dividing out an x on both sides of the equation ; such a cancellation excludes
          $x = 0$ as a solution.)
    $x = \frac{4}{3}$ (Of course, you still obtain one of the solutions but lose the other solution).

Example 3   Solve by factoring: $9x^2 = 36$

Solution $9x^2 = 36$
Step 1: $9x^2 - 36 = 0$

Step 2:   Factor the left-hand side completely.
$$9(x^2 - 4) = 0$$
$$9(x + 2)(x - 2) = 0$$
Step 3:   Set each factor equal zero and solve for x: (Note that $9 \neq 0$; only factors with a variable $= 0$)

$$x + 2 = 0 \quad \text{or} \ x - 2 = 0$$
$$\underline{\quad -2 \ -2\quad} \qquad \underline{\ +2 \ +2\ }$$
$$x = -2 \qquad\qquad x = 2$$

The solutions are 2 and -2.

Another method (The square root method)

$$9x^2 = 36$$
$$\frac{9x^2}{9} = \frac{36}{9}$$
$$x^2 = 4$$
$$x = \pm\sqrt{4}$$
$$x = \pm 2 \ (+2 \text{ or } -2)$$
Again, the solutions are 2 and -2.

Applications of the quadratic equation

Example   The width of a rectangle is 3 units less than the length. If the area of the rectangle is 28 sq. units, determine the dimensions of this rectangle.

Solution
Let the length of the rectangle $= x$.
Then, the width of this rectangle $= (x - 3)$
Area of a rectangle is given by LW (Where L is the length and W is the width).
The required equation: $x(x - 3) = 28$.
$$x^2 - 3x - 28 = 0$$
$$(x - 7)(x + 4) = 0$$
$$x = 7 \ \text{or} \ x = -4 \qquad\qquad \text{(Solving by factoring)}$$
We reject the negative value, -4, since the dimension of a rectangle cannot be negative.
When the length is 7 units, the width is 7 - 3 = 4 units.

(**Check**: If the length is 7 and the width is 4, the area is 7(4) = 28 sq. units; and also the width, 4, is 3 less than the length, 7. Therefore, the conditions in the original word problem have been satisfied.)
The length is 7 units and the width is 4 units.

Lesson 10 Exercises

A. Solve for x by factoring: **1.** $x^2 - 3x - 10 = 0$;   **2.** $2x^2 - 6x = 0$ **3.** $10x^2 = 15x$

Answers: **1.** - 2 and 5;   **2.** 0 and 3 ;   **3.** 0 and $\frac{3}{2}$

B. 1. Solve for x: $2x^2 = 32$;   **2.** Solve for x: $3x^2 - 75 = 0$.

3. The sum of two numbers is 20, and their product is 96. Find these numbers.

4. The length of a rectangle is 5 units more than the width. If the area of this rectangle is 126 sq, units, find the length and width of this rectangle.

Ans: **1.** 4 and -4;   **2.** 5 and -5;   **3.** 8 and 12;   **4.** Length = 14, width = 9.

Test # 9 -Student's Self-Test (Always, Test yourself before you are tested)

Attempt all questions on clean sheets of paper, **Do not write in the book** Show all necessary work.

---

1 Factor completely: $6x^2 - 13x + 6$.

2. Factor completely: $10x^2 - 21x + 2$.

---

3. Factor completely: $3x^2 + 13x + 4$.

4. Factor completely: $10x^2 - 13x$.

---

5. Factor completely: $2x^2 + 20x + 18$.

6. Factor completely: $4x^2 + 44x + 72$.

---

7. Factor completely: $8x^2 - 2x - 1$.

8. Factor completely: $8x^2 - 26x + 15$.

---

9. Factor completely: $18x^2 - 3x - 10$.

10. Factor completely: $15x^2 - 14x - 8$.

---

11. Factor completely: $3x^2 - 24x + 36$.

12. Factor completely: $2x^2 + 2x - 84$.

---

13. Factor completely: $7x^2 - 23x + 6$.

14. Solve for x by factoring: $4x^2 - 9 = 0$.

---

15. Solve for x by factoring: $4x^2 - 36 = 0$.

16. Factor completely: $24x^2 + 2x - 12$.

17. Solve for x by factoring: $-x^2 - 7x - 12 = 0$

18. Factor completely: $6x^2 - 19x + 10$.

19. Solve for x by factoring: $3x^2 - 6x = 0$.

20. Factor completely: $10x^2 - 29x + 21$.

21. Solve for x by factoring: $4x^2 = 36$.

22. Solve for x by factoring: $3x^2 - 75 = 0$.

23. Solve for x by factoring:
$x^2 - 3x - 10 = 0$.

24. The sum of two numbers is 24, and their product is 108. Find these numbers.

25. The length of a rectangle is 6 units more than the width. If the area of this rectangle is 112 sq. units, find the length and width of this rectangle.

Bonus:

$a^4 - 7a^2 + 12$;

$2y^4 + 5y^2 - 12$.

Answers: 1. $(2x - 3)(3x - 2)$; **2.** $(x - 2)(10x - 1)$; **3.** $(x + 4)(3x + 1)$; **4.** $x(10x - 13)$;
5. $2(x + 1)(x + 9)$; **6.** $4(x + 2)(x + 9)$; **7.** $(2x - 1)(4x + 1)$; **8.** $(2x - 5)(4x - 3)$;
9. $(6x - 5)(3x + 2)$; **10.** $(3x - 4)(5x + 2)$; **11.** $3(x - 6)(x - 2)$; **12.** $2(x - 6)(x + 7)$;
13. $(x - 3)(7x - 2)$; **14.** $x = \pm\frac{3}{2}$; **15.** $x = \pm 3$; **16.** $2(3x - 2)(4x + 3)$; **17.** $x = -3$, $x = -4$;
18. $(2x - 5)(3x - 2)$; **19.** $x = 0$, $x = 2$; **20.** $(5x - 7)(2x - 3)$; **21.** $x = \pm 3$; **22** $x = \pm 5$;
23. $x = 5$, $x = -2$; **24.** The numbers are 6 and 18;
25. The length is 14 units and the width is 8 units.
Bonus: (a) $(a^2 - 3)(a + 2)(a - 2)$; (b) $(2y^2 - 3)(y^2 + 4)$.

Lesson 11: Factoring the Sum and Difference between two Cubes.

50

CHAPTER 3B
More Factoring Polynomials
Lesson 11

Factoring the Sum and Difference between two Cubes;

Factoring the Sum of Two Cubes

$$\boxed{a^3 + b^3 = (a + b)(a^2 - ab + b^2)}\ \ldots\ldots\ldots\ldots\ldots\ldots(1)$$

Similarly, $x^3 + y^3 = (x + y)(x^2 - xy + y^2)$

Example Factor completely: $8x^3 + y^3$

$$\begin{aligned}
&\ 8x^3 + y^3 \\
&= (2x)^3 + y^3 \\
&= (2x + y)((2x)^2 - 2x(y) + y^2) \qquad \text{(Replace } a \text{ by } 2x;\ b \text{ by } y \text{ in the factoring formula (1) above.} \\
&= (2x + y)(4x^2 - 2xy + y^2)
\end{aligned}$$

Factoring the Difference between (of) Two Cubes

$$\boxed{a^3 - b^3 = (a - b)(a^2 + ab + b^2)}\ \ldots\ldots\ldots\ldots(2)$$

Similarly, $x^3 - y^3 = (x - y)(x^2 + xy + y^2)$

Example Factor completely: $x^3 - 8y^3$

$$\begin{aligned}
&\ x^3 - 8y^3 \\
&= x^3 - (2y)^3 \\
&= (x - 2y)(x^2 + x(2y) + (2y)^2) \qquad \text{(Replace } a \text{ by } x;\ \text{and } b \text{ by } 2y \text{ in the factoring formula (2) above)} \\
&= (x - 2y)(x^2 + 2xy + 4y^2)
\end{aligned}$$

It is good practice to check the factoring by multiplying the factors.
In the future, imitate the substitution (replacement) technique when factoring the sum of or the difference between two cubes.

Mnemonic device:
Find a way to remember and distinguish between the basic factoring formulas as in (1) and (2) above: You may observe that in the factors on the right-hand side, there is only one minus, "-", involved; and the other signs are all "+" signs.

$a^3 + b^3 = \boxed{(a + b)(a^2 - ab + b^2)}$. <-- the minus sign precedes the ab-term. Also there is only one minus sign.

$a^3 - b^3 = \boxed{(a - b)(a^2 + ab + b^2)}$ <--The minus sign precedes the b-term . Also there is only one minus sign

To make sure you have the correct factorization, check by multiplying the binomial and the trinomial.

Lesson 11 Exercises

A Factor completely: $y^3 + 8$; **2.** $x^3 + 27$; **3.** $x^3 - 27$; **4.** $8x^3 - 27$; **5.** $3x^3 - 81$

Answers: **1.** $(y + 2)(y^2 - 2y + 4)$; **2.** $(x + 3)(x^2 - 3x + 9)$; **3.** $(x - 3)(x^2 + 3x + 9)$; **4.** $(2x - 3)(4x^2 + 6x + 9)$
  5. $3(x - 3)(x^2 + 3x + 9)$

Introductory Theme for Chapters 4A & 4B
(Next two chapters)
Straight Line
Theme: Two points

1. Why two points? Two points, because given or knowing two points, a straight line can be drawn by connecting the two points, using a straight edge and pencil.

2. Why two points? Two points, because given (knowing) two points; the slope, m, of the line segment connecting the two points $P_1(x_1, y_1)$ and $P_2(x_2, y_2)$ can be found by applying $m = \dfrac{y_2 - y_1}{x_2 - x_1}$

3a. Why two points? Two points, because if we know the **slope, *m*,** and the **y-intercept, *b*,** of the line, we can obtain two points and draw the graph of the line

3b Note: y–intercept, b implies the point $(0, b)$, By choosing a point (x, y) on a line, we have two points, and the slope (as well as an equation) of the line connecting the two pints $(0, b)$ and (x, y)

is given by $m = \dfrac{y - b}{x - 0}$ <--**slope = slope**

$mx = y - b$ or $\boxed{y = mx + b}$ <------**slope-intercept form** of the equation of a line

4. Why two points? Two points, because given or knowing two points an equation of the line segment connecting the two points $P_1(x_1, y_1)$ and $P_2(x_2, y_2)$ can be found by

applying $\boxed{y - y_1 = \left(\dfrac{y_2 - y_1}{x_2 - x_1}\right)(x - x_1)}$ (from $\dfrac{y - y_1}{x - x_1} = \dfrac{y_2 - y_1}{x_2 - x_1}$ <-- **is slope = slope**

or $\boxed{y - y_1 = m(x - x_1)}$ <------**point-slope form,** where $m = \dfrac{y_2 - y_1}{x_2 - x_1}$

5. Why two points? Two points, because given or knowing the two-intercept points $(a, 0)$, $(0, b)$ an equation of the line segment connecting the two points $P_1(a, 0)$ and $P_2(0, b)$ can be found by

applying $y - y_1 = \left(\dfrac{y_2 - y_1}{x_2 - x_1}\right)(x - x_1)$ to obtain $y - 0 = \dfrac{b - 0}{0 - a}(x - a)$; from $\dfrac{y - 0}{x - a} = \dfrac{b - 0}{0 - a}$)

or $y = \dfrac{b}{-a}(x - a)$ or $y = \dfrac{b}{-a}x + \left(\dfrac{b}{-a}\right)(-a)$ or $\boxed{y = -\dfrac{b}{a}x + b}$ also $\dfrac{x}{a} + \dfrac{y}{b} = 1$ (Two intercept form)

6. Why two points? Two points, because given the graph (picture) of a line , we are given infinitely many points from which we can read the coordinates of any two points on the line and write an equation of a line by applying **3, 4** or **5** above. A picture is worth a thousand words

7. Why two points? Two points, because given or knowing two points; the **midpoint** of the line segment connecting the points $P_1(x_1, y_1)$ and $P_2(x_2, y_2)$ is given by x-coordinate, $x_m = \dfrac{x_1 + x_2}{2}$,

and the y-coordinate, $y_m = \dfrac{y_1 + y_2}{2}$

8. Why two points? Two points, because given or knowing) two points; $P_1(x_1, y_1)$, $P_2(x_2, y_2)$, the distance, d, between the two points on a line **in a plane** is given by

$d = \sqrt{(x_2 - x_1)^2 + (y_2 - y_1)^2}$

9. Why two points? Two points, because given or knowing) two points, $P_1(x_1, y_1)$, $P_2(x_2, y_2)$ on each of two lines, the slopes m_1, m_2 as well as parallelism or perpendicularity can be determined.
The above theme summarizes next two chapters.

CHAPTER 4A

Lesson 12A: **Rectangular Coordinate System**
Lesson 12B: **Drawing the Graph of a Straight Line**
Lesson 13: **Points on a Line; Slopes of lines; Intercepts**
Lesson 14: **Equations of Straight Lines**

Lesson 12A

Graphing in the Rectangular Coordinate System of Axes

We begin by means of an example. Let us consider the equation $y = 3x + 2$ (1)
If we let $x = 1$ in equation (1) then $y = 3(1) + 2$
$$= 5$$
Thus, when $x = 1$, $y = 5$
If we let $x = 2$ in equation (1) m then $y = 3(2) + 2$
$$= 8$$
Similarly, when $x = 2$, $y = 8$.

From the above example, we can say that for each value of x we choose, there is a corresponding value of y. By agreement, we can write the $x-$ and $y-$values values as an ordered pair (x, y), where x is the first component and y is the second component.

Then for the solutions to $y = 3x + 2$, when $x = 1$, $y = 5$, we can represent the solution as the ordered pair $(1, 5)$.

For $x = 2$, $y = 8$, we can write the ordered pair $(2, 8)$

Similarly, when $x = 4$, $y = 14$, giving us the ordered pair $(4, 14)$.

The equation $y = 3x + 2$ has infinitely many solutions. The solution set consists of all ordered pairs.

If an ordered pair is a solution of a given equation, then these numbers when substituted in the equation for the unknowns, should make both sides of the equation equal. We should note that we could have chosen values for y and then calculate the corresponding values of x. In fact, in some instances, it may be more convenient to choose y and calculate x

Rectangular Coordinate System of Axes (Cartesian Coordinate System)

Consider a horizontal number line (scale) labeled as in Figure. On this line, all numbers to the right of the origin are positive, and all numbers to the left of the origin are negative.

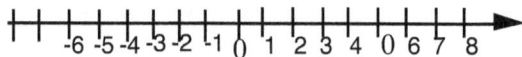

Figure 1: Horizontal real number line

Consider also a vertical line **(Figure 2)**. On this line, all numbers above the origin are positive, and all numbers below the origin are negative.

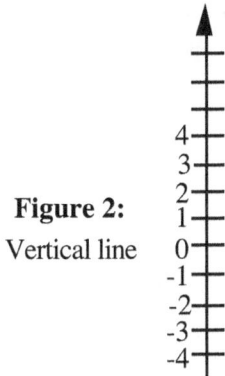

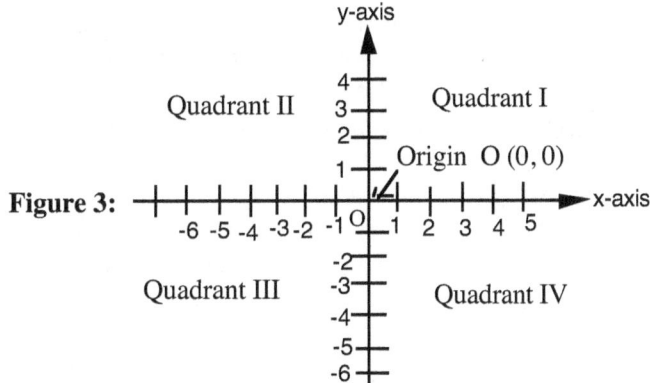

Figure 2:
Vertical line

Figure 3:

If we combine the horizontal and vertical lines at right angles so that the zero points (the origins) coincide, then we obtain what is called the rectangular coordinate system of axes. **(Figure 3).** We call the intersection of these lines, the origin and we label it $O(0,0)$. We call the horizontal number scale the x–axis, and we call the vertical number scale the y–axis. The two axes divide the plane (area) into four quadrants. These quadrants are numbered counterclockwise as quadrants I, II, III, and IV (i.e., first quadrant, second quadrant, third quadrant and fourth quadrant).

Graphing Ordered Pairs in a Rectangular Coordinate Plane

Geometrically, the first component (element) of each ordered pair is called the x-coordinate or the abscissa of the point. The second component (element) of the ordered pair is called the y-coordinate or ordinate of the point. The x-coordinate is the directed (positive or negative) distance from the y–axis and the y-coordinate is the directed distance from the x–axis. In quadrant I, both the x- and y-coordinates are positive. In quadrant II, the x–coordinate is negative but the y–coordinate is positive, In quadrant III, both coordinates are negative, In quadrant IV, the x–coordinate is positive, but the y–coordinate is negative.

Geometrically, each ordered pair represents a **point** in an x–y coordinate system of axes.

Plotting Points

In plotting points, we will be guided as follows:
1. We always count from the origin $O(0,0)$.
2. If the x–coordinate is positive, we count horizontally to the right from the origin, but if the x–coordinate is negative, we count horizontally to the left from the origin.
3. If the y–coordinate is positive, we count vertically upwards from the origin, but if the y–coordinate is negative, we count vertically downwards from the origin. The counting is done visually, and we do not make any marks on the graph paper as we count.

Example .Graph the following ordered pairs (points) in a rectangular coordinate system of   5 4
axes. $A(1,5)$. $B(-2,4)$, $P_1(-3,-4)$. $P_2(1,-2)$.

Solution

For $A(1,5)$: From the origin, count horizontally one unit to the right ($x=1$) and stop.
From this point, count 5 units ($y=5$) vertically upwards and stop. Place a dot here and label this point as $A(1,5)$.

For $B(-2,4)$: From the origin, count horizontally 2 units to the left ($x=-2$) and stop,
From this point, count 4 units ($y=4$) vertically upwards and stop. Place a dot here and label this point as $B(-2,4)$.

For $P_1(-3,-4)$: From the origin, count horizontally 3 units to the left ($x=-3$) and
stop. From this point, count 4 units vertically downwards ($y=-4$) and stop. Place a dot here and label this point as $P_1(-3,-4)$.

Similarly, for $P_2(1,-2)$, count horizontally 1 unit to the right ($x=1$) and 2 units vertically downwards ($y=-2$) and stop. Place a dot here and label this point as $P_1(-3,-4)$.

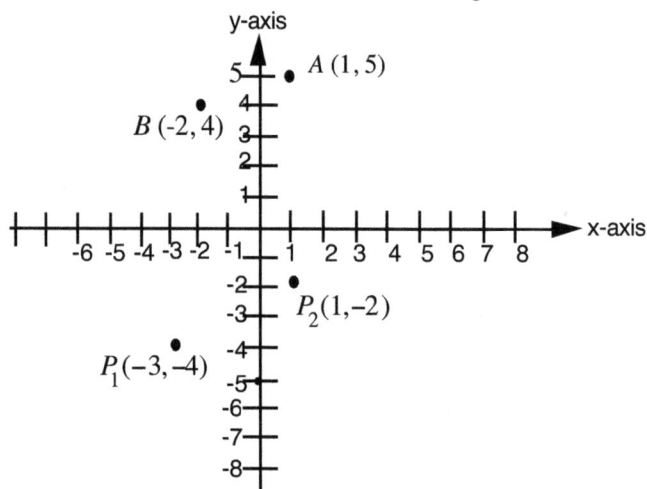

Figure 4

Lesson 12A Exercises

Graph the following ordered pairs in a system of rectangular axes.
$A(2,3)$, $B(-2,-3)$, $C(-4,5)$, $D(5,-4)$

Lesson 12B
Drawing the Graph of a Straight Line

Example Draw the graph of the line whose equation is given by : $y = 3x - 5$

We will cover three methods, namely a general method; the x- and y-intercepts method; and the intercept-slope method.

Method 1: General method

Step 1: Choose three convenient x-values and calculate the corresponding y-values to obtain ordered pairs.
(Actually, two ordered pairs will be sufficient; the third pair is used as a check: all **three** points must be in line)

choosing $x = 1$, $y = 3(1) - 5$ ordered pairs

$$y = 3 - 5 \quad\Big\} \quad \text{------->} (1,-2)$$
$$y = -2$$

choosing $x = 0$, $y = 3(0) - 5 \quad \Big\}\text{------->} (0, -5)$
$$y = -5$$

choosing $x = -2$, $y = 3(-2) - 5$

$$y = -6 - 5 \quad \Big\}\text{------->} (-2,-11)$$
$$y = -11$$

Step 2: Plot the points $(1,-2)$, $(0,-5)$ and $(-2,-11)$ on a rectangular coordinate system of axes
(on graph paper) and connect the points by a straight line.

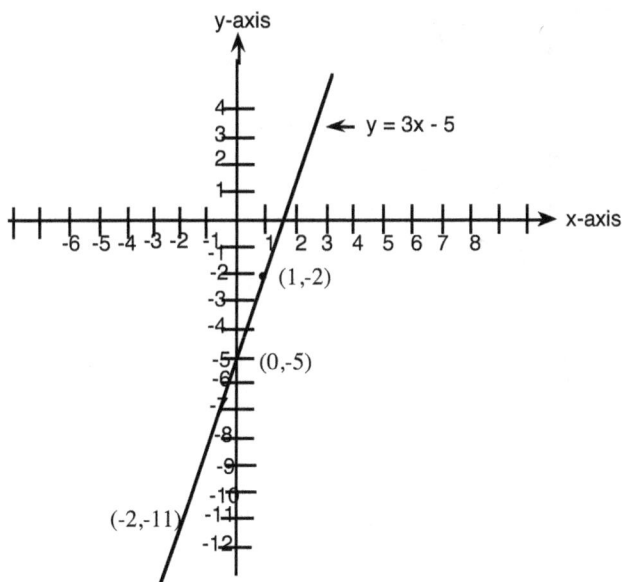

Note that from the given equation, the line has a positive slope and therefore the direction (see also page 67) of the
line (it **leans to the right**) in the figure is appropriate.

Note also that the above general method can be extended to draw the graphs of nonlinear equations such as $y = x^2$.

Method 2: The *x*- and *y*-intercepts method

We need two points to plot and connect by a straight line. The coordinates of the *x*-intercept yield one point and the coordinates of the *y*-intercept yield a second point..

Step 1: Find the *x*-intercept by letting $y = 0$ in the equation, $y = 3x - 5$, and solving for *x*.

Then $0 = 3x - 5$ and from which $x = \frac{5}{3}$. This step yields the point $(\frac{5}{3}, 0)$.

Step 2: Find the *y*-intercept by letting $x = 0$ in the equation and solving for *y*.

$$y = 3(0) - 5$$
$$y = -5$$
This yields the point $(0, -5)$

Step 3: Plot these two points from Steps 1 and 2 and connect them by a straight line.

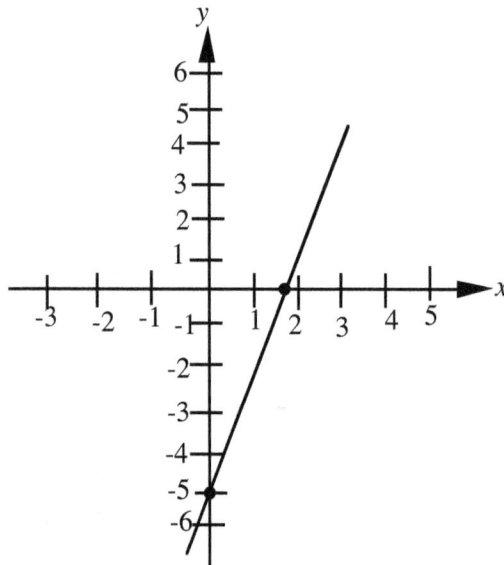

Figure: Graph of $y = 3x - 5$

Note that in Method 2, if the *y*-intercept is 0 (i.e., the equation is of the form $y = mx$), then we need to obtain another point by choosing another *x*-value (other than 0), calculate the corresponding *y*-value to obtain an ordered pair which we plot to locate a second point.

An example of such a case is given by the line $y = 3x$. Note also that if the *y*-intercept is 0, the line passes through the origin.

Method 3: Slope-Intercept method

Here also, we need two points to plot and connect by a straight line. The *y*-intercept locates one point, and we use the slope to locate a second point.

Step 1: Solve the equation for *y*, and read the *y*-intercept.

Since the equation has already been solved for *y*, we read the intercept, -5.

This step locates the point (0,−5).

Step 2: Graph the point (0, -5).

Step 3: Since the slope = +3 = $\frac{3}{1}$, the vertical change is +3 and the horizontal change is +1.

From the point (0, -5), the *y*-intercept, count 3 units vertically upwards and 1 unit horizontally to the right, stop and place a dot here; this is a second point. Connect the two points by a straight line.

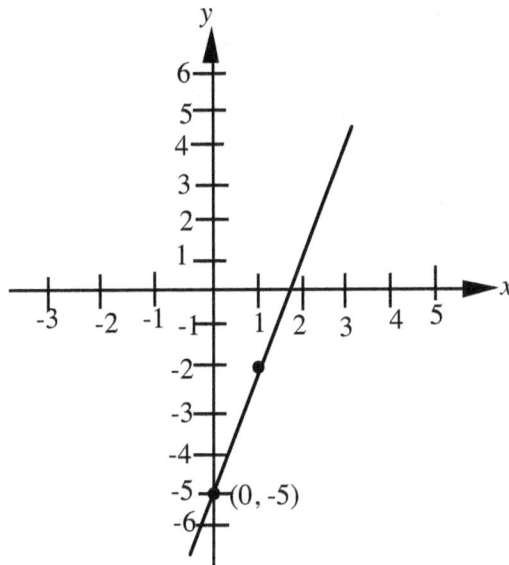

Figure: Graph of $y = 3x - 5$

Note In Step 3, we adopt the following convention.

1. Write the slope as a fraction if it is not a fraction. and if the slope is negative, give the minus sign to the numerator.

For example, a slope of -3 is expressed as $\frac{-3}{1}$; $-\frac{4}{5}$ is expressed as $\frac{-4}{5}$; $-\frac{1}{2}$ is expressed as $\frac{-1}{2}$

2. In counting from the *y*-intercept, if the numerator (change in *y*) is positive, we count vertically upwards, but if it is negative, we count vertically downwards.

For the denominator (change in *x*) we always count to the right since we have agreed to give the minus sign to the numerator.

Examples: For a slope of $\frac{-2}{3}$, starting from the *y*-intercept, we count 2 units vertically downwards (negative numerator) and 3 units horizontally to the right. For a slope of $\frac{-1}{2}$, starting from the *y*-intercept, we count

1 unit vertically downwards (negative numerator) and 2 units horizontally to the right. For a slope of $\frac{4}{5}$, starting from the *y*-intercept, we count 4 units upwards (positive numerator) and 5 units to the right

Lesson 12B Exercises

Draw the graphs of the following lines whose equations are given:

1. $y = -3x + 5$ **2.** $6y = 18 - 3x$ **3.** $2y - 6x = -3$

Answers: Graphs of the lines   **1.** $y = -3x + 5$ **2.** $6y = 18 - 3x$ **3.** $2y - 6x = -3$

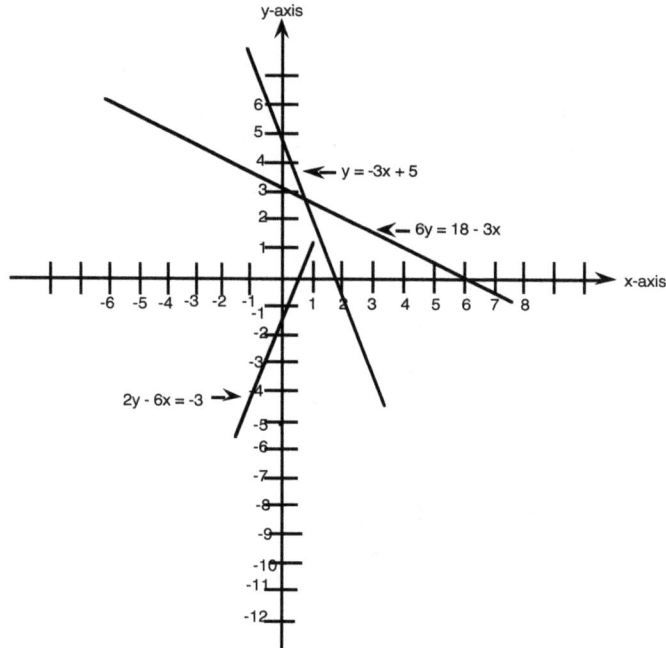

Lesson 13

Points on a Line; Slopes of Lines; x- and y-intercepts

Determining if a point is on a line whose equation is given (See also p.27)

Example   Which of the points $(5, 2)$ and $(3, 11)$ is on the line whose equation
is given by $y = 3x + 2$.

We will substitute each ordered pair in turn in the given equation. The ordered pair whose
substitution makes the left-hand side of the equation **equal** to the right-hand side **is on** the line.

Step 1:   Checking for the point (5,2)

Substitute the ordered pair (5,2) in

$$y = 3x + 2 \qquad\qquad (x = 5, y = 2)$$

The question mark "?" above the equality symbol

then, $2 \overset{?}{=} 3(5) + 2$ is there to show that at this step, we do not yet know
if the left-hand side and the right-hand side of the

$2 \overset{?}{=} 15 + 2$ equation are equal. In checking solutions, it is a good
practice to use the question mark.

$2 \overset{?}{=} 17$ No   (or $2 = 17$ is False)

The left-hand side is **not** equal to the right-hand side of the equation and therefore the
point (5,2) is **not** on the given line.

We also say (algebraically) that the ordered pair (5,2) is **not** a solution of the
the equation $y = 3x + 2$

Step 2: Checking for the point (3,11)

Substitute $(3, 11)$ in $y = 3x + 2$

then, $11 \overset{?}{=} 3(3) + 2$

$11 \overset{?}{=} 9 + 2$

$11 = 11$ True   (We no longer use the question mark since it is obvious that
the left- hand side equals the right-hand side of the equation)

Since the left-hand side **is equal to** the right-hand side of the equation,
the point $(3,11)$ **is on** the line $y = 3x + 2$.

Algebraically, we also say that
the ordered pair $(3,11)$ **is a solution** of the equation $y = 3x + 2$.

In fact, the last example (question) could have been posed as : Which of the following
ordered pairs is a solution of the equation $y = 3x + 2$? (a) $(5,2)$, (b) $(3,11)$

Solution: Proceed exactly as steps 1 and 2 of example above.

In calculations, the following have the same interchangeable implications:

1. A line passes through a given point.

2. A line contains a given point.

3. A given point is on a line.

4. The coordinates of a given point satisfy an equation of a line.

5. The coordinate pair of a given point is a solution of an equation of a line.

Example: The problem " find an equation of the line **passing through** the points.$(2, 3)$ and
$(6, 8)$" is **equivalent to** "find an equation of the line **containing** the points $(2, 3)$ and $(6, 8)$.

Finding the slope of a line given the coordinates of two points on the line   60

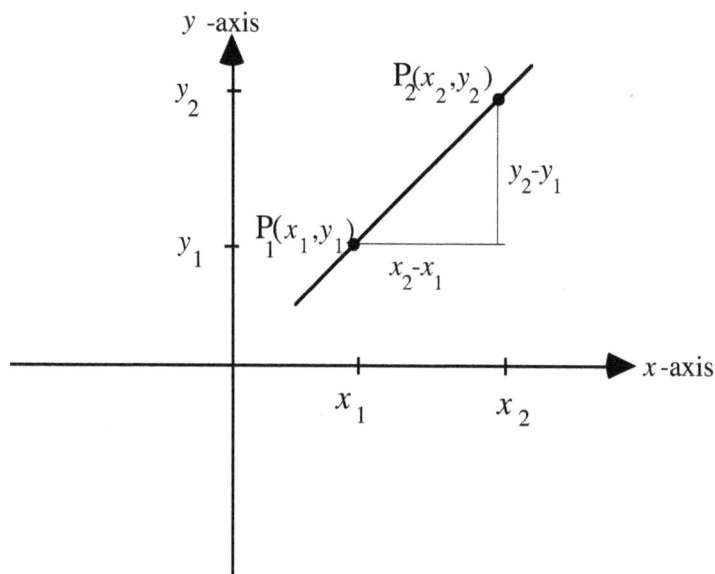

The slope, m, of the line segment connecting the points $P_1(x_1, y_1)$ and $P_2(x_2, y_2)$ is given by

$$m = \frac{y_2 - y_1}{x_2 - x_1}$$

Example   Find the slope of the line passing through the points (2,3) and (6,8).

Solution :   Identify the first point as $P_1(2,3)$ and the second point as $P_2(6,8)$

Then $x_1 = 2, y_1 = 3$; and $x_2 = 6$, $y_2 = 8$

Applying the slope formula, $m = \frac{y_2 - y_1}{x_2 - x_1}$

$$m = \frac{8 - 3}{6 - 2}$$

$$m = \frac{5}{4}$$

The slope is $\frac{5}{4}$.

Special Cases of the Slopes of Lines
The special cases are for horizontal and vertical lines

Slope of a horizontal Line
The **slope** of a **horizontal** line is **zero**, since the vertical change is zero. For example, the

slope, m, of the horizontal line in Figure **1**, below, by the slope formula is $m = \frac{3-3}{5-2} = \frac{0}{3} = 0.$

$\left(\text{Note that } m = \frac{\text{vertical change}}{\text{horizontal change}} = \frac{\text{change in } y}{\text{change in } x}\right)$

Slope of a Vertical Line

The **slope** of a **vertical** line is **undefined** since the horizontal change is zero. For example, the slope, m, of the vertical line in Figure **2**, below, by the slope formula is $m = \frac{2-(-3)}{4-4} = \frac{5}{0}$ is undefined.

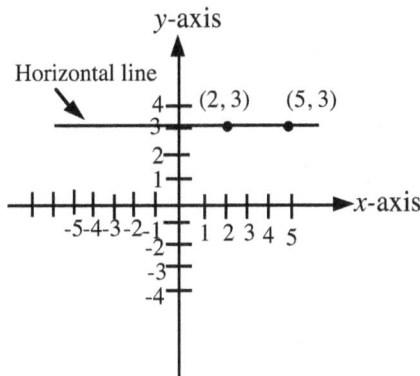

Figure 1

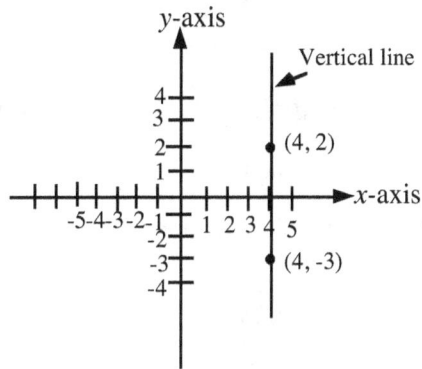

Figure 2

The *x*- and *y*-intercepts of a line

The *x***-intercept** of a line is the *x*-coordinate of the point where the line crosses or meets the *x*-axis. At this point, $y = 0$. (Geometrically, we should also say that the *x*-intercept is the point where the line intersects the *x*-axis, and in which case, we must specify two coordinates, with the y-coordinate always being zero.) See Figure 1 below.

The *y***-intercept** of a line is the *y*-coordinate of the point where the line crosses or meets the *y*-axis. At this point, $x = 0$. (Geometrically, we should also say that the *y*-intercept is the point where the line intersects the *y*-axis, and in which case, we must specify two coordinates, with the *x*-coordinate always being zero.) See Figure 1 below.

Example
In Figure 1 below, the *x*-intercept is -3; and the *y*-intercept is 2.
The *x*- and *y*--intercepts are at (-3,0) and (0,2) respectively.

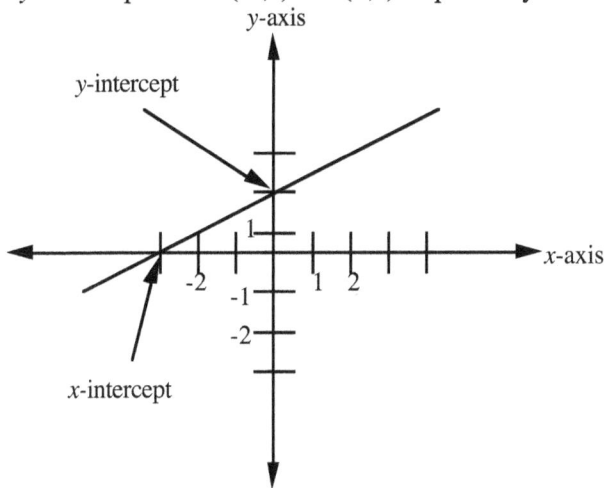

Figure 1

Example Find the x- and y-intercepts for the graph of $3x + 2y = 12$

For the x-intercept: Let $y = 0$ in $3x + 2y = 12$.
Then $3x + 2(0) = 12$; $3x = 12$; and from which $x = 4$.
The x-intercept is 4.

For the y-intercept: Let $x = 0$ in $3x + 2y = 12$.
Then $3(0) + 2y = 12$; $2y = 12$; and from which $y = 6$.
The y-intercept is 6.

Finding the slope and the y-intercept of a line, given the equation of the line

Example 1 Find the slope and the y-intercept of the line whose equation is given by $y = -5x + 6$.

The slope-intercept form of the equation of a straight line is given by $y = mx + b$, where m is the slope and b is the y-intercept.

Compare $y = -5x + 6$ to   **Note:** (The slope is the coefficient of the x- term
$y = mx + b$,   if the equation is in slope-intercept form.)

then, the slope, $m = -5$,
and the y-intercept $= 6$

Example 2   Find the slope and the y-intercept of the line whose equation is given by
$$5y + 4x = 20$$

Step 1: Solve the equation for y (that is, write the equation in slope-intercept form).
$$5y + 4x = 20$$
$$\underline{\quad -4x \qquad\qquad -4x \quad}$$
$$5y = -4x + 20$$
$$\frac{5}{5}y = \frac{-4}{5}x + \frac{20}{5}$$
$$y = -\frac{4}{5}x + 4$$

Step 2: By comparison with $y = mx + b$, the slope is $-\frac{4}{5}$ (the coefficient of the x-term).
and the y-intercept is +4 or 4.

Lesson 13 Exercises

A Determine which of the points (2,4), (-2,5), (-2,-16) are on the line whose equation is given by
$$y = 5x - 6$$

Answer: The points (2,4), and (-2,-16) are on the given line

B Find the slope of each line passing through the given points:

1. (2, 3) and (6, 4)   **2.** (3, -2) and (2, -4)   **3.** (5, 2) and (1, -1)

4. (2, 2) and (-3, 3)   **5.** (-4, 2) and (5, 2)   **6.** (-3, -5) and (-3, -6)

7. Find c so that the slope of the line passing through the points $(c, 2c)$ and (-6, 8) is 1.

Answers: 1. $\frac{1}{4}$;   **2.** 2 ;   **3.** $\frac{3}{4}$;   **4.** $-\frac{1}{5}$;   **5.** 0 ;   **6.** undefined;   **7.** $c = 14$

C **1..** Find the slope of the line passing through the points (3,2) and (5,9).

2. The points (2,2) and (-5,6) are on a line whose slope we want to determine. What is the slope of this line?

3. Find the slope of the line containing the points (1,6) and (2,8).

4. Find the slope of the line passing through the points (1,-1), (3,3) and (-2,-7).

Answers: 1. slope = $\frac{7}{2}$ **2.** slope = $-\frac{4}{7}$ **3.** slope = 2 ;   **4.** slope =2

D Find the slope and the y-intercept of the following:

1.. $y = -6x + 4$;   **2.** $-7 + 5x = y$; **3.** $y = -\frac{x}{2} + 8$

Answers: 1. slope = - 6, y-intercept = 4; **2.** slope = 5, y-intercept = -7;   **3.** slope = $-\frac{1}{2}$, y-intercept = 8

E Find the slope and y-intercept: **1.** $3x + 8y = 32$ **2.** $2y - 6x = 21$;   **3.** $y = -3x + 4$

4. $y = \frac{1}{2}x - 3$;   **5.** $3y = 6x + 12$;   **6.** $y = \frac{x}{3} + 5$;   **7.** $2y + 5x = 8$;   **8.** $4x + 3y = 12$

Ans 1. slope = $-\frac{3}{8}$, y -intercept = 4;   **2.** slope =3, y -intercept = $\frac{21}{2}$; **3.** slope = - 3 ; y -intercept = 4;

4. slope = $\frac{1}{2}$, y -intercept = -3 ;   **5.** slope = 2, y -intercept = 4;   **6.** slope = $\frac{1}{3}$, y -intercept = 5;

7. slope = $-\frac{5}{2}$, y -intercept = 4;   **8.** slope = $-\frac{4}{3}$, y -intercept = 4;

Lesson 14
Equations of Straight Lines

There are a number of approaches that we can use in **finding equations** of straight lines. Each approach depends upon what we are given.

Generally, we can easily write down an equation of a line if we know any of the pairs of properties or characteristics of (information about) a line in the formulas presented below.

These formulas are based on finding two different expressions for the slope of a line and equating these expressions to each other.

1. If we know the **slope, m,** and the **y-intercept, b,** of the line, we can apply the slope-intercept form, $y = mx + b$.

2. If we know the slope, m, and the coordinates (x_1, y_1) of one point on the line, we can apply the point-slope form, $(y - y_1) = m(x - x_1)$.

3. If we know the coordinates (x_1, y_1) and (x_2, y_2) of two points on the line,

    we can apply $(y - y_1) = \dfrac{y_2 - y_1}{x_2 - x_1}(x - x_1)$.

4. If we are given the graph (picture), we can determine any of the above pairs of properties (information) from the graph, and then apply the formulas in the above cases; however, if we want to memorize one more formula, we can apply the equation

    $y = -\dfrac{b}{a}x + b$, where a is the x-intercept and b is the y-intercept. (If $b = 0$ or $a = 0$ use the other methods.)

(This equation is another form of the two-intercept form: $\dfrac{x}{a} + \dfrac{y}{b} = 1$)

We will now cover the above **four** cases in detail with examples.

Case 1: Finding an equation of a line given the slope and the y-intercept of the line

Example 1   Find an equation of the line with slope 3 and y-intercept of 4.

Solution   We will apply the slope-intercept form of the equation

of a straight line, $y = mx + b$.

Substituting the slope, $m = 3$, and the y-intercept, $b = 4$ in this equation,

$y = 3x + 4$

Example 2   Find an equation of the line with slope - 2 and y-intercept of - 5.

Solution: Substituting $m = -2$, $b = -5$ in $y = mx + b$, we obtain

the equation $y = -2x - 5$

Case 2: **Finding an equation of a line given the slope of the line and the coordinates of one point on the line**

    Example 1 Find an equation of the line passing through the point $(3,-2)$ and having a slope 4.

Solution. We will cover two methods.

Method 1

Step 1: Find the y-intercept, b, by substituting $x_1 = 3$, $y_1 = -2$, $m = 4$ in
$$y = mx + b$$
$$\text{then, } -2 = 4(3) + b$$
$$-2 = 12 + b$$
$$-14 = b$$
Step 2: Now, since we know that $m = 4$, $b = -14$, we can apply $y = mx + b$ (as in Case 1, above)
and then, $y = 4x - 14$

Method 2

    We will use the point-slope form of the equation of a straight line which is given by

$$y - y_1 = m(x - x_1) \text{ <------point-slope form.} \tag{1}$$

Substituting the coordinates, $x_1 = 3$, $y_1 = -2$ and slope, $m = 4$ in equation (1), we obtain

$$y - (-2) = 4(x - 3)$$

$$y + 2 = 4(x - 3) \tag{2}$$

Equation (2) is the point-slope form of the required equation.

By solving equation (2) for y, we obtain

$$y = 4x - 14 \quad \text{ <----------- slope-intercept form} \tag{3}$$

Equation (3) is the slope-intercept form of the required equation.
In this form, we can, by inspection, determine the slope and the y-intercept.

The author recommends the slope-intercept form (for this course), unless otherwise specified.

Case 3: **Finding an equation of a line given the coordinates of two different points on the** **line**

Example   Find an equation of the line passing through the points (2,1) and (-3,-4).

Solution
    Step 1: Find the slope, m, with $x_1 = 2, y_1 = 1, x_2 = -3 \ y_2 = -4$

$$m = \frac{y_2 - y_1}{x_2 - x_1}$$

Scrapwork

$$m = \frac{-4 - 1}{-3 - 2}$$

$$\frac{-4 - 1}{-3 - 2} = \frac{-5}{-5} = 1$$

$$m = 1$$

Now, $m = 1, x_1 = 2, y_1 = 1, x_2 = -3, y_2 = -4,$ and we can apply the procedure in Case 2.

  Step 2: Applying Method 2 of Case 2 and
            Substituting $m = 1, x_1 = 2, y_1 = 1$ in $y - y_1 = m(x - x_1)$, we obtain
$$y - 1 = 1(x - 2)$$

$$y = x - 1$$

Also,   If we substitute $m = 1, x_2 = -3, y_2 = -4$ in $y - y_2 = m(x - x_2)$, we obtain

$$y - (-4) = 1(x - (-3))$$

$$y + 4 = x + 3$$
$$y = x - 1$$

Again, we obtain the same equation. An equation of the line is $y = x - 1$.
We conclude also that any of the two given points can be used in finding an equation of the
straight line.
From the above solution, we can state a general formula for Case 3 as

$$(y - y_1) = \frac{y_2 - y_1}{x_2 - x_1}(x - x_1)$$

Case 4: Finding an equation of a line given the graph (picture) of the line

If we are given the graph (picture), we can determine any of the above pairs of properties (information) from the graph (see p.64, case 4).

It is also useful to be able to tell immediately from the graph if the line has a positive slope, a negative slope, a zero slope, or an undefined slope.

The signs of the slopes of lines

The lines in Figure 1 have positive slopes.   The lines in Figure 2 have negative slopes.

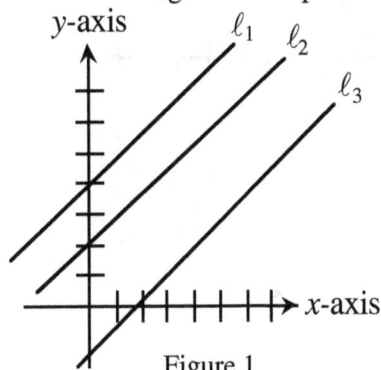

Figure 1

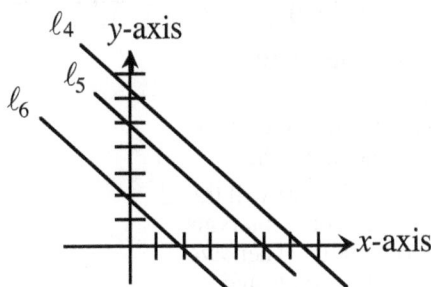

Figure 2

Lines ℓ_1, ℓ_2, and ℓ_3 have positive slopes.
(Simply, these lines **lean** to the **right** in the page; or these lines rise as one moves ones head from the left to the right in the page.)

Lines ℓ_4, ℓ_5 and ℓ_6 have negative slopes.
(Simply, these lines **lean** to the **left** in the page; or these lines fall as one moves ones head from the left to the right in the page.)

Note: The slope of a **horizontal line** is zero. The slope of a **vertical line** is undefined. (For the equations of horizontal and vertical lines, see p.70-71)

Example 1 Find an equation of the line whose graph is given below.

Method 1: Apply $y = -\dfrac{b}{a}x + b,$

where a is the x-intercept and b is the y-intercept. (see also p.61)

Step 1: From the graph, we read the values of a and b: $.a = 3, b = -4$

Step 2: Substitute $a = 3, b = -4$ in

$$y = -\frac{b}{a}x + b.$$

Then $y = -\dfrac{-4}{3}x + (-4)$

(Make sure you take into account the **minus sign** that comes with the formula.)

$y = +\dfrac{4}{3}x - 4$ (Two minus signs make

the x-term positive)

An equation of the line is $y = \dfrac{4}{3}x - 4.$

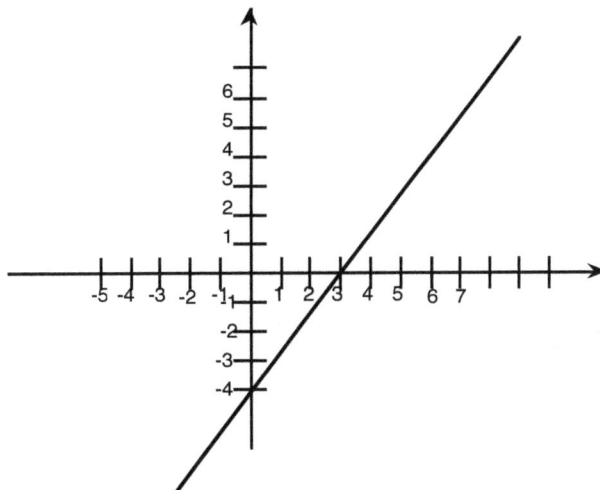

Method 2

Step 1: By inspection, the line has a positive slope. (See figure 1 above.)

Step 2: Slope, $m = \dfrac{\text{vertical change}}{\text{horizontal change}}$

$= \dfrac{\text{change in } y}{\text{change in } x}$

Pick any two convenient points on the line, say, $P_1(0,-4)$ and $P_2(3,0)$ (see figure 2, below).

Finding the vertical and horizontal changes by counting.

We will assume that we want to go from the point P_1 to the point P_2 , and count the number of equal intervals (units) we will travel vertically, and the number of equal intervals (units) we will travel horizontally. Then, we will go up vertically 4 equal intervals and horizontally to the right 3 equal intervals. Divide the vertical change, 4, by the

horizontal change, 3 , to obtain the slope $\dfrac{4}{3}$

Note that the slope is positive. (see p.67, Figure 1).

Step 3:From the graph , the y-intercept, $b = -4$

Step 4:With $m = \dfrac{4}{3},\ b = -4,$ apply $y = mx + b$

then, $y = \dfrac{4}{3} x - 4$

Therefore, an equation of the line is

$y = \dfrac{4}{3} x - 4$

In fact, once we have been able to specify the points $P_1(0,-4)$, $P_2(3,0)$, we

can apply $m = \dfrac{y_2 - y_1}{x_2 - x_1} = \dfrac{0 - (-4)}{3 - 0} = \dfrac{+4}{3}$

to find the slope $m = \dfrac{4}{3}$,

and then apply $y - y_1 = m(x - x_1)$ to obtain

$y + 4 = \dfrac{4}{3} (x - 0) \qquad (x_1 = 0, y_1 = -4)$

$y + 4 = \dfrac{4}{3} x - 0$

$y = \dfrac{4}{3} x - 4 <$ -------slope-intercept form of the

equation of a straight line.

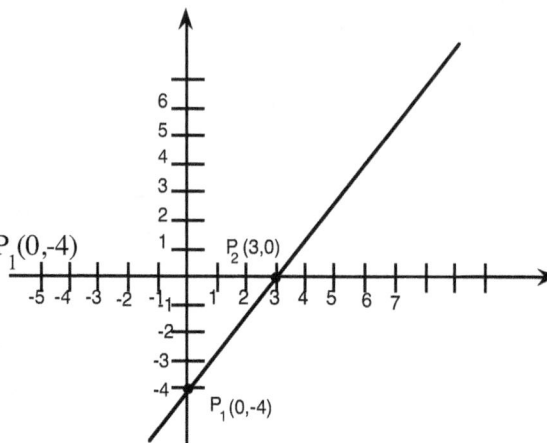

Figure 2

Example 2 Find an equation of the line whose graph is given below.

Method 1 Apply $y = -\dfrac{b}{a}x + b$ where a is the x-intercept and b is the y-intercept.

Step 1: From the graph, $a = 3, b = 6$

Step 2: Substitute $a = 3, b = 6$ in $y = -\dfrac{b}{a}x + b$

Then $y = -\dfrac{6}{3}x + 6$

$y = -2x + 6$ < ---slope-intercept form of the equation of a straight line.

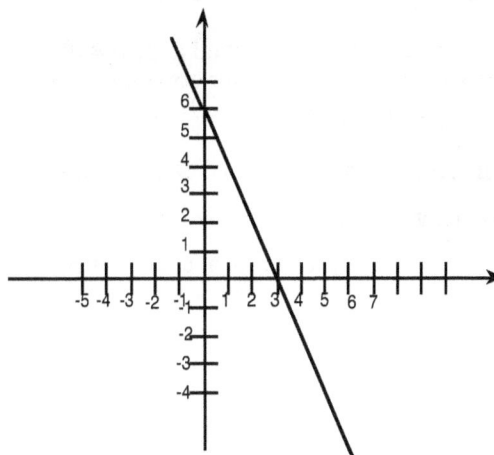

Figure

Method 2
Step 1: Note that the slope is negative.
(See p.67, Figure 2.)
Step 2: Pick any two convenient points on the line, say P_1 and P_2
Step 3: Assume that we want to travel (vertically and horizontally only) from P_1 to P_2.
Then, we will go down 6 units and then to the right 3 units.

The slope, $m = -\dfrac{6}{3} = -2$

(The slope is negative since the line leans to the left. See page 67, Fig..2)

Step 4: Read the y-intercept, $b = 6$
Step 5: Apply $y = mx + b$, with $m = -2, b = 6$

Then, an equation of the given graph is
$y = -2x + 6$

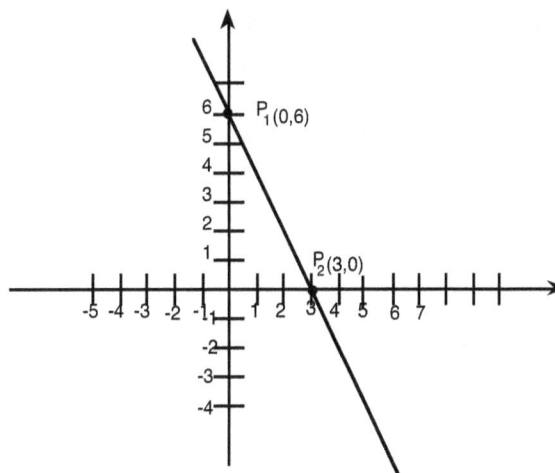

Figure

Method 3 From $P_1 (0,6)$ and $P_2(3,0)$

Step 1: $m = \dfrac{y_2 - y_1}{x_2 - x_1}$ $(x_1 = 0, y_1 = 6)$

$= \dfrac{0 - 6}{3 - 0} = \dfrac{-6}{3} = -2$

$m = -2$

Step 2: Apply $y - y_1 = m(x - x_1)$
$y - 6 = -2(x - 0)$
$y - 6 = -2x + 0$
$y = -2x + 6$
An equation of the line is $y = -2x + 6$

Note that, in **Method 3**, the negative sign results solely from the calculation and we do not have to know in advance the sign of the slope)

In **Case 4**, the author recommends **Method 1**.

Special Cases of the Equations of Straight Lines and their Graphs   70

The equation $y = mx + b$, (with m defined and, $m \neq 0$) geometrically represents oblique lines (i.e., lines which are neither vertical nor horizontal). The special cases of the equation of a straight line are for horizontal and vertical lines. In these cases, either the x- or the y-term is missing.

The Equation and Graph of a Horizontal Line

The equation of a horizontal line is of the form $y = b$, where b is the y-intercept.

   The slope, $m = 0$, since the vertical change is zero.

Substituting $m = 0$, in $y = mx + b$,

$$y = 0x + b$$

$$y = b \qquad\qquad (1)$$

Equation (1) means that as x varies, y remains unchanged.

Examples: Sketch the graphs of the following lines:

   1. $y = 3$,   **2.** $y = -4$.   **3.** $y = 0$.

Solution: The line $y = 3$ is the horizontal line passing through $(0,3)$. See Figure 1, below.

The line $y = -4$ is the horizontal line passing through $(0,-4)$.

The line $y = 0$ is the horizontal line along the x-axis (i.e., the line $y = 0$ is the x-axis)

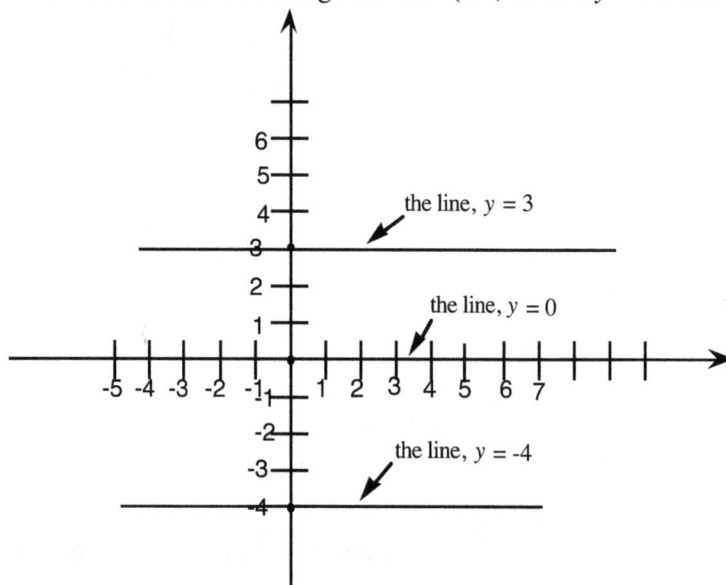

Figure 1

The Equation and Graph of a Vertical Line

The slope of a vertical line is undefined since the horizontal change is zero.
The equation of a vertical line is of the form $x = a$, where a is the x-intercept. This form of the equation means that as y varies, x remains unchanged.

Examples Sketch the lines with the following equations:

       1. $x = 2$,   **2.** $x = 0$,   **3.** $x = -4$

Solution: See Figure 2, below.

   1. The line $x = 2$ is the vertical line passing the point (2,0).

   2. The line $x = 0$ is the vertical line along the y-axis (i.e., the line $x = 0$ is the y-axis).

   3. The line $x = -4$ is the vertical line passing through the point (-4,0).

   4. The line $x = -2$ is the vertical line passing through the point (-2,0).

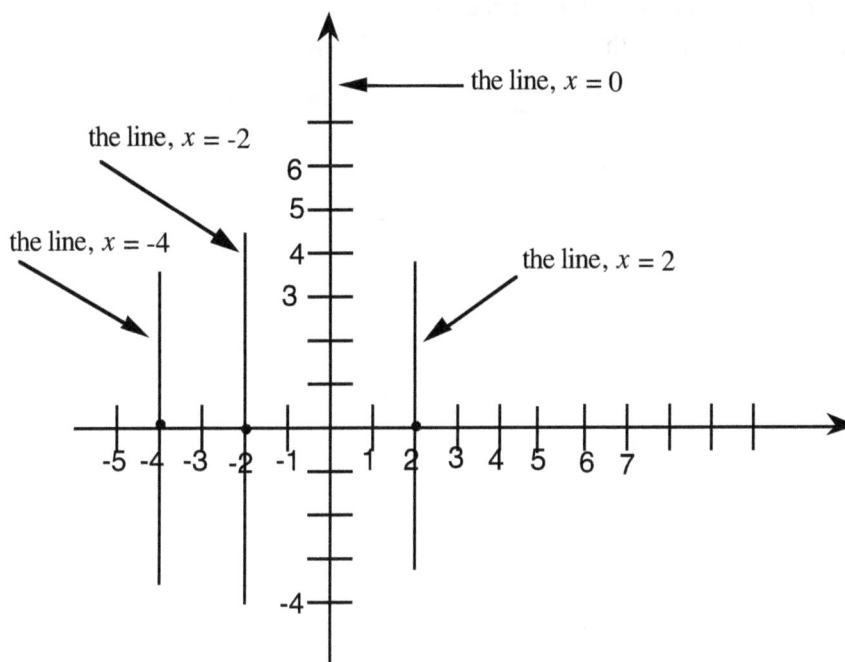

Figure 2

Lesson 14 Exercises

A **1.** Find an equation of the line with slope 7 and y-intercept -2.

  2. Find an equation of the line with slope $\frac{2}{3}$ and y-intercept 5.

  3. A line has a y-intercept -3 and a slope of 6. Find an equation for this line.

Answers: 1. $y = 7x - 2$;   2. $y = \frac{2}{3}x + 5$;   3. $y = 6x - 3$

B **1.** Find an equation of the line with slope -4 and passing through the point (3,-2).

2. A line passes through the point (-1,-7) and has a slope of 5. Find an equation for this line.

Answers: **1.** $y = -4x + 10$;   2. $y = 5x - 2$

C   **1.** Find an equation of the line passing through the points (2,2) and (-5,6)

    2. If the points (2,-5) and (-3,1) are on a certain line, find an equation for this line.

    3. Find an equation of the line with slope -3 and y-intercept 8.

    4. Find an equation of the line with slope 2 and passing through the point (1,6)

Answers: **1.** $y = -\frac{4}{7}x + \frac{22}{7}$;   **2.** $y = -\frac{6}{5}x - \frac{13}{5}$;   **3.** $y = -3x + 8$;   **4.** $y = 2x + 4$.

D Find the equations (slope-intercept forms) of the lines ℓ_1, ℓ_2, ℓ_3, ℓ_4

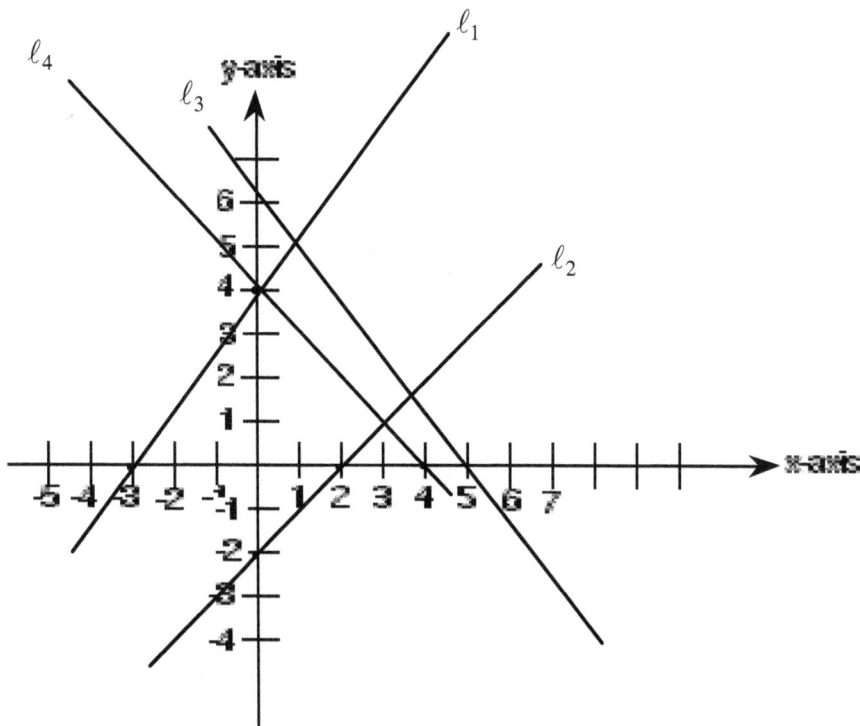

Answers: For ℓ_1: $y = \frac{4}{3}x + 4$. For ℓ_2: $y = x - 2$. For ℓ_3: $y = -\frac{5}{4}x + \frac{25}{4}$ For ℓ_4: $y = -x + 4$

E 1. Find an equation of the horizontal line passing through the point (3,-4).

  2. Sketch the graph the line in Problem 1.

  3. Does the line in Problem 1 have a y-intercept ? If yes, find it.

  4. Sketch the graph of the line $y = -5$.

  5. Sketch the graph of the line $y = 2$.

Answers: **1.** $y = -4$; **2.** See Fig 1, above and imitate. ; **3.** Yes. y-intercept = - 4 ; **4. & 5.** Imitate Fig. 1 above

F 1. Find an equation of the vertical line passing through the point (2,-3).

  2. Sketch the graph the line in Problem **1**.

  3. Does the line in Problem **1** have a y-intercept ? If yes, find it.

Answers: **1.** $x = 2$; **2.** See Fig. 2 and imitate **3.** No.

G 1. Draw the graph of the line $x = 4$. **2.** Draw the graph of the line $x = -1$.

Hint: See Fig. 2, above, and imitate.

H **1.** Find an equation of the line with slope -5 and y-intercept 2.

  2. Find an equation of the line whose y-intercept is -3 and whose slope is $\frac{1}{2}$.

  3. Find an equation of the line whose slope is -2 and which passes through the point (4, -5).

  4. A line passes through the point (-3, 2) and has slope 4. Find an equation for this line.

  5. Find an equation of the line passing through the points (5, 4) and (8, 6).

Answers: **1.** $y = -5x + 2$; **2.** $y = \frac{1}{2}x - 3$; **3.** $y = -2x + 3$; **4.** $y = 4x + 14$; **5.** $y = \frac{2}{3}x + \frac{2}{3}$

CHAPTER 4B

Lesson 15A: **Equations of Parallel lines and Perpendicular Lines**
Lesson 15B: **Midpoint of a Line; Distance between Points;** Distance Formula
Lesson 15C: **Applications: 1. Distance from a point to a line whose equation is given**
                                 2. Perpendicular Bisector of a Line

Lesson 15A

Parallel Lines: Slopes and Equations

Note: If two lines are parallel, then they have the same slope.

If the slope of one line is m_1, then the slope of the other parallel line is also m_1.

Example Find an equation of the line passing through the point (2,-5), and parallel to the
        line $y = -3x + 2$.

Solution

The slope of the line whose equation is given = -3. (comparing $y = -3x + 2$. with $y = mx + b$. $m = -3$)
The slope of the line whose equation we want to find = -3 (since two parallel lines have the same slope).
Applying $y - y_1 = m(x - x_1)$; with $m = -3, x_1 = 2, y_1 = -5$

$$y - (-5) = -3(x - 2)$$
$$y + 5 = -3x + 6$$
$$y = -3x + 1 \quad \text{<------slope-intercept form.}$$

Perpendicular Lines: Slopes and Equations

If two lines are perpendicular, then their slopes are negative reciprocals of each other; or simply,

if m_1 and m_2 are their slopes, then $m_1 = -\dfrac{1}{m_2}$ or $m_2 = -\dfrac{1}{m_1}$ or $m_1 \cdot m_2 = -1$

Example Find an equation of the line passing through the point (-2,4) and
           perpendicular to the line $y = 3x - 5$.

Solution

Step 1: Determine the slope of the line whose equation we want to find.

The slope of $y = 3x - 5$ is 3

The slope of the line whose equation we are to find is $-\dfrac{1}{3}$ (Since the two lines are perpendicular).

Step 2:   Apply $y - y_1 = m(x - x_1)$

               with $m = -\dfrac{1}{3}$ and $x_1 = -2, y_1 = 4$

               Then, $y - 4 = -\dfrac{1}{3}(x - (-2))$

$$y - 4 = -\frac{1}{3}(x + 2)$$
$$y - 4 = -\frac{1}{3}x - \frac{2}{3}$$
$$y = -\frac{1}{3}x - \frac{2}{3} + 4$$
$$y = -\frac{1}{3}x + \frac{10}{3} \quad \text{<---------slope-intercept form.}$$

Lesson 15A: Equations of Parallel lines and Perpendicular Lines

A **note** about finding the slope in the above problem:
The slope of the given line is 3
To find the slope of the other (perpendicular line) line,

Step 1. Invert 3 to obtain $\frac{1}{3}$

Step 2. Change the sign. Since $\frac{1}{3}$ has a plus sign, after the change, we obtain $-\frac{1}{3}$

Similarly, if the slope of one line were -5, the slope of the other perpendicular line would be $+\frac{1}{5}$.

If the slope of one line were $-\frac{1}{4}$, the slope of the other perpendicular line would be +4.

Summary: Invert and change the sign, or change the sign and invert.

Distinction between the point $x = a$ and the line $x = a$

We must distinguish between, for example, the point $x = 2$ and the line $x = 2$.
We use a dot to locate the point $x = 2$ on a number line.

For the graph of the line $x = 2$, we draw a vertical line through the point $(2, 0)$

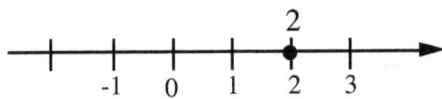

Figure: The graph of the point $x = 2$.

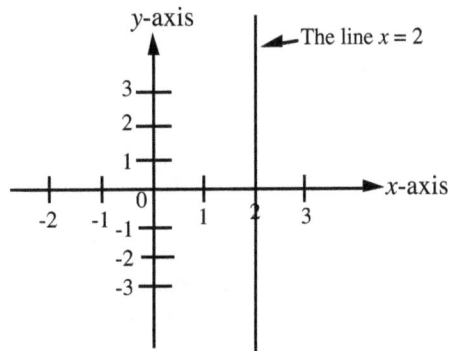

Figure: The graph of the line $x = 2$.

Similarly, the point $y = 2$ and the line (horizontal line) $y = 2$ are shown in Figures .. and respectively.

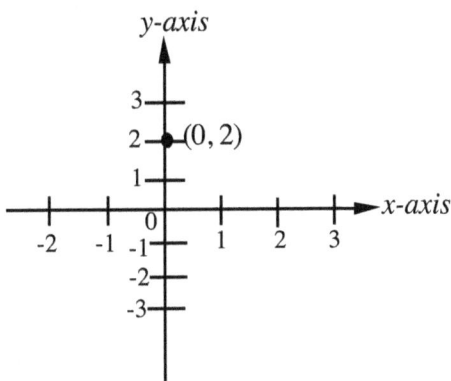

Figure: The graph of the point $y = 2$.

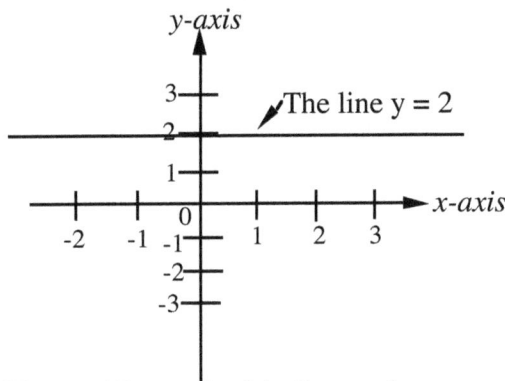

Figure: The graph of the line $y = 2$.

Lesson 15A Exercises

A 1.Find an equation of the line through the point $(3,-4)$ and perpendicular to the line $y = 2x + 5$.

2. Find an equation of the line with the same y–intercept as the line $2y = 3x + 8$ and parallel to the line $y = -4x + 2$.

Answers: **1.** $y = -\frac{1}{2}x - \frac{5}{2}$;   **2.** $y = -4x + 4$

B   **1. By** graphing, distinguish between the point $x = 2$ on the number line and the line $x = 2$.

  2. By graphing, distinguish between the point y = 2 on the y-axis and the line $y = 2$.

1.

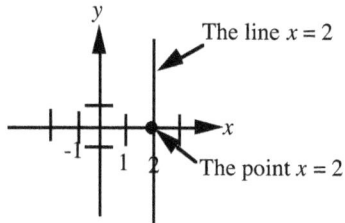

2.

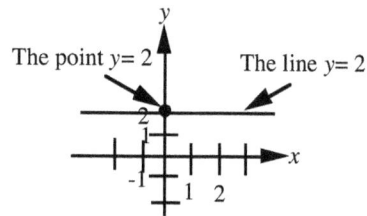

Lesson 15B

Midpoint of a Line; Distance between Points; Distance Formula

Midpoint of a Line

The **midpoint** of (Figure below), the line, P_1P_2, connecting the points $P_1(x_1, y_1)$ and $P_2(x_2, y_2)$ has the coordinates given by the following formulas:

The x-coordinate, x_m, of the midpoint is given by $x_m = \dfrac{x_1 + x_2}{2}$

The y-coordinate, y_m of the midpoint is given by $y_m = \dfrac{y_1 + y_2}{2}$

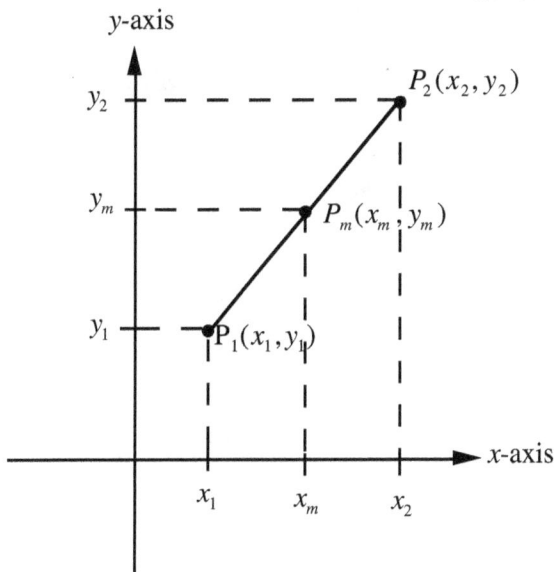

Example Find the coordinates of the mid-point of the line connecting the points (3, 2) and (-9, 4).

Solution

The x-coordinate of the mid-point $= \dfrac{x_1 + x_2}{2} = \dfrac{3 + (-9)}{2}$ ($x_1 = 3, x_2 = -9$)

$$= \dfrac{-6}{2}$$

$$= -3$$

The y-coordinate of the mid-point $= \dfrac{y_1 + y_2}{2} = \dfrac{2 + 4}{2}$ ($y_1 = 2, y_2 = 4$)

$$= \dfrac{6}{2}$$

$$= 3$$

Therefore $(x_m, y_m) = (-3, 3)$ where x_m and y_m are the coordinates of the mid-point.
The mid-point is at (-3, 3).

Distance Between two Points on a Line

The distance between two points A and B on a line is equal to the number of units (equal intervals) between A and B. If we denote the distance between A and B by $d(A, B)$. Then, we can write

$$d(A, B) = |A - B|$$

that is, the distance between A and B is the absolute value of the difference between A and B.

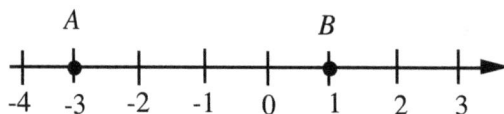

Figure: Distance on a horizontal line

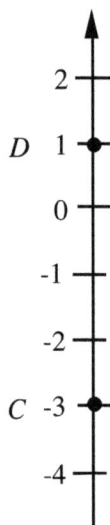

Figure: Distance on a vertical line

Example

$$d(-3, 1) = |-3 - 1|$$
$$= |-4|$$
$$= 4$$

We conclude also that the distance between two points on the same **horizontal line** or the same **vertical line** can be found algebraically.

However, the distance between two points in a plane cannot be found algebraically, but must be found geometrically as discussed in the next section.

Distance Between two Points on a Line in a Plane: Distance Formula

The distance, d, between the points $P_1(x_1, y_1)$ and $P_2(x_2, y_2)$ on a line **in a plane** (Fig.1) is given by

$$d = \sqrt{(x_2 - x_1)^2 + (y_2 - y_1)^2}$$ (By applying the Pythagorean theorem to the right triangle, and solving for d)

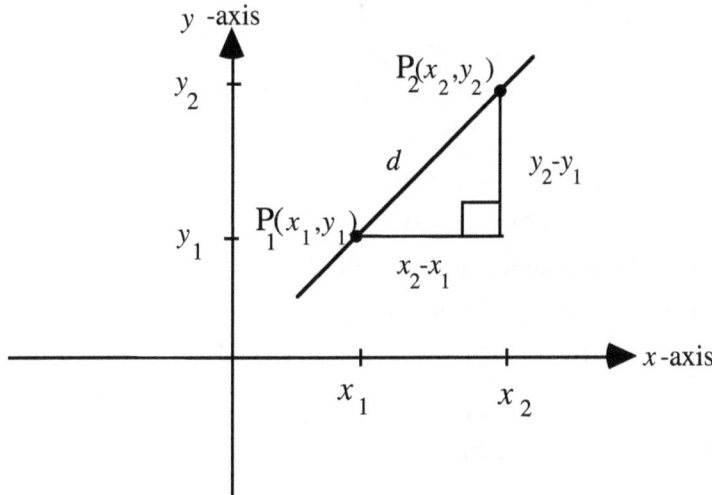

Figure 1

Example Find the distance between the points (-2,3) and (4, -5) .

Solution Apply the distance formula:

$$d = \sqrt{(x_2 - x_1)^2 + (y_2 - y_1)^2}$$ (1)

(where d is the distance between the points $P_1(x_1, y_1)$ and $P_2(x_2, y_2)$).

Substituting $x_1 = -2, y_1 = 3, x_2 = 4, y_2 = -5$ in equation (1) above,

$$d = \sqrt{(4 - (-2))^2 + (-5 - 3)^2}$$
$$d = \sqrt{(4 + 2)^2 + (-8)^2}$$
$$= \sqrt{(6)^2 + (-8)^2}$$
$$= \sqrt{36 + 64}$$
$$= \sqrt{100}$$
$$= 10$$

$\therefore$ the distance between the given points is 10 units.

Lesson 15B Exercises

A   **1.** Find the distance between the points (3,4) and (5, -1) .

2. Find the distance between the points (-4, 2) and (-6, -3)

Answers: **1.** $\sqrt{29}$; **2.** $\sqrt{29}$

B   Find the distance between each given pair of points.

1. (2, 3) and (−7, 4);   **2.** (−4, 3) and (6, 2);   **3.** (−3, 5) and (7, 10)

4. (a, b), and (c, d);   **5.** (−3, 5), and (4, 7)

6. (a) Find the directed distance from (4,−5) to (4,−7);

(b) Find the distance from (4,−5) to (4,−7)

Find the coordinates of the mid–point of each of the following points :

7. (2,−4), and (4, 0);   **8.** (3, 2), and (4, 6).

9. (−4, − 4), and (5, 3);   **10.** (5, 12), and (0, 0).

Answers: **1.** $\sqrt{82}$; **2.** $\sqrt{101}$; **3.** $5\sqrt{5}$; **4.** $\sqrt{a^2+b^2+c^2+d^2-2ac-2bd}$; **5.** $\sqrt{53}$; **6.** (a) −2 ; (b) 2.
7. $(3,-2)$; **8.** $(\frac{7}{2}, 4)$; **9.** $(\frac{1}{2},-\frac{1}{2})$; **10.** $(\frac{5}{2}, 6)$.

Lesson 15C
Applications
1. Distance from a point to a line whose equation is given
2. Perpendicular Bisector of a Line

Distance from a point to a line whose equation is given

Find a formula for the distance, d, from the point $A(x_0, y_0)$ to the line $y = mx + b$.

Solution
Method 1

Given: The point $A(x_0, y_0)$ and the line $y = mx + b$.

Required: To find a formula for the distance, d, between the point $A(x_0, y_0)$ and the line $y = mx + b$.

Construction: Draw the perpendicular line from $A(x_0, y_0)$ to meet the line $y = mx + b$ (line ℓ_2) at E.

Also, draw $\overline{AC} \perp$ to the x-axis.

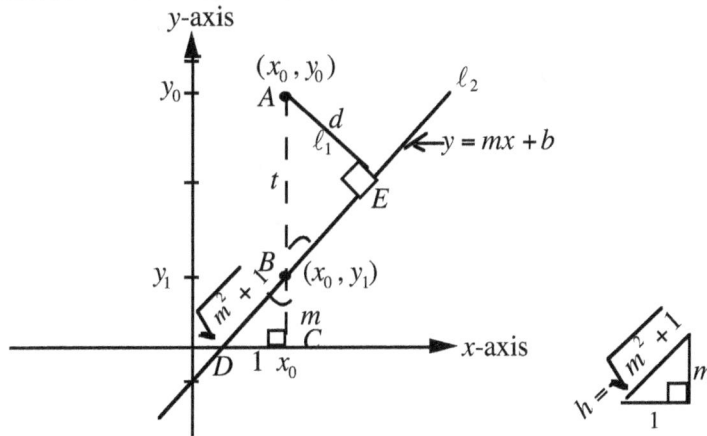

Step 1: Let $h = DB$. If the slope of $y = mx + b$ is $m = \frac{m}{1}$, then $BC = m$, $DC = 1$, and $h = \sqrt{m^2 + 1}$.

ΔABE and ΔDBC are similar: (Two angles of ΔABE are congruent to two angles of ΔDBC)

$\therefore \dfrac{AE}{DC} = \dfrac{AB}{DB}$ (Corresponding sides of similar triangles are in proportion)

$\dfrac{d}{1} = \dfrac{t}{\sqrt{m^2 + 1}}$ (1) ($AE = d$, $AB = t$, $DC = 1$, $DB = \sqrt{m^2 + 1}$)

Step 2: $t = y_0 - y_1$

$t = y_0 - (mx_0 + b)$ ($y_1 = mx_0 + b$ is obtained by substituting x_0 in $y = mx + b$)

$t = y_0 - mx_0 - b$

Step 3: Substitute for t in equation (1)

$\dfrac{d}{1} = \dfrac{y_0 - mx_0 - b}{\sqrt{1 + m^2}}$

$d = \dfrac{y_0 - mx_0 - b}{\sqrt{1 + m^2}}$

$\therefore d = \dfrac{|y_0 - mx_0 - b|}{\sqrt{m^2 + 1}}$

Note: If given $Ax + By + C = 0$, $y = -\frac{A}{B}x - \frac{C}{B}$

Then $d = \dfrac{|Ax_0 + By_0 + C|}{\sqrt{A^2 + B^2}}$ by substitution.

(We take the absolute value of the numerator in case the numerator is negative)

Method 2

Given: The point $A(x_0, y_0)$ and the line $y = mx + b$.

Required: To find a formula for the distance, d, between the point $A(x_0, y_0)$ and the line $y = mx + b$.

Construction: Draw the perpendicular line from the point $A(x_0, y_0)$ to meet the line $y = mx + b$ ℓ_2 at E.

Step 1: Let lines ℓ_1, and ℓ_2 meet at (r, s). (Figure below)

Let the distance between the points (x_0, y_0) and $(r, s) = d$.

Then $d^2 = (x_0 - r)^2 + (y_0 - r)^2$ or $d = \sqrt{(x_0 - r)^2 + (y_0 - r)^2}$

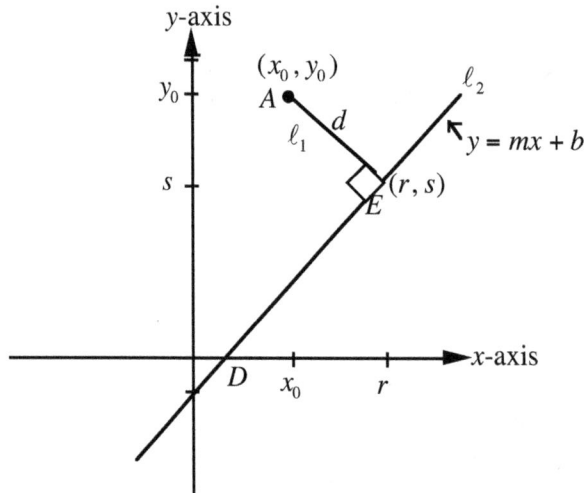

Step 2: We shall now express d in terms of x_0, y_0, m, and b only.

(That is we shall eliminate r, and s from the formula)

The slope of line $\ell_1 = \dfrac{y_0 - s}{x_0 - r}$.   (1)

The slope of line $\ell_2 = m$. (from $y = mx + b$)

Therefore slope of $\ell_1 = -\dfrac{1}{m}$ (2)   (Since ℓ_1 and ℓ_2 meet at right angles at E, the slopes are negative reciprocals of each other)

Equating right-hand-sides of (1) and (2) to each other, we obtain

$\dfrac{y_0 - s}{x_0 - r} = -\dfrac{1}{m}$

$my_0 - ms = -x_0 + r$ (cross-multiplying)

$r = my_0 + x_0 - ms$ (3)   (solving for r)

Since (r, s) is on the line $y = mx + b$

$s = mr + b$ (4)   (substituting r for x and s for y.)

Step 3: We now solve equations (3) and (4) simultaneously for r and s.

Substitute for s from (4) in (3): $my_0 + x_0 - m(mr + b) = r$

Now, we solve for r: $my_0 + x_0 - m^2r - mb = r$

$my_0 + x_0 - mb = m^2r + r$

$my_0 + x_0 - mb = r(m^2 + 1)$

$$r = \frac{my_0 + x_0 - mb}{m^2 + 1} \qquad (5)$$

Substitute for r from (5) in (4). Then we obtain

$$s = m\left[\frac{my_0 + x_0 - mb}{m^2 + 1}\right] + b$$

$$= \frac{m^2 y_0 + mx_0 - m^2 b + m^2 b + b}{m^2 + 1}$$

$$s = \frac{m^2 y_0 + mx_0 + b}{m^2 + 1} \qquad (6)$$

Step 4: We now substitute right-hand sides of equations (5) and (6) for r and s respectively in the formula $d^2 = (x_0 - r)^2 + (y_0 - r)^2$. Then we obtain

$$d^2 = \left\{x_0 - \left[\frac{my_0 + x_0 - mb}{m^2 + 1}\right]\right\}^2 + \left\{y_0 - \left[\frac{m^2 y_0 + mx_0 + b}{m^2 + 1}\right]\right\}$$

$$= \frac{\{m^2 x_0 + x_0 - my_0 - x_0 + mb\}^2}{(m^2 + 1)^2} + \frac{\{m^2 y_0 + y_0 - m^2 y_0 - mx_0 - b\}^2}{(m^2 + 1)^2}$$

$$= \frac{\{m^2 x_0 - my_0 + mb\}^2}{(m^2 + 1)^2} + \frac{\{y_0 - mx_0 - b\}^2}{(m^2 + 1)^2}$$

$$= \frac{\{-m(-mx_0 + y_0 - b)\}^2}{(m^2 + 1)^2} + \frac{\{y_0 - mx_0 - b\}^2}{(m^2 + 1)^2}$$

$$= \frac{\{(-m)^2(-mx_0 + y_0 - b)\}^2}{(m^2 + 1)^2} + \frac{\{y_0 - mx_0 - b\}^2}{(m^2 + 1)^2}$$

$$= \frac{m^2(y_0 - mx_0 - b)^2}{(m^2 + 1)^2} + \frac{(y_0 - mx_0 - b)^2}{(m^2 + 1)^2}$$

$$= \frac{m^2(y_0 - mx_0 - b)^2 + (y_0 - mx_0 - b)^2}{(m^2 + 1)^2}$$

$$= \frac{(y_0 - mx_0 - b)^2(m^2 + 1)}{(m^2 + 1)^2}$$

$$= \frac{(y_0 - mx_0 - b)^2(m^2 + 1)}{(m^2 + 1)^2}$$

$$d^2 = \frac{(y_0 - mx_0 - b)^2}{(m^2 + 1)}$$

$$d = \sqrt{\frac{(y_0 - mx_0 - b)^2}{(m^2 + 1)}}$$

$$= \frac{\sqrt{(y_0 - mx_0 - b)^2}}{\sqrt{m^2 + 1}}$$

$$\therefore d = \frac{|y_0 - mx_0 - b|}{\sqrt{m^2 + 1}} \qquad \text{(noting that } \sqrt{a^2} = |a|)$$

and the derivation is complete.

Application 2: Perpendicular Bisector of a Line

Example

Find the slope-intercept form of the equation of line L_1 (Figure below) which passes through the mid-point C of line L_2 given that line L_2 passes through the two points A (1 ,4) and B (7, 8)

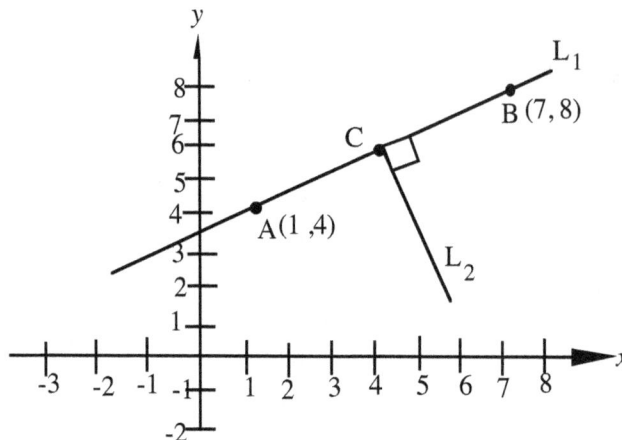

Solution

Step 1 Find the slope of line L_2

$$m = \frac{y_2 - y_1}{x_2 - x_1}$$

$$= \frac{8 - 4}{7 - 1}$$

$$= \frac{4}{6}$$

$$m = \frac{2}{3}$$

Step 2: Find the slope of L_1

Since L_1 is perpendicular to L_2, the slope of L_2 is $-\frac{3}{2}$ ($m_2 = -\frac{1}{m_1}$. The lines are perpendicular)

Step 3 : Find the coordinates of the midpoint C

$$x_m = \frac{1 + 7}{2} = 4 \quad y_m = \frac{4 + 8}{2} = 6$$

The mid-point is at (4, 6)

Step 4: now find the slope-intercept form of the equation of the line with slope $-\frac{3}{2}$ and passing through the point (4, 6).

Applying $y - y_1 = m(x - x_1)$, we obtain

$y - 6 = -\frac{3}{2}(x - 4)$ <-------- point-slope form

$y - 6 = -\frac{3}{2}x + 6$

$y = -\frac{3}{2}x + 6 + 6$

$y = -\frac{3}{2}x + 12$ <-----slope-intercept form

Test # 10 –Student's Self-Test (**Always, Test yourself before you are tested**)

Attempt all questions on clean sheets of paper, **Do not write in the book** Show all necessary work.

---

1. Draw the graphs of the following lines whose equations are given:

    (a) $y = -2x + 5$

    (b) $3y = 18 - 3x$

    (c) $2y - 6x = -8$

2. Determine which of the points $(2, 3)$, $(-2, 5)$, $(-2, -16)$ are on the line whose equation is given by $y = 5x - 6$.

---

3. Find the slope of each line passing through the given points:

    (a) $(5, 6)$ and $(3, 2)$;

    (b) $(2, -4)$ and $(3, -2)$.

4. Find the slope of each line passing through the given points:

    (a) $(1, -1)$; and $(5, 2)$.

    (b) $(-3, 3)$ and $(2, 2)$.

---

5. Find the slope of each line passing through the given points:

    (a) $(5, 2)$ and $(-4, 2)$;

    (b) $(-3, -6)$ and $(-3, -5)$.

6. Find the slope of the line passing through the points $(5, 9)$ and $(3, 2)$.

---

7. The points $(-5, 6)$ and $(2, 2)$ are on a line whose slope we want to determine. What is the slope of this line?

8. Find the slope of the line containing the points $(2, 8)$ and $(1, 6)$.

---

9. Find the slope of the line passing through the points $(1, -1)$, $(3, 3)$ and $(-2, -7)$.

10. Find the slope and the y-intercept of the following:

    (a) $y = -6x + 4$;   (b) $-7 + 5x = y$;

    (c) $y = -\frac{x}{2} + 8$

---

11. Find the slope and the y-intercept:

    (a) $4x + 8y = 32$

    (b) $5y - 6x = 21$

12. Find the slope and the y-intercept:

    (a) $y = -2x + 4$

    (b) $y = \frac{1}{5}x - 3$

---

13. Find the slope and the y-intercept:

    (a) $4y = 6x + 12$

    (b) $y = \frac{x}{3} + 5$

15. Find an equation of the line with slope 5 and y-intercept -3.

14. Find the slope and the y-intercept:

    (a) $3y + 5x = 8$

    (b) $5x + 4y = 12$

16. Find an equation of the line with slope $\frac{2}{3}$ and y-intercept 5.

17. A line has a y-intercept -2 and a slope of 6. Find an equation for this line.	**18.** Find an equation of the line with slope -3 and passing through the point $(5, -2)$.
19. A line passes through the point $(-1, -7)$ and has a slope of 5. Find an equation for this line.	**20.** Find an equation of the line passing through the points $(2, 2)$ and $(-5, 6)$
21 (a) If the points $(2, -5)$ and $(-3, 1)$ are on a certain line, find an equation for this line. (b) Find an equation of the line with slope -3 and y-intercept 8.	**22.** Find equations for lines m and l 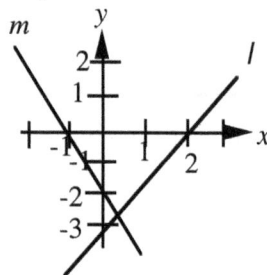
23. (a) Find an equation of the line with slope 2 and passing through the point $(1, 6)$ (b) Sketch the graph of the line $y = 2$ (c) Draw the graph of the line $x = 4$. (d) Draw the graph of the line $x = -1$.	**24.** (a) Find an equation of the horizontal line passing through the point $(3, -4)$. (b) Find an equation of the vertical line passing through the point $(2, -3)$. (c) Does the line in (b) have a y-intercept? If yes, find it.
25. (a) Does the line in Problem **23** (b) have a y-intercept? If yes, find it. **(b)** Sketch the graph of the line $y = -5$.	**Bonus:** (a) Sketch the graph of the line $y = -4$. (b) Sketch the graph of the line $y = 1$.

Answers: **2.** The point $(-2, -16)$; **3.** (a) 2; (b) 2; **4.** (a) $\frac{3}{4}$; (b) $-\frac{1}{5}$; **5.** (a) 0;

(b) undefined; **6.** $\frac{7}{2}$; **7.** $-\frac{4}{7}$; **8.** 2; **9.** 2; **10.** (a) Slope: -6, y-intercept: 4;

(b) Slope: 5, y-intercept:: -7; (c) Slope: $-\frac{1}{2}$, y-intercept: 8; **11.** (a) slope: $-\frac{1}{2}$, y-intercept: 4;

(b) Slope: $\frac{6}{5}$, y-intercept: $\frac{21}{5}$; **12.** (a) Slope: -2, y-intercept: 4; (b) Slope: $\frac{1}{5}$, y-intercept: -3;

13. (a) Slope: $\frac{3}{2}$, y-intercept: 3; (b) Slope: $\frac{1}{3}$, y-intercept: 5; **14.** (a) Slope: $-\frac{5}{3}$, y-intercept: $\frac{8}{3}$;

(b) Slope: $-\frac{5}{4}$, y-intercept: 3; **15.** $y = 5x - 3$; **16.** $y = \frac{2}{3}x + 5$; **17.** $y = 6x - 2$;

18. $y = -3x + 13$; **19.** $y = 5x - 2$; **20.** $y = -\frac{4}{7}x + \frac{22}{7}$ or $4x + 7y = 22$;

21. (a) $y = -\frac{6}{5}x - \frac{13}{5}$ or $6x + 5y = -13$; (b) $y = -3x + 8$; **22.** For the line m: $y = -2x - 2$,

for the line l: $y = \frac{3}{2}x - 3$; **23.** (a) $y = 2x + 4$; **24.** (a) $y = -4$; (b) $x = 2$; (c) No;

25. (a) Yes. y-intercept: 2.

CHAPTER 5A
Exponents and Radicals

Lesson 16: **Review of Exponents and Radicals**
Lesson 17: **Definitions and Simplification of Radicals**
Lesson 18: **Addition and Multiplication of Radicals**

Lesson 16
Review of Exponents and Radicals (see also Lesson 2)

Mnemonic Examples for the Rules of Exponents and Radicals

1. $x^2 x^3 = x^5$

2. $\dfrac{x^7}{x^4} = x^{7-4} = x^3$

3. $\left(x^4\right)^2 = x^8$

4. $(xy)^5 = x^5 y^5$

5. $\left(\dfrac{x}{y}\right)^6 = \dfrac{x^6}{y^6}$

6. $x^{-2} = \dfrac{1}{x^2}$

7. $x^0 = 1$

8. $\dfrac{x^4}{x^7} = x^{4-7} = x^{-3} = \dfrac{1}{x^3}$

9. $9^{1/2} = \sqrt{9}$

10. $\sqrt[3]{xy} = \sqrt[3]{x}\,\sqrt[3]{y}$

11. $8^{1/3} = \sqrt[3]{8} = 2$

12. $8^{2/3} = \left(\sqrt[3]{8}\right)^2 = \sqrt[3]{8^2} = 4$

13. $\left(\sqrt[4]{8}\right)^4 = \left(\sqrt[4]{8^4}\right) = 8$

14. $\sqrt[3]{\dfrac{x}{y}} = \dfrac{\sqrt[3]{x}}{\sqrt[3]{y}}$

Extra: Show that $x^{-2} = \dfrac{1}{x^2} \cdot \dfrac{x^7}{x^3} = x^{7-4} = x^3$

Note: $x^{-2} = \dfrac{x^{-2}}{1} \cdot \dfrac{x^2}{x^2} = \dfrac{x^{-2+2}}{x^2} = \dfrac{x^0}{x^2} = \dfrac{1}{x^2}$.

Example 1 Evaluate 7^{-2}

Step 1: Express with positive exponents.
$$7^{-2} = \dfrac{1}{7^2}$$

Step 2: Simplify. $\dfrac{1}{7^2} = \dfrac{1}{49}$

Example 2 Simplify, leaving answer with only positive exponents

$$\dfrac{x^{-3}y^2 z^6}{x^2 y^{-3} z}$$

Method 1 (Strictly, using the rules of exponents)

$$\frac{x^{-3}y^2z^6}{x^2y^{-3}z} = x^{-3-2}y^{2+3}z^{6-1}$$

$$= x^{-5}y^5z^5$$

$$= \frac{y^5z^5}{x^5}$$

Method 2 Taking powers across the division bar, and changing the signs of exponents, followed by cancellation

$$\frac{x^{-3}y^2z^6}{x^2y^{-3}z} = \frac{y^2y^3\cancel{z^6}\,z^5}{x^2x^3\cancel{z}}$$

(change x^{-3} to x^3 and write it in the denominator. Change y^{-3} to y^3 and write it in the numerator)
If the exponent is positive, leave the power where it is. The changes are for the powers with negative exponents).

$$= \frac{y^5z^5}{x^5}$$

 A note about cancellation in Method 2 above:
In order to apply cancellation (as done in arithmetic), the exponents of the powers involved in the cancellation must have the same sign; preferably, the exponents must be positive (or made positive by following the instructions outlined above) .

Example 3   Simplify $(4x)^0 - 4\,x^0$

Solution

$(4x)^0 - 4x^0 = 4^0x^0 - 4x^0$ **Note:** $x^0 = 1, (4)^0 = 1$

$\qquad\qquad = (1)(1) - 4(1)$

$\qquad\qquad = 1 - 4$
$\qquad\qquad = -3$

Example 4   Multiply $\left(\frac{-8a^2b}{5}\right)\left(\frac{9c}{3abc}\right)\left(\frac{-15}{27c}\right)$

Solution $\dfrac{-8(9)(-15)a^2bc}{5(3)(27)abc^2}$

$\qquad = +\,\dfrac{8(9)(15)a^2bc}{5(3)(27)abc^2}$

$\qquad = \dfrac{8a}{3c}$ (After canceling the common factors in the numerator and the denominator)

Lesson 16 Exercises

A Simplify the following:

1. $16^{1/2}$

2. $\sqrt[3]{x^6 y^{12}}$

3. $64^{1/3}$

4. $27^{2/3}$

5. $(\sqrt[4]{16})^3$

6. $\sqrt[3]{\dfrac{x^6}{y^{15}}}$

Answers: 1. 4 ; **2.** $x^2 y^4$; **3.** 4 ; **4.** 9; **5.** 8; **6.** $\dfrac{x^2}{y^5}$

B Simplify the following:

1. 3^{-3}

3. $\left(\dfrac{4}{9}\right)^{-3/2}$

4. $\dfrac{x^{-4} y^3 z^2}{x^4 y^{-2} z}$

5. $(6x)^0 - 6x^0$

2. $100^{-1/2}$

6. $\left(\dfrac{-4a^3 b}{3b}\right)\left(\dfrac{9c}{2bc}\right)\left(\dfrac{-12}{18c}\right)$

7. $81^{-1/4}$

Answers: 1. $\dfrac{1}{27}$; **2.** $\dfrac{1}{10}$; **3.** $\dfrac{27}{8}$; **4.** $\dfrac{y^5 z}{x^8}$; **5.** -5; **6.** $\dfrac{4a^3}{bc}$; **7.** $\dfrac{1}{3}$

Lesson 17

Definitions and Simplification of Radicals

Definitions of **Rational and Irrational Numbers**

Rational number : A rational number (a fraction) is a real number which **can** be written as the ratio of two integers. The word **rational** pertains to the word **ratio.**

Examples are (a) $\frac{2}{3}$; (b) $\frac{1}{5}$; (c) 4 (since $4 = \frac{4}{1}$)

(d) 0 (since $0 = \frac{0}{7} = \frac{0}{3}$ ... or $0 = \frac{0}{b}$, where b is an integer and b $\neq$ 0)

(e) $\sqrt{4}$ (because $\sqrt{4} = 2 = \frac{2}{1}$)

Irrational number: An irrational number is a real number which **cannot** be written as the ratio of two integers. However, we can approximate irrational numbers as closely as we wish by rational numbers or decimals.

Examples of irrational numbers are $\sqrt{2}, \sqrt[3]{4}$, and π (pi). We can for example, approximate $\sqrt{2}$ by 1.414, and π by $\frac{22}{7}$ or 3.142. Except otherwise instructed, in simplifying radical expressions, we prefer to leave the irrational numbers in **radical** forms.

Radicals and Roots

Consider two real numbers r and A. We denote the **principal nth root** of A (n being a positive integer) by $\sqrt[n]{A}$. Also $\sqrt[n]{A}$ is called a **radical.** We define the principal nth root of A as follows:

$\sqrt[n]{A} = r$ if $r^n = A$ (i.e., the nth root of A = r if r^n = A) with the following qualifications:

1. Any root of zero is zero (i.e. $\sqrt[n]{0} = 0$)
2. If A is a positive number, then the principal nth root of A is the positive nth root of A.
3. If A is a negative number, then the principal nth root of A is the negative nth root of A.

Note from above that a **radical** is an expression used for indicating the root of a number.

We call n the index of the radical (n indicating the type of root being considered); A is called the radicand (the number of which a root is being taken). The radicand is thus the expression under the symbol " $\sqrt[n]{\ }$ " . Thus, the radical consists of the index n with root symbol " $\sqrt[n]{\ }$ " and the radicand.

If $n = 2$, we obtain $\sqrt[2]{A}$, the square root of A, We usually omit the "2" and write $\sqrt{A}$.

Therefore, $\sqrt[2]{A} = \sqrt{A}$. However, if explicit indication helps you to understand a problem, write the 2. When $n = 3$, we obtain $\sqrt[3]{A}$, the cube root A, but in this case, we have to write the "3". Similarly, for $n = 4$ and higher indices, we must always write the index. Some higher orders of the root of A are $\sqrt[4]{A}, \sqrt[5]{A}$, and $\sqrt[8]{A}$.

Some specific radicals are $\sqrt{2}, \sqrt[3]{4}$, $\sqrt{10}$, and $\sqrt{5}$.

Example 1 Name the parts of the radical $\sqrt{32}$

Solution   The index is understood to be 2, since $\sqrt{32} = \sqrt[2]{32}$
The radicand is 32.

Example 2   Name the parts of the radical $\sqrt[3]{16}$

Solution   The index is 3
The radicand is 16

Square Roots

Example $\sqrt{9}$ or $\sqrt[2]{9}$

This radical is read as the square root of 9. The square root of 9 is 3 (because $3^2 = 9$).

The following definition is useful in finding the **square root** of a number:

The **principal square root** of a number (nonzero number) is one of the two equal positive factors of that number.

Thus, if we can "break up" a number into two equal positive factors, then, one of the positive factors is the square root.(Note that we exclude negative roots).

Square root of zero: $\sqrt{0} = 0$ (that is, the square root of zero is zero).
because $0^2 = 0$

Examples   (a) $\sqrt{9} = \sqrt{(3)(3)} = 3$;   (b) $\sqrt{64} = \sqrt{(8)(8)} = 8$ (one of (8)(8) is 8)

Cube Roots

Example $\sqrt[3]{8}$

The above radical consists of the index 3, the radical sign and the radicand 8.
This radical is read " the cube root of 8". The cube root of 8 is 2. (because $2^3 = 8$)

The following definition is also useful in finding the cube root of a number:

The **cube root** of a number is **one of the three equal** factors of that number. Here, we do not specify positive root, since the cube root may be positive or negative, depending on whether the given number is positive or negative.

Examples (a) $\sqrt[3]{8} = \sqrt[3]{(2)(2)(2)} = 2$ **(b)** $\sqrt[3]{-8} = \sqrt[3]{(-2)(-2)(-2)} = -2$

Check: $2^3 = 8$ Check: $(-2)^3 = -8$

Note: For **even roots** such as the square root of a , $(\sqrt{a})$, the fourth root of a, $(\sqrt[4]{a})$, the sixth root

of a, $(\sqrt[6]{a})$, $\sqrt[n]{a}$ implies the positive root.

For **odd roots** such as the cube root $(\sqrt[3]{a})$, the fifth root $(\sqrt[5]{a})$, or the seventh root $(\sqrt[7]{a})$, the root may be positive or negative according to whether the given number is positive or negative.

More Examples

(a) $\sqrt[4]{16} = \mathbf{2}$ or $\sqrt[4]{16} = \sqrt[4]{(2)(2)(2)(2)} = \mathbf{2}$
because $2^4 = 16$

(b) $\sqrt[5]{32} = 2$ or $\sqrt[5]{32} = \sqrt[5]{(2)(2)(2)(2)(2)} = \mathbf{2}$
because $2^5 = 32$

(c) $\sqrt[5]{-32} = -2$ or $\sqrt[5]{-32} = \sqrt[5]{(-2)(-2)(-2)(-2)(-2)} = \mathbf{-2}$
because $(-2)^5 = -32$

Simplifying Radicals

Two approaches are considered: One approach depends on guessing correctly a perfect power which is a factor of the radicand. The other approach is the application of prime factorization, and the grouping of the factors into n equal factors: for example, for square root, we would like to have two equal **positive** factors. For cube root, we would like to have three equal factors.

Example Simplify the following: **(a)** $\sqrt{32}$; **(b)** $\sqrt{\dfrac{49}{64}}$; **(c)** $\sqrt[3]{80}$.

Solutions

(a) Method 1

$$\sqrt{32} = \sqrt{(16)(2)}$$
$$= \sqrt{16}\sqrt{2}$$
$$= 4\sqrt{2} \quad (\sqrt{16} \text{ is } \textbf{rational but } \sqrt{2} \text{ is } \textbf{irrational})$$

Method 2

$$\sqrt{32} = \sqrt{(2)(2)(2)(2)(2)}$$
$$= \sqrt{(4)(4)(2)} \quad (\textbf{Rule}: \sqrt[n]{ab} = \sqrt{a}\sqrt{b})$$
$$= 4\sqrt{2}$$

(b) $\sqrt{\dfrac{49}{64}} = \dfrac{\sqrt{49}}{\sqrt{64}} = \dfrac{7}{8}$ $\qquad \left(\textbf{Rule}: \sqrt[n]{\dfrac{a}{b}} = \dfrac{\sqrt[n]{a}}{\sqrt[n]{b}}\right)$

(c) $\sqrt[3]{80} = \sqrt[3]{(2)(2)(2)(2)(5)}$

$\qquad\qquad = \sqrt[3]{(2)(2)(2)} \sqrt[3]{10}$

$\qquad\qquad = \sqrt[3]{8} \sqrt[3]{10}$

$\qquad\qquad = 2\sqrt[3]{10}$ (We leave the product of the "extra" 2 and the 5 under the radical sign as 10)

Scrapwork:

$80 = 2 \times 2 \times 2 \times 2 \times 5$

Simplify the following: **1.** $\sqrt{x^3 y^2}$; **2.** $\sqrt{27}$; **3.** $\sqrt{18x^3 y^4}$; **4.** $\sqrt{64x^7 y^{14} z^3}$; **5.** $\sqrt{\dfrac{8}{9}}$

Solutions

1. $\sqrt{x^3 y^2} = \sqrt{x^2 y^2 x} = xy\sqrt{x}$

2. $\sqrt{27} = \sqrt{(9)(3)} = \sqrt{9}\sqrt{3} = 3\sqrt{3}$

3. $\sqrt{18x^3 y^4} = \sqrt{(9)(2)x^2 x y^4} = \sqrt{9x^2 y^4 \cdot 2x} = 3xy^2 \sqrt{2x}$

4. $\sqrt{64x^7 y^{14} z^3} = \sqrt{64x^6 y^{14} z^2 \cdot xz} = 8x^3 y^7 z\sqrt{xz};$

5. $\sqrt{\dfrac{8}{9}} = \dfrac{\sqrt{8}}{\sqrt{9}} = \dfrac{2\sqrt{2}}{3}$

Lesson 17 Exercises

A Simplify:

1. $\sqrt{18}$;   **2.** $\sqrt{50}$;   **3.** $\sqrt{24}$;   **4.** $\sqrt{72}$;   **5.** $\sqrt{288}$;   **6.** $\sqrt{48}$;   **7.** $\sqrt{\dfrac{81}{100}}$

Answers: **1.** $3\sqrt{2}$; **2.** $5\sqrt{2}$; **3.** $2\sqrt{6}$; **4.** $6\sqrt{2}$; **5.** $12\sqrt{2}$; **6.** $4\sqrt{3}$; **7.** $\dfrac{9}{10}$

B Simplify the following: **1.** $\sqrt{16x^6}$ **2.** $\sqrt{9x^8y^2}$ **3.** $\sqrt{27x^9y^4}$

4. $\sqrt{18x^3y^5}$;   **5.** $\sqrt{32x^8y}$;   **6.** $\sqrt{8x^4y^2z^3}$

Answers: **1.** $4x^3$;   **2.** $3x^4y$;   **3.** $3x^4y^2\sqrt{3x}$;   **4.** $3xy^2\sqrt{2xy}$;   **5.** $4x^4\sqrt{2y}$; **6.** $2x^2yz\sqrt{2z}$

Lesson 18
Addition and Multiplication of Radicals

Addition of Radicals (Like Radicals or Similar Radicals)

Like radicals have the same index **and** the same radicand.

Examples: **1.** $4\sqrt{3}$ and $\sqrt{3}$ are like radicals
(index = 2 , the square root index; radicand = 3).

2. $5\sqrt[3]{4}$ and $2\sqrt[3]{4}$ are like radicals
(index = 3, radicand = 4).

Unlike radicals: Unlike radicals have different indices **or** radicands or both.

Examples   **1.** $\sqrt{3}$ and $\sqrt{2}$ are unlike radicals. (They have different radicands.)

2. $\sqrt[3]{2}$ and $\sqrt[4]{2}$ are unlike radicals (They have different indices.)

Like radicals can be combined into a single radical (i.e., added or subtracted in much the same way as like terms of polynomials are added or subtracted).

Unlike radicals cannot be combined into a single radical (i.e., cannot be added). However, in some cases, by simplifying the given radical(s), like radicals may be obtained, and which then may be added.

To **add like radicals**, add the non-radical parts (coefficients) and keep the radical part.

Example 1   Add: $3\sqrt{11} + 5\sqrt{11}$ (same index = 2; same radicand =11)

Solution $3\sqrt{11} + 5\sqrt{11}$
$= (3 + 5)\,\sqrt{11}$
$= 8\sqrt{11}$

Note that $3\sqrt{11}$ means 3 times $\sqrt{11}$; but $\sqrt[3]{11}$ means the cube root of 11.

Example 2   Simplify : $3\sqrt{19} + 8\sqrt{10} + 6\sqrt{19} - 2\sqrt{10}$

Solution $3\sqrt{19} + 8\sqrt{10} + 6\sqrt{19} - 2\sqrt{10}$

$= 3\sqrt{19} + 6\sqrt{19} + 8\sqrt{10} - 2\sqrt{10}$ <-------- You may skip this step.

$= (3 + 6)\sqrt{19} + (8 - 2)\sqrt{10}$
$= 9\sqrt{19} + 6\sqrt{10}$

Example 3 Add : $\sqrt{50} + \sqrt{72}$

Solution Step 1: Simplify the radicals first to see if there are any like radicals

$\sqrt{50} = \sqrt{(25)(2)} =$ or $\sqrt{50} = \sqrt{(5)(5)(2)}$
$\sqrt{50} = 5\sqrt{2}$ $= 5\sqrt{2}$

$\sqrt{72} = \sqrt{(36)(2)}$ or $=\sqrt{(3)(3)(2)(2)(2)}$
$\sqrt{72} = 6\sqrt{2}$ $= (3)(2)\sqrt{2} = 6\sqrt{2}$

Step 2: Now, we have like radicals and therefore, we can add.

$\sqrt{50} + \sqrt{72}$
$= 5\sqrt{2} + 6\sqrt{2}$
$= (5 + 6)\sqrt{2}$
$= 11\sqrt{2}$

In the future, Step 1 could be considered as scrapwork and show only Step 2.

Example 4   Simplify: $2\sqrt{75} - 4\sqrt{27}$
Solution

$$= 2\sqrt{75} - 4\sqrt{27}$$
$$= 2\sqrt{(25)(3)} - 4\sqrt{(9)(3)}$$
$$= 2(5)\sqrt{3} - 4(3)\sqrt{3}$$
$$= 10\sqrt{3} - 12\sqrt{3}$$
$$= (10 - 12)\sqrt{3}$$
$$= -2\sqrt{3}$$

Example 5   Simplify: $4x\sqrt{25y} - 6x\sqrt{16y}$.
Solution

$$4x\sqrt{25y} - 6x\sqrt{16y}$$
$$= 4x(5)\sqrt{y} - 6x(4)\sqrt{y} \qquad \text{Scrapwork: } \sqrt{25} = 5; \sqrt{16} = 4$$
$$= 20x\sqrt{y} - 24x\sqrt{y}$$
$$= (20x - 24x)\sqrt{y} \qquad \text{(Adding the coefficients of } \sqrt{y}.)$$
$$= -4x\sqrt{y}$$

Multiplication of Radicals

Compared to the conditions in the addition of radicals, the only condition here is that the radicals to be multiplied must have the **same index**. Thus, the radicands may be different or the same, but the index must be the same. Note however that radicals with different indices can always be changed to radicals with a common index by using fractional exponents (see p.102 Examples 6 & 7, bottom of page) and then multiplying the resulting radicals (The common index will be the LCM of the different indices.)

Example 1   Multiply $\sqrt{3}$ and $\sqrt{5}$
Solution

$$(\sqrt{3})(\sqrt{5}$$
$$= \sqrt{(3)(5)}$$
$$= \sqrt{15}$$

Example 2 Multiply $7\sqrt{3}$ and $9\sqrt{5}$
Solution
Procedure: First multiply the non-radical parts (coefficients) and then multiply the radical parts and simplify if possible.

$$(7\sqrt{3})(9\sqrt{5})$$
$$= (7)(9)\sqrt{(3)(5)}$$
$$= 63\sqrt{15}$$

Example 3   Multiply $4\sqrt{3}$ and $5\sqrt{3}$
Solution

$$(4\sqrt{3})(5\sqrt{3}) = (4)(5)\sqrt{(3)(3)}$$
$$= 20\sqrt{9}$$
$$= 20(3)$$
$$= 60$$

Example 4 Find the product of $2\sqrt{3}$ and $5\sqrt{2} - 4\sqrt{3}$

Solution $2\sqrt{3}\,(5\sqrt{2} - 4\sqrt{3})$<----------The parentheses are important.

$= (2\sqrt{3})(5\sqrt{2}) - (2\sqrt{3})(4\sqrt{3})$ <--------Application of the distributive rule.

$= (2)(5)\sqrt{(3)(2)} - (2)(4)\sqrt{(3)(3)}$

$= 10\sqrt{6} - 8\sqrt{9}$

$= 10\sqrt{6} - 8(3)$

$= 10\sqrt{6} - 24$

$= -24 + 10\sqrt{6}$

Note: In Example 4, you may skip lines 2 and 3.

Example 5 Simplify: $\sqrt{8}\sqrt{24}$

Method 1

$\sqrt{8}\sqrt{24} = \sqrt{(8)(24)}$

$= \sqrt{(2)(2)(2)(3)(2)(2)(2)}$

$= \sqrt{(2)(2)(2)\,(2)(2)(2)\,(3)}$

$= \sqrt{(8)(8)3}$

$= 8\sqrt{3}$

Method 2 $\sqrt{8}\sqrt{24}$

Step 1: Simplify each radical first.

$\sqrt{8}\sqrt{24}$

$= (2\sqrt{2})(2\sqrt{6})$

Step 2: Multiply the radicals.

$= (2)(2)(\sqrt{2})(\sqrt{6})$

$= 4\sqrt{12}$

$= 4\sqrt{4}\sqrt{3}$

$= 4(2)\sqrt{3}$

$= 8\sqrt{3}$

Scrapwork:
1. $\sqrt{8} = 2\sqrt{2}$
2. $\sqrt{24} = 2\sqrt{6}$

Example 6 $(\sqrt{5})(\sqrt{5}) = \sqrt{25} = 5$

Example 7 $(\sqrt{6})(\sqrt{6}) = \sqrt{36} = 6$

Lesson 18 Exercises

A Add or subtract:

1. $2\sqrt{3} + 4\sqrt{3}$

2. $3\sqrt{5} + \sqrt{98} + \sqrt{20} + 8\sqrt{2}$;

3. $4\sqrt{48} - \sqrt{12} + \sqrt{75}$

4. $2x\sqrt{9y} - 5x\sqrt{4y}$

Answers: **1.** $6\sqrt{3}$;   **2.** $5\sqrt{5} + 15\sqrt{2}$;   **3.** $19\sqrt{3}$;   **4.** $-4x\sqrt{y}$

B Multiply and simplify:

1. $2\sqrt{5}$ and $3\sqrt{2}$;   **2.** $3\sqrt{5}$ and $4\sqrt{10}$;   **3.** $4\sqrt{2}$ and $5\sqrt{2}$

4. $4\sqrt{3}$ and $4\sqrt{3}$;   **5.** $2x\sqrt{3y}$ and $\sqrt{6y}$

Answers: **1.** $6\sqrt{10}$;   **2.** $60\sqrt{2}$;   **3.** 40;   **4.** 48;   **5.** $6xy\sqrt{2}$

CHAPTER 5B
More Radicals

Lesson 19: **Division of Radicals; Rationalization of Denominators**
Lesson 20: **Fractional Exponents and Reduction of Indices**

Lesson 19

Division of Radicals; Rationalization of Denominators,

Division of a Radical by a Rational Number

Example   Simplify: $\dfrac{6-\sqrt{12}}{2}$

Solution $\dfrac{6-\sqrt{12}}{2}$

$= \dfrac{6}{2} - \dfrac{\sqrt{12}}{2}$

$= 3 - \dfrac{\overset{1}{\cancel{2}}\sqrt{3}}{\cancel{2}_1}$

$= 3 - \sqrt{3}$

Scrapwork: $\sqrt{12} = \sqrt{(2)(2)(3)} = = 2\sqrt{3}$

Rationalization of Denominators

To rationalize a denominator, we change a given "fraction" to an "equivalent fraction" so that there are no radicals in the denominator. The equivalent fraction may have radicals in the numerator (but not in the denominator).

Example 1 Rationalize the denominator: $\dfrac{3}{\sqrt{6}}$

We will multiply both the denominator and the numerator by a radical such that the radicand in the denominator becomes a perfect square.

We will multiply both the denominator and the numerator by $\sqrt{6}$

$\dfrac{3}{\sqrt{6}} = \dfrac{3}{\sqrt{6}}\dfrac{\sqrt{6}}{\sqrt{6}}$

$= \dfrac{3\sqrt{6}}{\sqrt{36}}$

$= \dfrac{3\sqrt{6}}{6}$

$= \dfrac{\overset{1}{\cancel{3}}\sqrt{6}}{\cancel{6}_2}$

$= \dfrac{\sqrt{6}}{2}$

A motivation for rationalizing denominators:

Note above that in practical applications, $\dfrac{\sqrt{6}}{2}$ is more convenient to use than $\dfrac{3}{\sqrt{6}}$; for example, it is easier

to divide the decimal approximation of $\sqrt{6}$ by 2 than to divide 3 by the decimal approximation of $\sqrt{6}$.

Example 2   Rationalize the denominator: $\sqrt{\frac{7}{3}}$ (one-term denominator)

Solution   We shall multiply both the denominator and the numerator by $\sqrt{3}$.

$$\sqrt{\frac{7}{3}} = \sqrt{\frac{7 \cdot 3}{3 \cdot 3}} \quad \text{or} \quad \frac{\sqrt{7}}{\sqrt{3}} \cdot \frac{\sqrt{3}}{\sqrt{3}}$$

$$= \frac{\sqrt{21}}{\sqrt{9}} \quad \text{or} \quad \frac{\sqrt{21}}{\sqrt{9}}$$

$$= \frac{\sqrt{21}}{3} \quad \text{or} \quad \frac{\sqrt{21}}{3}$$

Example 3   Rationalize the denominator : $\sqrt[3]{\frac{5}{4}}$

Solution

$$\sqrt[3]{\frac{5}{4}} = \sqrt[3]{\frac{5}{2 \cdot 2}}$$

$$= \sqrt[3]{\frac{5 \cdot 2}{2 \cdot 2 \cdot 2}} \quad \text{<-------One more 2 will make the denominator a perfect cube.}$$

$$= \frac{\sqrt[3]{10}}{\sqrt[3]{8}}$$

$$= \frac{\sqrt[3]{10}}{2} \; .$$

Example 4 Rationalize the denominator: $\sqrt[3]{\frac{8}{9}}$

Method 1 $\sqrt[3]{\frac{8}{9}} = \sqrt[3]{\frac{(2)(2)(2)}{(3)(3)}}$

$$= \sqrt[3]{\frac{(2)(2)(2)(3)}{(3)(3)(3)}} \quad \text{<--- one more 3 will make the denominator a perfect cube}$$

$$= \frac{2\sqrt[3]{3}}{3}$$

Method 2 $\sqrt[3]{\frac{8}{9}} = \frac{\sqrt[3]{8}}{\sqrt[3]{9}}$

$$= \frac{2}{\sqrt[3]{9}} \qquad (\sqrt[3]{8} = 2)$$

$$= \frac{2\sqrt[3]{3}}{\sqrt[3]{9} \, \sqrt[3]{3}}$$

$$= \frac{2\sqrt[3]{3}}{\sqrt[3]{27}}$$

$$= \frac{2\sqrt[3]{3}}{3}$$

Lesson 19: Division of Radicals; Rationalization of Denominators

Method 3 In the following approach, simplifying the numerator becomes more involved than Methods 1 & 2:

$$\sqrt[3]{\frac{8}{9}} = \frac{\sqrt[3]{8}}{\sqrt[3]{9}} \cdot \frac{\sqrt[3]{9}(\sqrt[3]{9})}{\sqrt[3]{9}(\sqrt[3]{9})} = \frac{\sqrt[3]{8 \cdot 9 \cdot 9}}{\sqrt[3]{9 \cdot 9 \cdot 9}} = \frac{\sqrt[3]{8 \cdot 9 \cdot 9}}{9}$$

$$= \frac{\sqrt[3]{2 \cdot 2 \cdot 2 \cdot 3 \cdot 3 \cdot 3 \cdot 3}}{9} = \frac{2 \cdot 3\sqrt[3]{3}}{9} = \frac{2\sqrt[3]{3}}{3}$$

Example 5 Rationalize the denominator: $\dfrac{\sqrt{5}}{\sqrt[3]{2}}$

Solution $\dfrac{\sqrt{5}}{\sqrt[3]{2}} = \dfrac{\sqrt{5}}{\sqrt[3]{2}} \cdot \dfrac{\sqrt[3]{2}(\sqrt[3]{2})}{\sqrt[3]{2}(\sqrt[3]{2})} = \dfrac{\sqrt{5} \cdot \sqrt[3]{4}}{2}$ or $\dfrac{\sqrt[6]{2000}}{2}$ (See also page 102 Examples 6 & 7, bottom of page)

Definition : The **conjugate** of a given binomial is another binomial that differs from the given binomial only in the sign of one of the terms. The conjugate of $a + b$ is $a - b$; and the conjugate of $a - b$ is $a + b$.

Example 6 Rationalize the denominator $\dfrac{5}{\sqrt{3} + \sqrt{2}}$ (two-term denominator).

Procedure: Multiply both the denominator and the denominator by the conjugate of $\sqrt{3} + \sqrt{2}$.

The conjugate of $\sqrt{3} + \sqrt{2}$ is $\sqrt{3} - \sqrt{2}.$ (You may also multiply by $-\sqrt{3} + \sqrt{2}$; try it later on.)

Step 1: $\dfrac{5}{\sqrt{3} + \sqrt{2}} = \dfrac{5 \;(\sqrt{3} - \sqrt{2})}{(\sqrt{3} + \sqrt{2})(\sqrt{3} - \sqrt{2})}$

Step 2: $= \dfrac{5(\sqrt{3} - \sqrt{2})}{(\sqrt{3})(\sqrt{3}) - (\sqrt{3})(\sqrt{2}) + (\sqrt{2})(\sqrt{3}) - (\sqrt{2})(\sqrt{2})}$

$= \dfrac{5(\sqrt{3} - \sqrt{2})}{\sqrt{9} + 0 - \sqrt{4}}.$ Note: $(-\sqrt{3})(\sqrt{2}) + (\sqrt{2})(\sqrt{3}) = 0$

$= \dfrac{5(\sqrt{3} - \sqrt{2})}{3 - 2}$

$= \dfrac{5(\sqrt{3} - \sqrt{2})}{1}$

$= 5(\sqrt{3} - \sqrt{2})$ or $5\sqrt{3} - 5\sqrt{2}$

Note above: As was suggested in the procedure, you could also have multiplied by $-\sqrt{3} + \sqrt{2}$ and obtained the same result. Therefore, it is not critical which of the terms differ in sign.

Example 7 Rationalize the denominator : $\dfrac{1}{\sqrt{8} - 2}$

Procedure: Multiply both the denominator and the numerator by the conjugate of $\sqrt{8}$ - 2.
The conjugate of $\sqrt{8}$ - 2 is $\sqrt{8} + 2$ (You may also multiply by $-\sqrt{8}$ - 2; try it later on.)

$$\dfrac{1}{\sqrt{8} - 2} = \dfrac{1}{(\sqrt{8} - 2)} \dfrac{(\sqrt{8} + 2)}{(\sqrt{8} + 2)}$$

$$= \dfrac{\sqrt{8} + 2}{(\sqrt{8})(\sqrt{8}) + 2\sqrt{8} - 2\sqrt{8} - 4}$$ <----After some practice, you may skip writing this step.

$$= \dfrac{\sqrt{8} + 2}{\sqrt{64} + 0 - 4}$$ $(2\sqrt{8} - 2\sqrt{8} = 0)$

$$= \dfrac{\sqrt{8} + 2}{8 - 4}$$

$$= \dfrac{2\sqrt{2} + 2}{4}$$

$$= \dfrac{2(\sqrt{2} + 1)}{4}$$

$$= \dfrac{\sqrt{2} + 1}{2} \quad \text{or} \quad \dfrac{1}{2} + \dfrac{\sqrt{2}}{2}$$

Note above that $\dfrac{1}{\sqrt{8} - 2} = \dfrac{1}{-2 + \sqrt{8}}$

* Perhaps, it would be better to say "a" conjugate instead of "the" conjugate since it does not matter which terms differ in sign, so far as the rationalization of denominators is concerned.

Lesson 19 Exercises

A 1. Simplify: $\dfrac{8 + \sqrt{48}}{4}$ 2. $\dfrac{12 - \sqrt{18}}{3}$

Answers: 1. $2 + \sqrt{3}$;   2. $4 - \sqrt{2}$

B Rationalize the denominators:

1. $\dfrac{1}{\sqrt{3}}$;   2. $\dfrac{2}{\sqrt{5}}$;   3. $\dfrac{5}{\sqrt{2}}$;   4. $\dfrac{6}{\sqrt{5} - \sqrt{3}}$;   5. $\dfrac{\sqrt{3}}{4 + \sqrt{3}}$;

6. $\sqrt[3]{\dfrac{4}{9}}$;   7. $\sqrt[3]{\dfrac{8}{9}}$.   8. Is $\dfrac{\sqrt{3}}{\sqrt{2}} = \dfrac{3}{\sqrt{6}}$?

Answers: 1. $\dfrac{\sqrt{3}}{3}$; 2. $\dfrac{2\sqrt{5}}{5}$; 3. $\dfrac{5\sqrt{2}}{2}$; 4. $3\sqrt{5} + 3\sqrt{3}$; 5. $\dfrac{-3 + 4\sqrt{3}}{13}$; 6. $\dfrac{\sqrt[3]{12}}{3}$; 7. $\dfrac{2\sqrt[3]{3}}{3}$; 8. Yes

Lesson 20

Fractional Exponents and Reduction of Indices

Fractional Exponents

Algebraic operations involving radicals can sometimes be made easy by changing the radicals involved to their equivalent exponential forms, and then applying the laws of exponents. The advice is that when a radical is easy to simplify as is, simplify accordingly; but if it is too complicated and not easy to simplify, consider changing to exponential form and then simplifying, followed by conversion back to radical form.

Interconversion between radicals and exponential expressions

Examples

1. $25^{1/2} = \sqrt[2]{25} = \sqrt{25} = 5$

2. $9^{1/2} = \sqrt{9} = 3$

3. $8^{1/3} = \sqrt[3]{8} = 2$

4. $8^{2/3} = (8^{1/3})^2 = (\sqrt[3]{8})^2 = 2^2 = 4$

5. $16^{3/4} = (16^{1/4})^3 = (\sqrt[4]{16})^3 = 2^3 = 8$

Rules: **1** . $\sqrt[n]{a} = a^{1/n}$

2. $\sqrt[n]{a^m} = a^{m/n}$

3. $\sqrt[n]{a^m} = (\sqrt[n]{a})^m$

Example 1   Simplify the following:

1. $a^{1/3} \cdot a^{2/3}$; **2.** $b^{2/3} \cdot b^{1/2}$;   **3.** $4^{-1/2}$;   **4.** $2^{2/5} \cdot 2^{1/4}$ **5.** $\sqrt[5]{x^2} \cdot \sqrt[5]{x^3}$

6. $\sqrt[5]{3^2} \cdot \sqrt[4]{3}$;   **7.** $\sqrt{x} \cdot \sqrt[3]{x^2}$

Solutions

1. $a^{1/3} \cdot a^{2/3} = a^{1/3 + 2/3} = a^1 = a$

2. $b^{2/3} \cdot b^{1/2} = b^{2/3 + 1/2} = b^{7/6}$ $(2/3 + 1/2 = 7/6)$

3. $4^{-1/2} = \dfrac{1}{4^{1/2}} = \dfrac{1}{\sqrt{4}} = \dfrac{1}{2}$

4. $2^{2/5} \cdot 2^{1/4} = 2^{2/5 + 1/4} = 2^{13/20}$

5. $\sqrt[5]{x^2} \cdot \sqrt[5]{x^3} = x^{2/5} x^{3/5} = x^{5/5} = x^1 = x$ or $\sqrt[5]{x^2} \cdot \sqrt[5]{x^3} = \sqrt[5]{x^2 \cdot x^3} = \sqrt[5]{x^5} = x$

6. $\sqrt[5]{3^2} \cdot \sqrt[4]{3} = 3^{2/5} \cdot 3^{1/4} = 3^{13/20} = \sqrt[20]{3^{13}}$ <---(Multiplying radicals with different indices)

7. $\sqrt{x} \cdot \sqrt[3]{x^2} = x^{1/2} \cdot x^{2/3} = x^{7/6} = \sqrt[6]{x^7} = x\sqrt[6]{x}$ <---(Multiplying radicals with different indices)

Example 2   Evaluate. $1000^{-1/3}$

Step 1: Express with positive exponents.

$$1000^{-1/3} = \frac{1}{1000^{1/3}}$$

Step 2:   Change the exponential form to radical form.

then, $\dfrac{1}{1000^{1/3}} = \dfrac{1}{\sqrt[3]{1000}} = \dfrac{1}{10}$

Example 3   Evaluate. $81^{-1/2}$

$$81^{-1/2} = \frac{1}{81^{1/2}}$$

$$= \frac{1}{\sqrt{81}}$$

$$= \frac{1}{9}$$

Example 4   Evaluate $\left(-\dfrac{8}{27}\right)^{-5/3}$

We will cover two methods.
Method 1

$\left(-\dfrac{8}{27}\right)^{-5/3}$ $= \left(\dfrac{-8}{27}\right)^{-5/3}$ <--- (We can give the minus sign to either the "8"or the"27")

$$= \frac{1}{\left(\dfrac{-8}{27}\right)^{5/3}}$$

$$= \frac{1}{\left(\sqrt[3]{\dfrac{-8}{27}}\right)^{5}}$$

$$= \frac{1}{\dfrac{\left(\sqrt[3]{-8}\right)^{5}}{\left(\sqrt[3]{27}\right)^{5}}}$$

$$= \frac{1}{\dfrac{-32}{243}} \quad \text{or} \quad 1 \div \frac{-32}{243}$$

$$= -\frac{243}{32}$$

Scrapwork: **1.** $\sqrt[3]{-8} = -2$

2. $(-2)^{5} = \mathbf{-32}$

3. $\sqrt[3]{27} = 3$

4 . $3^{5} = \mathbf{243}$

Method 2 (Much faster than Method 1)

$$\left(-\frac{8}{27}\right)^{-5/3} = \left(\frac{-8}{27}\right)^{-5/3}$$

$$= \left(\frac{27}{-8}\right)^{5/3} \quad \text{(Interchange the numerator and the denominator and change the sign of the exponent)}$$

$$= \frac{(\sqrt[3]{27})^5}{(\sqrt[3]{-8})^5} = \frac{(3)^5}{(-2)^5} = \frac{243}{-32}$$

$$= -\frac{243}{32}$$

Reduction of Indices

Example 1   Reduce the index: $\sqrt[4]{36}$

Step 1:   Change the radical to exponential form (power form).

Then, $\sqrt[4]{36} = \sqrt[4]{6^2} = 6^{2/4}$

Step 2: Reduce the exponent $\frac{2}{4}$ to lowest terms.

then $\frac{2}{4} = \frac{1}{2}$

Step 3: $6^{2/4} = 6^{1/2}$

Step 4:   Change back to the radical form.

then $6^{1/2} = \sqrt{6}$ **Note:** $6^{1/2} = \sqrt[2]{6} = \sqrt{6}$

Perfect powers

Examples: **1.** 9 is a perfect square.
   2. 8 is a perfect cube .
   3. $16y^2$ is a perfect square .
   4. $\frac{4}{9}$ is a perfect square .

What is meant by to simplify a radical?

1. It may mean remove all perfect powers from the radicand (page 92); or

2. It may mean rationalize the denominator (page 98)

3. It may mean reduce the index of the radical (page 104)

4. It may mean perform a combination of the above.

Lesson 20 Exercises

A   Simplify the following:

1. $x^{3/4} \cdot x^{1/4}$ **2.** $x^{2/3} \cdot x^{3/4}$ **3.** $3^{1/2} \cdot 3^{2/3}$

Answers: **1.** x ;   **2.** $x^{17/12}$;   **3.** $3^{7/6}$

B   Evaluate the following:

1. $100^{-1/2}$;   **2.** $25^{-1/2}$;   **3.** $\left(\dfrac{16}{81}\right)^{-1/4}$

Answers: **1.** $\dfrac{1}{10}$;   **2.** $\dfrac{1}{5}$;   **3.** $\dfrac{3}{2}$

C Reduce the index:

1. $\sqrt[4]{x^2}$;   **2.** $\sqrt[6]{x^8}$;   **3.** $\sqrt[4]{25}$;   **4.** $\sqrt[4]{4}$

Answers: **1.** $\sqrt{x}$; **2.** $\sqrt[3]{x^4}$; **3.** $\sqrt{5}$; **4.** $\sqrt{2}$.

CHAPTER 6

ALGEBRAIC FRACTIONS

Lesson 21: **Reduction, Multiplication, Division of Algebraic Fractions**
Lesson 22: **Addition of Fractions; Simplification of Complex Fractions**

Lesson 21

Reduction, Multiplication, and Division of Algebraic Fractions

Reduction of Fraction to Lowest Terms

The most important (and perhaps the only) prerequisite in being able to reduce algebraic fractions to lowest terms is **knowing how to factor polynomials.** Therefore, go back and review the factoring of polynomials.

Example 1   Reduce to lowest terms: $\dfrac{x^2 - 5x}{x^2 - 4x - 5}$

Step 1   Factor both the numerator and the denominator

$$\dfrac{x^2 - 5x}{x^2 - 4x - 5}$$

$$= \dfrac{x(x - 5)}{(x + 1)(x - 5)}$$

Step 2:   Cancel (divide out) the common factor in the numerator and the denominator.

$$\dfrac{x\cancel{(x-5)}}{(x + 1)(\cancel{x - 5})}$$

$$= \dfrac{x}{x + 1}$$

Example 2   Reduce to lowest terms

$$\dfrac{x^2 + 10x + 24}{3x + 12}$$

Step 1: Factor both the numerator and the denominator.

$$\dfrac{x^2 + 10x + 24}{3x + 12} = \dfrac{(x + 4)(x + 6)}{3(x + 4)}$$

Step 2: Cancel the common factors in the numerator and the denominator

$$\dfrac{x^2 + 10x + 24}{3x + 12} = \dfrac{\cancel{(x + 4)}(x + 6)}{3\cancel{(x + 4)}}$$

$$= \dfrac{x + 6}{3}$$

Example 3 Reduce to lowest terms $\dfrac{x^2 + 2x - 24}{12 - 3x}$

Solution

Step 1: Factor the numerator, factor the denominator:

Then, $\dfrac{x^2 + 2x - 24}{12 - 3x} = \dfrac{(x - 4)(x + 6)}{3(4 - x)}$

Step 2: $\qquad\qquad = \dfrac{(x - 4)(x + 6)}{-3(x - 4)}$ Note that $4 - x = -(-4 + x) = -(x - 4)$

Step 3: $\qquad\qquad = \dfrac{\cancel{(x - 4)}(x + 6)}{-3\cancel{(x - 4)}}$

Step 4: $\qquad\qquad = \dfrac{x + 6}{-3}$

Step 5: $\qquad\qquad = -\dfrac{x + 6}{3}$

A note about cancellation (dividing out): When may we cancel?
We may cancel when the following two conditions have been satisfied simultaneously.

1. All quantities (factors) in the numerator are multiplying each other, and all the quantities (factors) in the denominator are multiplying each other.

2. There are common quantities (factors) in both the numerator and the denominator.

Also, we consider each expression (such as a binomial) within the parentheses as one quantity.

Example in which we **can** cancel: $\dfrac{2x(x + 2)(x - 1)(x + 3)}{x(x - 1)(x + 3)} = \dfrac{2\cancel{x}(x + 2)\cancel{(x - 1)}\cancel{(x + 3)}}{\cancel{x}\cancel{(x - 1)}\cancel{(x + 3)}} = 2(x + 2)$

Example in which we **cannot** cancel as is: $\dfrac{2x(x + 2) + (x - 1)(x + 3)}{x(x - 1)(x - 3)}$

because of the "+" sign between the parentheses in the numerator.

Multiplication and Division of Fractions

As it was in the case of reduction of fractions to lowest terms, the most important (and perhaps the only) prerequisite in being able to multiply and divide algebraic fractions is **knowing how to factor polynomials.** Therefore, go back and review the factoring of polynomials.

Perform the indicated operations and simplify.

Example 1 $\dfrac{x}{x - 2} \cdot \dfrac{3x - 6}{6x}$

Solution: Factor and divide out (cancel) the common factors in the numerators and the denominators

$\dfrac{x}{x - 2} \cdot \dfrac{3x - 6}{6x} = \dfrac{x}{x - 2} \cdot \dfrac{3(x - 2)}{6x}$ Note: $3x - 6 = 3(x - 2)$

Consider $(x - 2)$ as a single quantity

$\qquad\qquad = \dfrac{\overset{1}{\cancel{x}}}{\underset{1}{\cancel{x - 2}}} \cdot \dfrac{\overset{1}{\cancel{3(x - 2)}}}{\underset{2}{\cancel{6x}}_1}$

$\qquad\qquad = \dfrac{1}{2}$

Example 2   Divide: $\dfrac{3y^2}{y+2} \div \dfrac{6y}{y^2-4}$

$\dfrac{3y^2}{y+2} \div \dfrac{6y}{y^2-4} \quad = \dfrac{3y^2}{y+2} \cdot \dfrac{y^2-4}{6y}$ (Changing to multiplication and inverting the divisor)

$= \dfrac{\cancel{3y^2}\,y}{\cancel{y+2}} \cdot \dfrac{(y+2)(y-2)}{\cancel{6y}\,2}$ Scrapwork: $y^2-4=(y+2)(y-2)$

$= \dfrac{y(y-2)}{2}$ or $\dfrac{y^2-2y}{2}$

Example 3   Simplify: $\dfrac{x^2-49}{x^4} \cdot \dfrac{x^2}{3x+21}$

Solution: Factor and cancel the common factors.

$\dfrac{x^2-49}{x^4} \cdot \dfrac{x^2}{3x+21} = \dfrac{(x+7)(x-7)}{x^4} \cdot \dfrac{x^2}{3(x+7)}$

$= \dfrac{\cancel{(x+7)}(x-7)}{\cancel{x^4}\,x^2} \cdot \dfrac{\cancel{x^2}\,1}{3\cancel{(x+7)}\,1}$

$= \dfrac{x-7}{3x^2}$

Lesson 21 Exercises

A Reduce to lowest terms:

1. $\dfrac{x+4}{x^2+7x+12}$;   2. $\dfrac{x^2-8x+15}{x^2-x-20}$ 3. $\dfrac{6-x}{x^2-9x+18}$;   4. $\dfrac{2x^2-32}{x^2-11x+28}$

Answers: 1. $\dfrac{1}{x+3}$; 2 $\dfrac{x-3}{x+4}$ 3. $-\dfrac{1}{x-3}$ 4. $\dfrac{2(x+4)}{x-7}$

B Reduce the following to lowest terms:

1. $\dfrac{2x-6}{x-3}$;   2. $\dfrac{x^2+x-6}{x-2}$;   3. $\dfrac{4-x}{x^2+x-20}$;   4. $\dfrac{x^2+x-30}{x^2-x-20}$

5. $\dfrac{x^2+4x-21}{x^2-49}$;   6. $\dfrac{x^2-81}{x^2-11x+18}$;   7. $\dfrac{x^2-8x+12}{4x^2-32x+48}$ 8. $\dfrac{2x^2-32}{2x^2+2x-40}$

Answers: 1. 2 ; 2. $x+3$; 3. $-\dfrac{1}{x+5}$; 4. $\dfrac{x+6}{x+4}$; 5. $\dfrac{x-3}{x-7}$; 6. $\dfrac{x+9}{x-2}$; 7. $\dfrac{1}{4}$; 8. $\dfrac{x+4}{x+5}$

C Perform the indicated operations:

1. $\dfrac{x}{x-3} \cdot \dfrac{4x-12}{x+3}$;

2. $\dfrac{10x^2}{x-5} \div \dfrac{5x}{x^2-25}$;

3. $\dfrac{x^2-81}{x^6} \cdot \dfrac{x^4}{x^2+6x-27}$

Answers: **1.** $\dfrac{4x}{x+3}$;   **2.** $2x^2 + 10x$;   **3.** $\dfrac{x-9}{x^2(x-3)}$

D Simplify the following:

1. $\dfrac{6x^4}{7y^2} \cdot \dfrac{21y}{3x}$;

2. $\dfrac{x^2}{x+3} \div \dfrac{x}{2x+6}$;

3. $\dfrac{8-x}{x^2-64} \cdot \dfrac{3x+24}{x+3}$

4. $\dfrac{x^2-9}{4x^2-36} \cdot \dfrac{2x^2-50}{3x+15}$;

5. $\dfrac{8-x}{x^8} \div \dfrac{x^2-4x-32}{x^6+4x^5}$

Answers: **1.** $\dfrac{6x^3}{y}$; **2.** $2x$; **3.** $-\dfrac{3}{x+3}$; **4.** $\dfrac{x-5}{6}$; **5.** $-\dfrac{1}{x^3}$

Lesson 22

Addition of Fractions; Simplification of Complex Fractions

Addition and Subtraction of Fractions

Like Fractions (Fractions with the same denominator)

Example Add: $\dfrac{2x}{x-3} + \dfrac{4}{x-3}$

Solution Since the above fractions already have a common denominator, we add the numerators and keep the common denominator.

$$\dfrac{2x}{x-3} + \dfrac{4}{x-3}$$

$$= \dfrac{2x+4}{x-3}$$

Unlike Fractions (Fractions with different denominators)

In this case, we have to change the fractions to (equivalent) like fractions, add the numerators and then reduce the fraction to lowest terms if possible.

Example 1 Add: $\dfrac{5}{x^2 + 7x + 12} + \dfrac{x+2}{x+3}$

Step 1: Factor the denominators if factorable.

$$\dfrac{5}{(x+3)(x+4)} + \dfrac{x+2}{x+3}$$

Step 2: We will make the denominators the same by looking for missing factors. The factors of the denominator of the first fraction are $(x+3)$ and $(x+4)$, but the denominator of the second fraction has the factor $x+3$. To make the denominators of the two fractions identical, we multiply both the denominator and the numerator of the second fraction by $x+4$, since the denominator of the second fraction is "missing" $x+4$, compared to the denominator of the first fraction.

Note that any time we multiply the denominator by any quantity, we must also multiply the numerator by the same quantity (in order **not** to change the value of the original fraction.).

$$= \dfrac{5}{(x+3)(x+4)} + \dfrac{(x+2)(x+4)}{(x+3)(x+4)}$$

Step 3: Since the denominators are now the same, add the numerators. (We have also produced the LCD)

$$= \dfrac{5 + (x+2)(x+4)}{(x+3)(x+4)}$$ The LCD we produced (in Step 1) is $(x+3)(x+4)$

$$= \dfrac{5 + x^2 + 4x + 2x + 8}{(x+3)(x+4)}$$

$$= \dfrac{x^2 + 6x + 13}{(x+3)(x+4)}$$ (Simplifying the numerator)

Note that if the numerator were factorable, we would have factored it to determine if there are common factors in the numerator and the denominator, and in which case, we would have canceled the common factors. In this example, the numerator is not factorable. It is preferable to leave the denominator in the factored form.

Example 2 Add: $\dfrac{3}{x+2} + \dfrac{2}{x^2-4}$

Step 1: Factor the factorable denominator(s).

Then, $\dfrac{3}{x+2} + \dfrac{2}{x^2-4}$

$= \dfrac{3}{x+2} + \dfrac{2}{(x+2)(x-2)}$

Step 2: The denominator of the first fraction is "missing" $(x-2)$ compared to the denominator of the second fraction. Therefore, multiply both denominator and numerator of the first fraction by $(x-2)$; add the numerators of the resulting like fractions and reduce if possible.

$= \dfrac{3(x-2)}{(x+2)(x-2)} + \dfrac{2}{(x+2)(x-2)}$ (We have produced the LCD which is $(x+2)(x-2)$)

$= \dfrac{3(x-2)+2}{(x+2)(x-2)}$

$= \dfrac{3x-6+2}{(x+2)(x-2)}$

$= \dfrac{3x-4}{(x+2)(x-2)}$

Example 3 Combine into a single fraction: $\dfrac{3x}{5y} + \dfrac{x}{2y}$

Solution

Step 1: Make the denominators identical by "looking for missing factors" and multiplying accordingly. (Multiply the first fraction by 2 and the second fraction by 5.)

$\dfrac{3x}{5y} + \dfrac{x}{2y}$

$= \dfrac{3x(2)}{5y(2)} + \dfrac{x(5)}{2y(5)}$

$= \dfrac{6x}{10y} + \dfrac{5x}{10y}$ <-------Like fractions   (The LCD produced is $10y$)

Step 2: Add the numerators since we have like fractions.

Then, $\dfrac{6x}{10y} + \dfrac{5x}{10y}$

$= \dfrac{11x}{10y}$

Example 4 Add: $\dfrac{7}{5} + \dfrac{3}{4n}$

Solution

Step 1: Make the denominators the same by looking for missing factors and multiplying accordingly. (Multiply the first fraction by 4n, and multiply the second fraction by 5)

$$\frac{7}{5} + \frac{3}{4n}$$

$$= \frac{7(4n)}{5(4n)} + \frac{3(5)}{4n(5)}$$

$$= \frac{28n}{20n} + \frac{15}{20n} \quad \text{<--------Like fractions (The LCD produced is } 20n)$$

Step 2: Add the numerators.

$$\text{Then, } \frac{28n}{20n} + \frac{15}{20n}$$

$$= \frac{28n + 15}{20n}$$

Example 5 Combine into a single fraction: $\dfrac{4}{x} + \dfrac{1}{2x} + \dfrac{2}{5}$

$$\frac{4}{x} + \frac{1}{2x} + \frac{2}{5}$$

Step 1: Multiply accordingly by 2, 5, and x.

$$= \frac{4(2)(5)}{x(2)(5)} + \frac{1(5)}{2x(5)} + \frac{2(x)(2)}{5(x)(2)} \qquad \text{(Making the denominators the same)}$$

$$= \frac{40}{10x} + \frac{5}{10x} + \frac{4x}{10x} \text{<----Like fractions}$$

Step 2: Add the numerators, since we have like fractions.

$$\text{Then, } \frac{40}{10x} + \frac{5}{10x} + \frac{4x}{10x}$$

$$= \frac{4x + 45}{10x}$$

Example 6. Add: $\dfrac{5}{x+4} - \dfrac{3x}{x-4} + \dfrac{2}{x^2 - 16}$

$$= \dfrac{5}{x+4} - \dfrac{3x}{x-4} + \dfrac{2}{(x+4)(x-4)}$$

$$= \dfrac{5(x-4)}{(x+4)(x-4)} - \dfrac{3x(x+4)}{(x-4)(x+4)} + \dfrac{2}{(x+4)(x-4)}$$

Now, since we have like fractions (all the denominators are the same), we add the numerators.

$$= \dfrac{5(x-4) - 3x(x+4) + 2}{(x+4)(x-4)}$$

$$= \dfrac{5x - 20 - 3x^2 - 12x + 2}{(x+4)(x-4)}$$

$$= \dfrac{-3x^2 - 7x - 18}{(x+4)(x-4)}$$

$$= -\dfrac{3x^2 + 7x + 18}{(x+4)(x-4)}$$

Example 7 Simplify: $\dfrac{3x+4}{x-3} - \dfrac{x-1}{x+1}$

Solution

$$= \dfrac{3x+4}{x-3} - \dfrac{x-1}{x+1}$$

$$= \dfrac{(x+1)(3x+4)}{(x+1)(x-3)} - \dfrac{(x-1)(x-3)}{(x+1)(x-3)} \text{ (Making the denominators the same and producing the LCD)}$$

$$= \dfrac{3x^2 + 4x + 3x + 4 - (x^2 - 3x - x + 3)}{(x+1)(x-3)} \quad \text{(The minus sign before the parentheses is important)}$$

$$= \dfrac{3x^2 + 7x + 4 - (x^2 - 4x + 3)}{(x+1)(x-3)} \quad \text{(The minus sign before the parentheses is important)}$$

$$= \dfrac{3x^2 + 7x + 4 - x^2 + 4x - 3}{(x+1)(x-3)}$$

$$= \dfrac{2x^2 + 11x + 1}{(x+1)(x-3)}$$

Example 8 Simplify : $3x + \dfrac{2x}{y}$

$$3x + \dfrac{2x}{y}$$

$$= \dfrac{3x(y)}{1y} + \dfrac{2x}{y}$$

$$= \dfrac{3xy}{y} + \dfrac{2x}{y}$$

$$= \dfrac{3xy + 2x}{y}$$

Example 9 Simplify: $x - 2 + \dfrac{3}{x - 4}$

$$= \dfrac{x - 2}{1} + \dfrac{3}{x - 4} \qquad \text{or} \qquad \dfrac{x}{1} - \dfrac{2}{1} + \dfrac{3}{x - 4}$$

$$= \dfrac{(x - 4)(x - 2)}{(x - 4)1} + \dfrac{3}{x - 4} \qquad \text{or} \qquad \dfrac{x(x - 4)}{1(x - 4)} - \dfrac{2(x - 4)}{1(x - 4)} + \dfrac{3}{x - 4}$$

$$= \dfrac{(x - 4)(x - 2) + 3}{x - 4} \qquad \text{or} \qquad \dfrac{x(x - 4) - 2(x - 4) + 3}{x - 4}$$

$$= \dfrac{x^2 - 2x - 4x + 8 + 3}{x - 4} \qquad \text{or} \qquad \dfrac{x^2 - 4x - 2x + 8 + 3}{x - 4}$$

$$= \dfrac{x^2 - 6x + 11}{x - 4} \qquad \text{or} \qquad \dfrac{x^2 - 6x + 11}{x - 4}$$

The above "or" means that you may follow the steps on the left or the steps on the right.

Example 10 Add: $x + 4 + \dfrac{3}{x - 2}$

$$= \dfrac{x}{1} + \dfrac{4}{1} + \dfrac{3}{x - 2}$$

$$= \dfrac{x(x - 2)}{1(x - 2)} + \dfrac{4(x - 2)}{1(x - 2)} + \dfrac{3}{x - 2}$$

$$= \dfrac{x(x - 2)}{(x - 2)} + \dfrac{4(x - 2)}{(x - 2)} + \dfrac{3}{x - 2}$$

(Now, we can add the numerators since we have like fractions, and write the denominator only once.)

$$= \dfrac{x(x - 2) + 4(x - 2) + 3}{x - 2}$$

$$= \dfrac{x^2 - 2x + 4x - 8 + 3}{x - 2}$$

$$= \dfrac{x^2 + 2x - 5}{x - 2}$$

Example 11   Add : $\dfrac{x^2 + 7x + 12}{x^2 + 8x + 15} + \dfrac{3}{x - 5}$

Solution

Step 1: Factor and reduce the first expression before proceeding to add. (Failure to reduce to lowest terms first, in this problem, will result in having to deal with higher powers of x.)

$$\dfrac{x^2 + 7x + 12}{x^2 + 8x + 15} + \dfrac{3}{x - 5}$$

$$= \dfrac{(x + 3)(x + 4)}{(x + 3)(x + 5)} + \dfrac{3}{x - 5}$$

$$= \dfrac{x + 4}{x + 5} + \dfrac{3}{x - 5}$$

Step 2: Now, proceed to add as before.

$$= \dfrac{(x + 4)(x - 5)}{(x + 5)(x - 5)} + \dfrac{3 \cdot (x + 5)}{(x - 5)(x + 5)}$$

$$= \dfrac{x^2 - 5x + 4x - 20 + 3x + 15}{(x + 5)(x - 5)}$$

$$= \dfrac{x^2 + 2x - 5}{(x + 5)(x - 5)}$$

Note above that sometimes a given fraction may **not** be in its lowest terms; and in this case, it is good practice to reduce to lowest terms before proceeding to add.

Simplification of Complex Fractions

Example 1   Simplify: $\dfrac{\dfrac{3x}{ax}}{\dfrac{5x}{bx}}$

Solution $\dfrac{\dfrac{3x}{ax}}{\dfrac{5x}{bx}} = \dfrac{3x}{ax} \div \dfrac{5x}{bx}$

$$= \dfrac{3x}{ax} \cdot \dfrac{bx}{5x} \qquad \text{(inverting the divisor, } \dfrac{5x}{bx} \text{)}.$$

$$= \dfrac{3b}{5a} \qquad\qquad \text{(canceling the common factors)}.$$

Example 2 Simplify:

$$\dfrac{\dfrac{y}{y^2 + 4y + 3}}{\dfrac{1}{y^2 - 1} + 1}$$

Solution   **Method 1** (Adding the denominator first)

$$\dfrac{\dfrac{y}{y^2 + 4y + 3}}{\dfrac{1}{y^2 - 1} + 1} = \dfrac{y}{y^2 + 4y + 3} \div \dfrac{1}{y^2 - 1} + 1$$

$$= \dfrac{y}{y^2 + 4y + 3} \div \dfrac{y^2}{(y + 1)(y - 1)}$$

$$= \dfrac{y}{y^2 + 4y + 3} \cdot \dfrac{(y + 1)(y - 1)}{y^2}$$

$$= \dfrac{\overset{1}{\cancel{y}}}{\cancel{(y + 1)}(y + 3)} \cdot \dfrac{\overset{1}{\cancel{(y + 1)}}(y - 1)}{\cancel{y^2}\,y}$$

$$= \dfrac{y - 1}{y(y + 3)}$$

Scrapwork

$$\dfrac{1}{(y + 1)(y - 1)} + \dfrac{1}{1}$$

$$= \dfrac{1 + (y + 1)(y - 1)}{(y + 1)(y - 1)}$$

$$= \dfrac{1 + y^2 - 1}{(y + 1)(y - 1)}$$

$$= \dfrac{y^2}{(y + 1)(y - 1)}$$

Method 2: Multiply both the numerator and denominator of the complex fraction by the LCD of all the fractions in the numerator and the denominator. The LCD $= (y + 1)(y - 1)(y + 3)$

$$\dfrac{\dfrac{y}{y^2 + 4y + 3}}{\dfrac{1}{y^2 - 1} + 1}$$

$$= \dfrac{\dfrac{y(y + 1)(y - 1)(y + 3)}{(y + 1)(y + 3)}}{\dfrac{(y + 1)(y - 1)(y + 3)}{(y + 1)(y - 1)} + \dfrac{1(y + 1)(y - 1)(y + 3)}{1}}$$

$$= \dfrac{\dfrac{y\cancel{(y + 1)}(y - 1)\cancel{(y + 3)}}{\cancel{(y + 1)}\cancel{(y + 3)}}}{\dfrac{\cancel{(y + 1)}\cancel{(y - 1)}(y + 3)}{\cancel{(y + 1)}\cancel{(y - 1)}} + \dfrac{(y + 1)(y - 1)(y + 3)}{1}}$$

$$= \dfrac{y(y - 1)}{(y + 3) + (y^2 - 1)(y + 3)} \qquad \text{Note :} y^2 - 1 = (y + 1)(y - 1)$$

$$= \dfrac{y(y - 1)}{(y + 3)[1 + y^2 - 1]}$$

$$= \dfrac{y(y - 1)}{(y + 3) y^2}$$

$$= \dfrac{\overset{1}{\cancel{y}}(y - 1)}{(y + 3) \cancel{y^2}\,y}$$

$$= \dfrac{(y - 1)}{y(y + 3)}$$

Lesson 22 Exercises

A Carry out the indicated operation and simplify:

1. $\dfrac{3}{x-4} + \dfrac{x+1}{x^2-16}$;

2. $\dfrac{2}{xy} - \dfrac{4}{y}$;

3. $\dfrac{3x}{x-4} + \dfrac{2}{x-4}$;

4. $x - 2 + \dfrac{2}{x-3}$

5. $\dfrac{5}{x^2-4} + \dfrac{2x}{x+2} - \dfrac{4}{x-3}$.

Answers: **1.** $\dfrac{4x+13}{(x-4)(x+4)}$; **2.** $\dfrac{-4x+2}{xy}$; **3.** $\dfrac{3x+2}{x-4}$; **4.** $\dfrac{x^2-5x+8}{x-3}$; **5.** $\dfrac{2x^3-14x^2+17x+1}{(x-3)(x-2)(x+2)}$;

B Carry out the indicated operation and simplify:

1. $\dfrac{x}{3} + \dfrac{4}{x+2}$

2. $\dfrac{2}{x-3} - \dfrac{x}{4}$

3. $\dfrac{x}{y} + \dfrac{2}{n}$

4. $\dfrac{2}{x} + \dfrac{3}{y}$;

5. $\dfrac{2}{x-5} + \dfrac{3}{x-5}$;

6. $\dfrac{x+1}{x+2} + \dfrac{x-3}{x-2}$;

7. $\dfrac{2x-1}{x+4} - \dfrac{x-5}{x-3}$

Answers: **1.** $\dfrac{x^2+2x+12}{3(x+2)}$; **2.** $\dfrac{-x^2+3x+8}{4(x-3)}$; **3.** $\dfrac{nx+2y}{ny}$; **4.** $\dfrac{3x+2y}{xy}$;

5. $\dfrac{5}{x-5}$; **6.** $\dfrac{2x^2-2x-8}{(x+2)(x-2)}$; **7.** $\dfrac{2x^2-6x+23}{(x-3)(x+4)}$

C Simplify the following:

1. $\dfrac{\dfrac{x}{x^2+7x+12}}{\dfrac{16}{x^2-16}+1}$;

2. $\dfrac{\dfrac{2ab}{cd}}{\dfrac{cd}{ab}}$;

3. $\dfrac{3-\dfrac{2}{x}}{9x}$;

4. $\dfrac{\dfrac{x}{y}+3}{\dfrac{y}{x}-3}$;

5. $\dfrac{\dfrac{5c^2y}{3bx}}{\dfrac{25cy}{6bx^2}}$

Answers: **1.** $\dfrac{x-4}{x(x+3)}$; **2.** $\dfrac{2a^2b^2}{c^2d^2}$; **3.** $\dfrac{3x-2}{9x^2}$; **4.** $\dfrac{x^2+3xy}{y(y-3x)}$; **5.** $\dfrac{2cx}{5}$

CHAPTER 7
Inequalities

Lesson 23: Set Notation, Set Operations, Interval Notation
Lesson 24: Linear Inequalities; Compound Inequalities; Word Problems

Lesson 23
Set Notation, Set Operations, Interval Notation

Sets, Set Notation

Set A set is a well-defined collection of objects or things. The objects or things are called the elements or members of the set. A set may contain many elements; it may contain only one element; or it may contain no element. We call the set which contains no element the empty set or the null set.

Representation of sets: We will cover the roster method and the set-builder notation.

Roster method: In the roster method, we list the elements of the set and enclose them by braces.
Example: We can denote the set of numbers $2, 3, 4$, as $\{2, 3, 4\}$.
We separate the elements by commas. Note that the order in which the elements are listed does not matter. We may **name** a set by a capital letter . For example, if we denote the set of numbers $2, 3, 4$ by A, then we write $A = \{2, 3, 4\}$.
We read this as "A is the set whose elements are $2, 3, 4$.

Set-builder notation: In the set builder notation, we state the conditions which the elements of the set must satisfy.
Example: Let E be the set of all odd integers between 4 and 12. Then if we use x to represent an arbitrary element of this set, we write $E = \{x \mid x$ is an odd integer between 4 and 12$\}$.
This is read " E is the set of numbers (elements) x such that x is an odd integer between 4 and 12".
We read the vertical line " l " as "such that".
(Note that by the roster method, the set E in this example would be represented by
$E = \{5, 7, 9, 11\}$

We denote the empty set by $\{\ \}$ or $\emptyset$.

We will use the **set-builder notation** in stating the solutions of inequalities.

Definition

An **inequality** is a mathematical statement that two expressions are not equal.

An inequality, like an equation, has three parts, namely the left-hand side, the inequality symbol which may be " $>$, $<$, $\geq$, $\leq$ or $\neq$", and the right-hand side.

Examples 1. $2x - 5 > 6x + 3$ read " $2x$ minus 5 **is greater than** $6x$ plus 3".

2. $2x + 3 \geq 6x + 7$ read " $2x$ plus 3 **is greater than or equal to** $6x$ plus 7".

3. $4x + 5 < 8x - 2$ read "$4x$ plus **five is less than** $8x$ minus 2".

4. $3x - 1 \leq 5x + 7$ read "$3x$ minus 1 **is less than or equal to** $5x$ plus 7".

5. $x > 6$ read " x is greater than 6"

Sense of an inequality (direction or order of an inequality)
The sense of an inequality refers to whether the inequality symbol is the greater than symbol " $>$" or the less than symbol "$<$".

Set Notation, Interval Notation and Graphs

Set notation Interval notation

$\downarrow$ $\quad\quad$ $\downarrow$

$\{x \mid x > a\} = (a, +\infty)$

$\{x \mid x \geq a\} = [a, +\infty)$

$\{x \mid x < b\} = (-\infty, b)$

$\{x \mid x \leq b\} = (-\infty, b]$

$\{x \mid a < x < b\} = (a, b)$

$\{x \mid a \leq x \leq b\} = [a, b]$

$\{x \mid a < x \leq b\} = (a, b]$

$\{x \mid a \leq x < b\} = [a, b)$

$\{x \mid x \text{ is a real number}\}$

$\quad = (-\infty, +\infty)$

Graphs

Example 1 Draw the graph for $\{x \mid x > 2\} = (2, +\infty)$

Solution

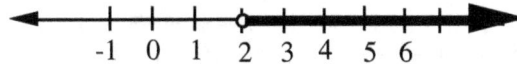

Graph for $\{x \mid x > 2\} = (2, +\infty)$

Note: The hollow circle at 2 indicates that 2 is **not** part of the solution set.

Example 2 Draw the graph for $\{x \mid x \geq 2\} = [2, +\infty)$

Solution

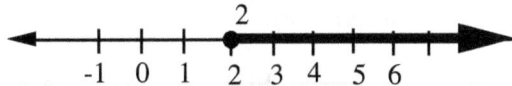

Graph for $\{x \mid x \geq 2\} = [2, +\infty)$

Note: The solid circle at 2 indicates that 2 is part of the solution set.

Example 3

$\{x \mid x \leq -8\} = (-\infty, -8]$

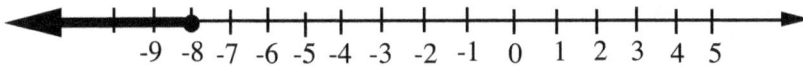

Graph for $\{x \mid x \leq -8\} = (-\infty, -8]$

Set Operations Involving Inequalities

Union of two sets

The union of two sets A and B, written $A \cup B$, is the set containing all the elements that belong to either set A or set B (or belong to both sets).

Symbolically, $A \cup B = \{x \mid x \in A, \text{ or } x \in B\}$

Example 1 If $A = \{x \mid x < -3\}$ and $B = \{x \mid x > 4\}$ then
$\quad\quad A \cup B = \{x \mid x < -3\} \cup \{x \mid x > 4\}$
$\quad\quad\quad = \{x \mid x < -3 \text{ or } x > 4\}$

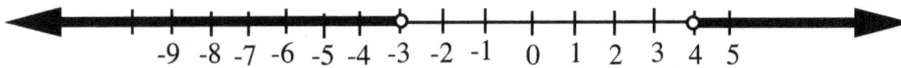

Graph for $\{x \mid x < -3 \text{ or } x > 4\}$

Intersection of two sets

The intersection of two sets A and B, written $A \cap B$ is the set containing the elements that are common to A and B. Thus the elements of the intersection are the elements that belong to both sets simultaneously.

Symbolically, $A \cap B = \{x \mid x \in A \text{ and } x \in B\}$

Example 2 If $A = \{x \mid x > -3\}$ and $B = \{x \mid x < 2\}$ then

$A \cap B = \{x \mid x > -3\} \cap \{x \mid x < 2\}$

$A \cap B = \{x \mid -3 < x < 2\}$

Example 3 If $A = \{x \mid x < -3\}$ and $B = \{x \mid x > 4\}$ then

$A \cap B = \{x \mid x < -3\} \cap \{x \mid x > 4\}$

$A \cap B = \{\ \}$, the empty set.

(Since a number cannot be less than -3 and greater than 4 at the same time.)

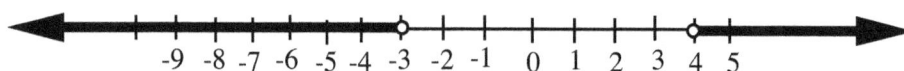

Compare the solutions to Examples 1 and 3, above, and note that even though they have identical graphs, Example 1 has a solution but Example 3 has no solution.

Lesson 23 Exercises

1. What is a set?
2. What is an inequality?
3.What is meant by the sense or direction or the order of an equality?
4. Draw graphs for the following:
a. $\{x \mid x > 3\}$; **b.** $\{x \mid x \geq 3\}$; **c.** $\{x \mid x < 3\}$; **d.** $\{x \mid x \leq 3\}$; **e.** $\{x \mid x < -5\}$

Answers: **4.**

(a)

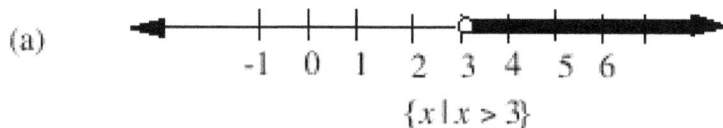

$\{x \mid x > 3\}$

(b)

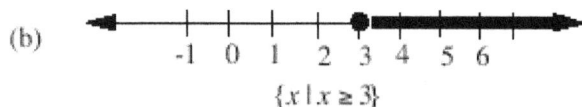

$\{x \mid x \geq 3\}$

(c)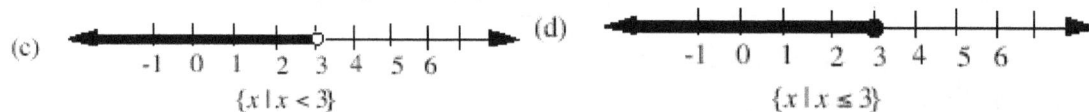

$\{x \mid x < 3\}$ $\{x \mid x \leq 3\}$

(e)

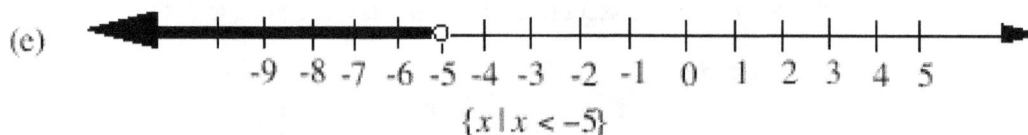

$\{x \mid x < -5\}$

Lesson 24

Solving Linear Inequalities; Compound Inequalities; Word Problems

Solving Linear Inequalities

The solution set of a linear inequality is the set of real numbers each of which when substituted for the variable makes the inequality true.

The techniques for solving linear inequalities are similar to the techniques for solving linear equations, except that when an inequality is divided or multiplied by a negative number, the sense of the inequality must be reversed as follows: The symbol " > " when reversed becomes " < "; the symbol " < " when reversed becomes " > "; the symbol " ≥ " when reversed becomes " ≤ "; and the symbol " ≤ " when reversed becomes " ≥ ".

Example 1 Solve and graph the solution set of the inequality,
$$3x - 4 > 0$$
Step 1: To undo the -4, add + 4 to both sides of the inequality.

$$
\begin{array}{r}
3x - 4 > 0 \\
+4 \;\; +4 \\
\hline
3x > 4
\end{array}
$$

Step 2: Divide both sides of the inequality by 3.

$$\frac{3x}{3} > \frac{4}{3} \qquad x > \frac{4}{3}$$

The solution set is $\{x \mid x > \frac{4}{3}\}$.   (The solution set is all real numbers greater than $\frac{4}{3}$)

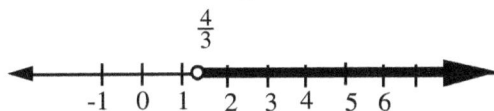

Graph for $x > \frac{4}{3}$

Note: The hollow circle at $\frac{4}{3}$ indicates that $\frac{4}{3}$ is **not** part of the solution set.

Example 2 Solve and graph the solution set of the inequality
$$3x - 4 \geq 0$$
Step 1: To undo the -4, add + 4 to both sides of the inequality.

$$
\begin{array}{r}
3x - 4 \geq 0 \\
+4 \;\; +4 \\
\hline
3x \geq 4
\end{array}
$$

Step 2: Divide both sides of the inequality by 3.

$$\frac{3x}{3} \geq \frac{4}{3}$$

$$x \geq \frac{4}{3}$$

The solution set is $\{x \mid x \geq \frac{4}{3}\}$. (The solution set is all real numbers greater than or equal to $\frac{4}{3}$)

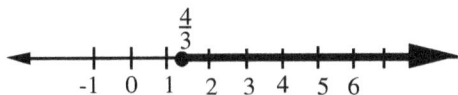

Graph for $x \geq \frac{4}{3}$ **Note**: The solid circle at $\frac{4}{3}$ indicates that $\frac{4}{3}$ is part of the solution set.

Example 3 $5x + 5 > 9x - 6$

Solution

Step 1: We undo the $9x$ by adding $-9x$ to both sides of the inequality.

$$5x + 5 > 9x - 6$$
$$\underline{-9x \qquad - 9x}$$
$$-4x + 5 > -6$$

Step 2: To undo the 5, add -5 to both sides of the inequality.

$$-4x + 5 > -6$$
$$\underline{\quad -5 \quad -5}$$
$$-4x > -11$$

Step 2: Divide both sides of the inequality by -4 and **reverse the sense** of the inequality.

$$\frac{-4x}{-4} < \frac{-11}{-4} \quad \text{(change this " > ' to that " < ")}$$

$$x < \frac{11}{4}$$

The solution set is $\{x \mid x < \frac{11}{4}\}$.   (The solution set is all real numbers less than $\frac{11}{4}$.)

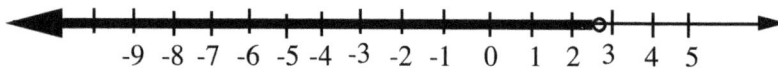

Graph for $x < \frac{11}{4}$

Example 4   Solve for x: $2(2x + 1) \geq 5(x + 2)$

Solution

$$2(2x + 1) \geq 5(x + 2)$$
$$4x + 2 \geq 5x + 10$$
$$\underline{-5x \qquad -5x}$$
$$-x + 2 \geq 10$$
$$\underline{\quad -2 \quad -2}$$
$$-x \geq 8$$
$$x \leq -8$$

(Multiplying or dividing both the left-hand side and the right-hand side by -1 and reversing the sense of the inequality or changing the signs of both sides and reversing the sense of the inequality)

The solution set is $\{x \mid x \leq -8\}$.

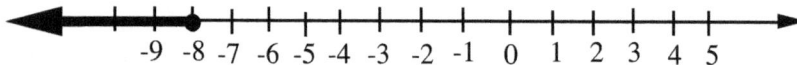

Graph for $x \leq -8$

Compound Inequalities

A compound inequality is formed by connecting two inequalities with the words "**and**" or " **or**".

Example We will call the following type an "**AND**" problem. (More formally, a **conjunction** problem)

$$3x - 2 < 4 \text{ and } 2x + 10 > 4$$

Because of the connective "and", we want a solution set (a set of numbers, if any) which is common to both solution sets of the two inequalities (i.e., the intersection of the solution sets). Any solution set must satisfy both inequalities simultaneously.

Example Solve for x: $3x - 2 < 4$ and $2x + 10 > 4$

Procedure: Solve each inequality separately and find the intersection (common set of numbers) of the solutions.

$$3x - 2 < 4 \quad \text{and} \quad 2x + 10 > 4$$
$$\underline{+2 \ +2} \qquad \qquad \underline{- 10 \ -10}$$
$$3x < 6 \qquad \qquad 2x > -6$$
$$x < 2 \quad \text{and} \quad x > -3$$

The solution is the intersection of $\{x \mid x < 2\}$ and $\{x \mid x > -3\}$ i.e., $\{x \mid x < 2\} \cap \{x \mid x > -3\}$
Graph the solutions to determine if the solutions intersect.

The solution is $\{x \mid -3 < x < 2\}$

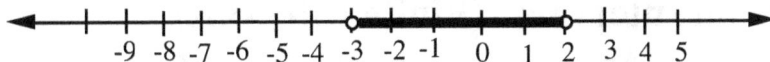

Graph for $-3 < x < 2$

Another "**AND**" problem

Example 2 Solve for x: $\quad -5 < 2x + 1 \le 7$

Solution: The above inequality is equivalent to the compound inequality
$$-5 < 2x+1 \text{ and } 2x+ 1 \le 7.$$

We can solve this compound inequality using the method used in Example 1 above, however, we will use a faster method called the "**Condensed**" or "**continued inequality**" method as follows:

$$-5 < 2x + 1 \le 7$$
$$\underline{-1 \qquad -1 \ -1} \qquad \text{(Adding -1 to all three sides of the inequality)}$$
$$-6 < 2x \qquad \le 6$$

$$\frac{-6}{2} < \frac{2x}{2} \le \frac{6}{2} \qquad \text{(Dividing each of all the three sides of the inequality by 2)}$$
$$-3 < x \le 3$$

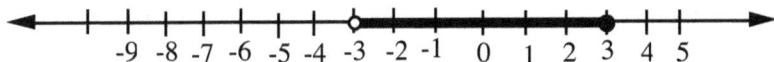

Graph for $-3 < x \le 3$ (Note the hollow circle at -3 and the solid circle at 3)

Example 3 Solve for x: $-6x > 12$ and $x + 1 > 5$

Solution: For the first inequality, divide by -6 and reverse the sense of the inequality.
For the second inequality, add -1 to both sides of the inequality.
$$-6x > 12 \text{ and } x + 1 > 5$$
$$x < -2 \quad \text{and} \quad x > 4$$
Graph the solution set for $x < -2$ and for $x > 4$ and determine if the two graphs intersect.
If they intersect, there is a solution, but if they do not intersect, then there is no solution.

$x < -2$ \qquad\qquad\qquad $x > 4$

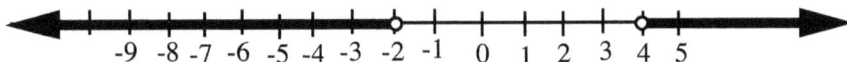

From the graph, the two solution sets do not intersect. (i.e., there are no numbers which belong to both solutions).
Therefore, there is no solution or the solution set is the empty set.

An "OR " Compound Inequality Problem (More formally, a **disjunction** problem) 124

Example 4 Solve for x: $2x - 3 > 5$ or $3x + 1 < -8$

Because of the connective "or" we want the solution set (if any) to be the union of the solution sets of the two inequalities. The solution is the set of real numbers which satisfies either inequality or both.

$$2x - 3 > 5 \text{ or } 3x + 1 < -8$$
$$\underline{+3 + 3} \qquad \underline{\quad -1 \quad -1}$$
$$2x > 8 \qquad \quad 3x < -9$$
$$x > 4 \quad \text{ or } \quad x < -3$$
Same as $x < -3$ or $x > 4$

Graph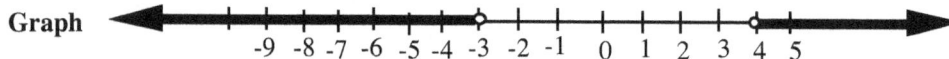

Solution set is $\{x \mid x > 4\} \cup \{x \mid x < -3\} = \{x \mid x < -3 \text{ or } x > 4\}$.

Comparison of Example 3 and Example 4

Note that even though the two problems have similar graphs, Example 3 has no solution because it is an "**and**" problem but Example 4 has a solution because it is an "**or**" problem".

More on Intervals   (Finite Intervals)

Open interval The **open** interval (Figure) from a to b , written (a, b) or $a < x < b$, (where x is real number between a and b) is the set of all numbers between a and b but excluding the end-points a and b. (" open" means the endpoints are **excluded**)

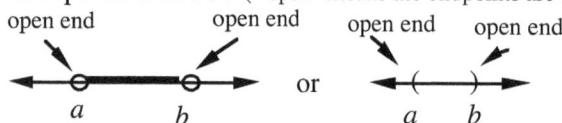

Closed interval The **closed** interval from a to b , written $[a, b]$ or $a \leq x \leq b$ (where x is real number between a and b) is the set of all numbers between a and b and including the end-points a and b. (" closed " means the endpoints are **included**)

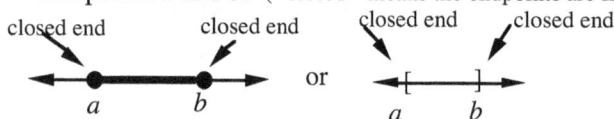

Half-closed and half-open Interval The half-closed and half-open interval from a to b is the set of all numbers between a and b, including either the endpoint a or the endpoint b but not both.

In Figure 1   the interval half-open on the left (or half-closed on the right) is the set of all numbers between a to b , including the endpoint b but excluding the endpoint a.. This interval is symbolized $(a, b]$ or $a < x \leq b$.

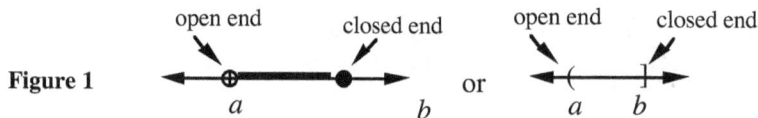

In Figure **2**, the interval half-open on the right (or half-closed on the left) is the set of all numbers between a to b , including the endpoint a but excluding the endpoint b.

This interval is symbolized $[a, b)$ or $a \leq x < b$.

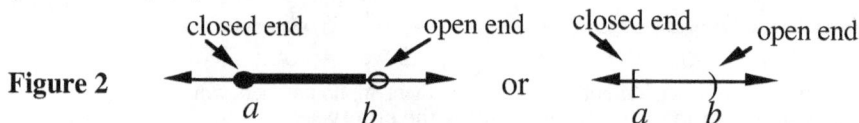

Note: Within any interval, there are infinitely many rational numbers and infinitely many irrational numbers.

Applications (Word Problems Using Inequalities)

Example During the semester James scored 85, 70, 75 on the first three tests. He wants his class average for the term to be at least 80. There is one more test to take. What is the score he needs on the last test ?

Solution Let x be the last score.

Then, the total score for the four tests will be $85 + 70 + 75 + x$.

$$\text{Average score} = \frac{\text{Total score}}{\text{Number of tests}} = \frac{85 + 70 + 75 + x}{4}$$

Now, we want this average score to be at least 80 (i.e., 80 or more).

$$\frac{85 + 70 + 75 + x}{4} \geq 80 \qquad \text{Scrapwork}$$

$$\frac{230 + x}{4} \geq 80 \qquad 85+70+75=230$$

We solve for x.

$$\frac{4(230 + x)}{4} \geq 80(4)$$

$$230 + x \geq 320$$

$$\underline{-230 \qquad\quad -230}$$

$$x \geq 90$$

∴ James needs at least a score of 90 on the last test. (i.e., 90 or more).

Some words to help translate inequality word problems

1. " is at least " translates to " $\geq$ ".

2. " is no less than" translates to " $\geq$ ".

3 " is greater than or equal to " translates to " $\geq$"

4 " is more than" translates to " $>$ ".

5. " is greater than" translates to " $>$ ".

6. " is over " translates to " $>$ "

7. " is no more than" translates to " $\leq$ ".

8." is at most " translates to " $\leq$ ".

9. " is less than or equal to " translates to " $\leq$ ".

10 " is less than " translates to " $<$ ".

11. " is under " translates to " $<$ "

Also: x is between a and b translates to " $a < x < b$ " .

x is greater than a but less than b translates to " $a < x < b$ " .

Note the difference between "**more than**" and "**is more than** " as in the following:

Example: 3 **more than** twice a number **is more than** 5 less than the number

If the number is x, the translation is $2x + 3 > x - 5$. ("**Is more than**" translates to " $>$")

Lesson 24 Exercises

A Solve and graph the solution set:

1. $3x - 6 > 0$;   2. $-4x + 1 > 13$;   3. $2x - 1 \leq 7x - 11$;   4. $2(x + 4) \geq 3(x - 2)$

5. $x + 4 > -5$;   6. $x + 4 < 6$;   7. $x + 4 \leq 6$;   8. $x + 4 \geq 6$;   9. $-4x > 12$

10. $-5x \leq -30$;   11. $\frac{-x}{4} \leq 9$;   12. $\frac{x}{4} - 6 > 14$;   13. $2(x - 4) + 3 \leq 17$

Answers; **1.** $\{x \mid x > 2\}$;   **2.** $\{x \mid x < -3\}$;   **3.** $\{x \mid x \geq 2\}$;   **4.** $\{x \mid x \leq 14\}$;   **5.** $\{x \mid x > -9\}$;   **6.** $\{x \mid x < 2\}$;
7. $\{x \mid x \leq 2\}$; **8.** $\{x \mid x \geq 2\}$; **9.** $\{x \mid x < -3\}$; **10.** $\{x \mid x \geq 6\}$; **11.** $\{x \mid x \geq -36\}$; **12.** $\{x \mid x > 80\}$; **13.** $\{x \mid x \leq 11\}$;

B Solve for: **1.** $2x - 3 < 1$ and $3x + 5 > 2$;   **2.** $-4 < 3x - 1 < 5$;   **3.** $2x - 3 < 1$ and $3x - 5 > 7$
4. $-1 \leq 2x + 5 < 3$;   **5.** $3x + 7 < 1$ and $2x - 2 < 5x + 7$

Answers: **1.** $\{x \mid -1 < x < 2\}$;   **2.** $\{x \mid -1 < x < 2\}$ **3.** No solution: $x < 2$ and $x > 4$ do not intersect.
      4. $\{x \mid -3 \leq x < -1\}$;   **5.** $\{x \mid -3 < x < -2\}$

C Solve for x:

1. $3x - 2 > 4$ or $3x - 1 > -7$;   **2.** $3x + 4 > 1$ or $2x + 1 < -9$;   **3.** $2x - 3 < 1$ or $3x - 5 > 7$
4. $2x + 3 < 5$ or $x + 1 > 4$;   **5.** $3x - 2 > 4$ or $x - 2 < 6$;   **6.** $x - 1 < \frac{1}{2}$ or $x + 3 > 5$

Answers: **1.** $\{x \mid x > -2\}$;   **2.** $\{x \mid x < -5$ or $x > -1\}$;   **3.** $\{x \mid x < 2$ or $x > 4\}$; **4.** $\{x \mid x < 1$ or $x > 3\}$;
   5. All real numbers since $x > 2$ or $x < 8$; **6.** $\{x \mid x < \frac{3}{2}$ or $x > 2\}$;

D During the semester, Betty scored 75 and 78 on the first two tests. She wants her class average for the term to be at least 80. There is one more test to take. What is the score she needs on the last test ?

      Answer: Betty needs at least a score of 87 on the last test

E Solve for x:

1. $\frac{x}{2} + 1 < x - 14$;   **2.** $3x - 2 < \frac{10 - x}{-2}$;   **3.** $\frac{x + 3}{2} \leq \frac{2 + x}{-2}$

      Answers **1.** $x > 30$;   **2.** $x < -\frac{6}{5}$;   **3.** $x \leq -\frac{5}{2}$

CHAPTER 8
Quadratic Equations

Lesson 25: **Solving by Factoring**
Lesson 26: **Solving by the Square Root Method**
Lesson 27: **Solving by Quadratic Formula**
Lesson 28: **Solving by Completing the Square**
Lesson 29: **Applications of the Quadratic Equation**

Standard form of the quadratic equation: $ax^2 + bx + c = 0$, where a, b, and c are constants and $a \neq 0$

Examples
1. $x^2 - 3x - 28 = 0$ <---------in standard form
2. $x^2 - 2x = 0$ <--------- in standard form
3. $x^2 = 5x + 14$ <---------**not** in standard form
4. $5x^2 = 8x$ <---------**not** in standard form

We shall consider four methods:

Method 1: **By factoring** (For easily recognizable factors and for cases in which the constant term is missing, i.e., $c = 0$)

Method 2: **By the square root method** (For cases in which the x-term is missing i.e., $b = 0$)

Method 3: **By the quadratic formula** (This always works)

Method 4: **By completing the square** (If asked to use this method, otherwise choose from Methods 1, 2, and 3 above).

Note that Method 4 also always works. **Note also** that the quadratic formula can be derived from the quadratic equation **by completing the square** and solving for x.

Lesson 25
Solving Quadratic Equations by Factoring

Principle of zero products: If $ab = 0$ then either $a = 0$, or $b = 0$ (or both $= 0$).

Example 1 Solve by factoring
$x^2 - 3x - 28 = 0$

Step 1: Factor the quadratic trinomial.

$(x + 4)(x - 7) = 0$

Step 2: Set each factor equal to zero and solve each equation for x.

$x + 4 = 0$ or $x - 7 = 0$
$\underline{\quad -4 \ -4}$ $\underline{\quad +7 \ +7}$
$x = -4$ $x = 7$

$\therefore \ x = -4$, or $x = 7$

The solutions are -4 and 7.

Example 2 Solve by factoring.

$$3x^2 - 4x = 0$$

Solution

Step 1: Factor. $3x^2 - 4x = 0$

$\qquad$ $x(3x - 4) = 0$ $\qquad$ (performing common monomial factoring).

Step 2: Set each factor equal to zero and solve for x. (Only each factor containing a variable
$\qquad$ is set to zero)

$$x = 0 \quad \text{or} \quad 3x - 4 = 0$$
$$\underline{+4 \quad +4}$$
$$\frac{3}{3}x = \frac{4}{3}$$

$$x = \frac{4}{3}$$

$$\therefore x = 0, \text{ or } x = \frac{4}{3} \qquad\qquad \textbf{Note} \text{ that 0 is also a solution .}$$

$\qquad$ The solutions are 0 and $\frac{4}{3}$

Note: A common **wrong** approach in the above problem:

$3x^2 - 4x = 0$

$\qquad 3x^2 = 4x$

$3x = 4$ (Canceling, that is dividing out an x on both sides of the equation ; such a cancellation excludes
$\qquad\qquad x = 0$ as a solution.)

$x = \frac{4}{3}$ (Of course, you still obtain one of the solutions but lose the other solution)

Example 3 Solve for x by factoring.

$$\frac{x^2}{2} + \frac{10x}{3} + 2 = 0$$

Step 1: Undo the denominators by multiplying the equation by the LCM of 2 and 3 which is 6.

$$(6)\frac{x^2}{2} + (6)\frac{10x}{3} + 2(6) = 6(0)$$

$$\overset{3}{\cancel{(6)}}\frac{x^2}{\cancel{2}_1} + \frac{\overset{2}{\cancel{(6)}}10x}{\cancel{3}_1} + 2(6) = 0$$

$$3x^2 + 20x + 12 = 0 \text{ ....(A)}$$

Step 2: Factor by the substitution method or otherwise

$$3(3x^2) + 20x(3) + 12(3) = 0$$
$$9x^2 + 20(3x) + 36 = 0$$
$$(3x)^2 + 20(3x) + 36 = 0 \text{ ....(B)}$$

Step 3: Let $3x = s$ in equation (B), and factor.

$$s^2 + 20s + 36 = 0$$
$$(s + 2)(s + 18) = 0\text{.....................................(C)}$$

Step 4: Replace s by $3x$ in equation (C)

$$(3x + 2)(3x + 18) = 0 \text{ ....(D)}$$

Step 5: Set each factor equal to zero and solve for x.

$$3x + 2 = 0 \qquad\qquad 3x + 18 = 0$$

$$\frac{-2 \quad -2}{} \qquad\qquad \frac{-18 \quad -18}{}$$

$$\frac{3}{3}x = \frac{-2}{3} \qquad\qquad \frac{3}{3}x = \frac{-18}{3}$$

$$x = -\frac{2}{3} \qquad\qquad\qquad x = -6$$

The solutions are $-\frac{2}{3}$ and -6. or solution set = $\{-\frac{2}{3}, -6\}$

Note that in the above problem , since it is an equation, it was not necessary to factor completely the left-hand side of equation (D) of Step 4.

Factorability of a quadratic trinomial

A quadratic trinomial is factorable if the discriminant $b^2 - 4ac$ is a perfect square. (see also p.41.)

Lesson 25 Exercises

A Solve by factoring: **1.** $x^2 - 11x + 18 = 0$; **2.** $x^2 - 5x - 36 = 0$;
3. $x^2 - 18x = 0$; **4.** $3x^2 = 24x$

Solutions **1.** {2, 9}; **2.** {- 4, 9}; **3.** {0,18}; **4.** {0,8}

B Solve by factoring: **1.** $x^2 - 4x - 21 = 0$; **2.** $x^2 + 4x - 45 = 0$; **3.** $x^2 - 25 = 0$

4. $9x^2 - 6x = 0$ **5.** $14t = 7t^2$ **6.** $x^2 = 11x - 18$

7. $2x^2 + 2x - 144 = 0$ **8.** $9x^2 - 36 = 0$ **9.** $ax^2 = -bx$

10. $x^2 - mx + nx - mn = 0$

Answers: **1.** { -3, 7} ; **2.** {-9, 5} ; **3.** {-5, 5} ; **4.** $\{0, \frac{2}{3}\}$; **5.** {0, 2}; **6.** {2, 9} ; **7.** {-9, 8} ; **8.** {-2, 2} ;

9. $\{0, -\frac{b}{a}\}$; **10.** {$m, -n$}

Lesson 26

Solving Quadratic Equations by the Square Root Method

(For cases in which $b = 0$, i.e., the x-term is missing)

Principle: If $x^2 = k$, then $x = \pm\sqrt{k}$ (i.e., $x = +\sqrt{k}$ or $x = -\sqrt{k}$)

Example 1 Solve by the square root method.

$$x^2 - 9 = 0$$

Solution

$$
\begin{aligned}
x^2 - 9 &= 0 \\
+9 \quad &+9 \\
\hline
x^2 &= 9 \\
x &= \pm\sqrt{9} \\
x &= \pm 3 \quad \text{(i.e. } x = 3 \text{ or } x = -3\text{)}
\end{aligned}
$$

The solutions are -3 and 3.

Example 2 Solve by the square root method.

$$9x^2 - 36 = 0$$

Solution

$$
\begin{aligned}
9x^2 - 36 &= 0 \\
+36 \quad &+36 \\
\hline
9x^2 &= 36 \\
\frac{9}{9}x^2 &= \frac{36}{9} \\
x^2 &= 4 \\
x &= \pm\sqrt{4} \\
x &= \pm 2
\end{aligned}
$$

The solutions are -2 and 2.

Lesson 26 Exercises

A Solve by the square root method:

1. $x^2 - 16 = 0$;   2. $4x^2 - 20 = 0$;   3. $x^2 - 49 = 0$;

Read page 195-196 before attempting Problem 4:   **4.** $x^2 + 49 = 0$

Answers: **1.** $\{-4, 4\}$;   **2.** $\{-\sqrt{5}, \sqrt{5}\}$;   **3.** $\{-7, 7\}$; 4. $\{-7i, 7i\}$

B Solve the following for x:

1. $x^2 - 9 = 0$ 2. $-8 + x^2 = 0$ 3. $ax^2 - c = 0$

4. $x^2 + 9 = 0$ 5. $s = ax^2$ 6. $x^2 - \frac{1}{4} = 0$

Answers: **1.** $\{-3, 3\}$;   **2.** $\{-2\sqrt{2}, 2\sqrt{2}\}$;   **3.** $\left\{-\sqrt{\frac{c}{a}}, \sqrt{\frac{c}{a}}\right\}$;   **4.** $\{-3i, 3i\}$;

5 $\left\{-\sqrt{\frac{s}{a}}, \sqrt{\frac{s}{a}}\right\}$;   **6.** $\left\{-\frac{1}{2}, \frac{1}{2}\right\}$

Lesson 27

Solving Quadratic Equations by Completing the Square

Example 1 Solve by completing the square.

$$x^2 - 12x + 8 = 0$$

Step 1: Eliminate the constant term (the "8")
from the left -hand side.   (Note that the 8 ends up on the right-hand side of the equation as -8).

$$
\begin{array}{r}
x^2 - 12x + 8 = 0 \\
\underline{-8 \quad -8} \\
x^2 - 12x = -8
\end{array}
$$

Step 2:   Add the square of half the coefficient of the x-term to both sides of the equation.

(i.e., add the square of $\frac{b}{2}$ to both sides of the equation)

$$x^2 - 12x + \left(\tfrac{-12}{2}\right)^2 = -8 + (-6)^2 \quad (b = -12, \ \tfrac{b}{2} = -6; \text{ and the square of } \tfrac{b}{2} = (-6)^2$$

$$x^2 - 12x + (-6)^2 = -8 + (-6)^2$$

Step 3:   Complete the square on the left-hand side of the equation

$$(x - 6)^2 = -8 + 36$$
$$(x - 6)^2 = 28$$

Step 4:   " Take the square root " of both sides of the equation.

$$x - 6 = \pm \sqrt{28}$$

Step 5: Solve for x and simplify right-hand side.

$$
\begin{array}{r}
x - 6 = \pm \sqrt{28} \\
\underline{+6 \quad +6}
\end{array}
$$
$$x = 6 \pm \sqrt{28}$$
$$\mathbf{x = 6 \pm 2\sqrt{7}}$$

Scrapwork:

$$\sqrt{28} = \sqrt{4}\sqrt{7} = 2\sqrt{7}$$

Lesson 27: Solving by Completing the Square

Example 2 Solve by completing the square.

$$3x^2 - 9x - 2 = 0$$

Step 1: $$3x^2 - 9x = 2$$

Step 2: Divide the equation by the coefficient of the x^2-term (we want this coefficient to be 1, for this method)

$$\frac{3x^2}{3} - \frac{9x}{3} = \frac{2}{3}$$

$$x^2 - 3x = \frac{2}{3}$$

Step 3: Add the square of the coefficient of the x-term to both sides of the equation, and complete the square.

$$x^2 - 3x + \left(\frac{-3}{2}\right)^2 = \frac{2}{3} + \left(\frac{-3}{2}\right)^2$$

Note: $\frac{b}{2} = \frac{-3}{2}$; $\left(\frac{b}{2}\right)^2 = \left(\frac{-3}{2}\right)^2$

$$\left(x - \frac{3}{2}\right)^2 = \frac{2}{3} + \frac{9}{4}$$

Multiply this out

$$\left(x - \frac{3}{2}\right)^2 = \frac{35}{12}$$

Scrapwork: $\frac{2}{3} + \frac{9}{4} = \frac{35}{12}$

$$x - \frac{3}{2} = \pm\sqrt{\frac{35}{12}}$$

Step 4: Solve for x and simplify right-hand side.

$$x - \frac{3}{2} = \pm\sqrt{\frac{35}{12}}$$
$$+\frac{3}{2} \qquad +\frac{3}{2}$$
_____

$$x = +\frac{3}{2} \pm \sqrt{\frac{35}{12}}$$

$$x = \frac{3}{2} \pm \frac{\sqrt{105}}{6}$$

Scrapwork: $\sqrt{\frac{35}{12}} = \sqrt{\frac{35 \cdot 3}{12 \cdot 3}}$

Solution is $\left\{\frac{3}{2} \pm \frac{\sqrt{105}}{6}\right\}$ or $\left\{\frac{3}{2} + \frac{\sqrt{105}}{6}, \frac{3}{2} - \frac{\sqrt{105}}{6}\right\}$

Lesson 27 Exercises

A Solve by completing the square:

1. $x^2 + 12x + 10 = 0$;

2. $x^2 - 4x + 6 = 0$;

3. $x^2 - 3x - 8 = 0$

4. $x^2 - 12x - 35 = 0$

Answers: **1.** $\{ -6 + \sqrt{26}, -6 - \sqrt{26}\}$; **2.** $\{2 + i\sqrt{2},\ 2 - i\sqrt{2}\}$; **3.** $\{\frac{3}{2} + \frac{\sqrt{41}}{2}, \frac{3}{2} - \frac{\sqrt{41}}{2}\}$;

4. $\{6 + \sqrt{71}, 6 - \sqrt{71}\}$

B Solve by completing the square:

1. $x^2 - 11x + 18 = 0$;

2. $2x^2 + 5x - 12 = 0$;

3. $x^2 - 4x - 1 = 0$;

4. $x^2 - 4x + 8 = 0$.

5. $4x^2 + 48x + 40 = 0$;

6. $2x^2 - 7x - 12 = 0$;

7. $\frac{1}{2}at^2 + bt + k = 0$;

8. $ax^2 + bx + c = 0$

Solutions: **1.** $\{2, 9\}$; **2.** $\{-4, \frac{3}{2}\}$; **3.** $\{2 + \sqrt{5}, 2 - \sqrt{5}\}$; **4.** $\{2 + 2i, 2 - 2i\}$; **5.** $\{-6 + \sqrt{26}, -6 - \sqrt{26}\}$;

6. $\{\frac{7}{4} + \frac{\sqrt{145}}{4}, \frac{7}{4} - \frac{\sqrt{145}}{4}\}$; **7.** $\{-\frac{b}{a} \pm \frac{\sqrt{b^2 - 2ak}}{a}\}$; **8.** $x = -\frac{b}{2a} \pm \frac{\sqrt{b^2 - 4ac}}{2a}; x = \frac{-b \pm \sqrt{b^2 - 4ac}}{2a}$

Lesson 28
Solving Quadratic Equations by the Quadratic Formula

Example 1   Solve for x by the quadratic formula

$$3x^2 - 6x - 2 = 0$$

Step 1: $a = 3, \ b = -6, \ c = -2$

Step 2: $x = \dfrac{-b \pm \sqrt{b^2 - 4ac}}{2a}$

Step 3: $x = \dfrac{-(-6) \pm \sqrt{(-6)^2 - 4(3)(-2)}}{2(3)}$ (Substituting for $a, b,$ and c)

$$= \dfrac{+6 \pm \sqrt{36 + 24}}{6}$$

$$= \dfrac{6 \pm \sqrt{60}}{6}$$

$$= \dfrac{6}{6} \pm \dfrac{\sqrt{60}}{6}$$

$$= 1 \pm \dfrac{2\sqrt{15}}{6}$$

$$x = 1 \pm \dfrac{\sqrt{15}}{3}$$

Solution set is $\left\{ 1 - \dfrac{\sqrt{15}}{3} , 1 + \dfrac{\sqrt{15}}{3} \right\}$

Example 2   Solve by the quadratic formula:

$$x^2 - 16 = 8 - 2x$$

Step 1:   Place the equation in standard form (i.e. rewrite the equation so that the only term on the right-hand side is zero)

$$
\begin{array}{rl}
x^2 - 16 & = 8 - 2x \\
+2x & +2x \\
\hline
x^2 + 2x - 16 & = 8 \\
-8 & -8 \\
\hline
x^2 + 2x - 24 & = 0 \;\text{<------------(Standard Form)}
\end{array}
$$

Step 2: $a = 1$, $b = 2$, $c = -24$

Step 3: $$x = \frac{-b \pm \sqrt{b^2 - 4\,ac}}{2a}$$ <----------The quadratic formula

Step 4 : $$x = \frac{-2 \pm \sqrt{(2)^2 - 4(1)(-24)}}{2(1)}$$ (Substituting for a, b, and c)

$$x = \frac{-2 \pm \sqrt{4 + 96}}{2}$$

$$x = \frac{-2 \pm \sqrt{100}}{2}$$

$$x = \frac{-2 \pm 10}{2}$$

$$x = \frac{-2 + 10}{2} \text{ , or } x = \frac{-2 - 10}{2}$$

$$x = \frac{8}{2}, \text{ or } x = \frac{-12}{2}$$

$$x = 4 \text{ or } x = -6$$

The solutions are 4 and -6.

Solving by any method

Example   Solve by any method: $x^2 = 6x$

$$x^2 = 6x$$

Step 1: Write the equation in standard form.

$$x^2 - 6x = 0$$

Step 2: We solve by factoring (since $c = 0$, the quadratic is easily factorable)

$$x(x - 6) = 0$$

Step 3: Set each factor equal to 0 and solve for x:

$$
\begin{array}{ll}
x = 0 \text{ , or } & x - 6 = 0 \\
& +6 \quad +6 \\
\hline
& x = +6
\end{array}
$$

$$x = 0, \text{ or } x = 6$$

The solutions are 0 and 6; or the solution set is {0,6}.

Discriminant of the quadratic equation: nature of roots and graphs of the equation   136

The expression $b^2 - 4ac$ is called the **discriminant** of the quadratic equation. The value of the discriminant determines the nature of the roots (or solutions) of the quadratic equation:

1. If $b^2 - 4ac > 0$, the equation $ax^2 + bx + c = 0$ has two real and unequal roots. Also if $b^2 - 4ac$ is a perfect square, the roots are rational. Graphically, the curve of the corresponding quadratic function, $f(x) = ax^2 + bx + c$, crosses the x-axis at two different points.

2. If $b^2 - 4ac = 0$, the quadratic equation has real equal roots (double root or repeated root). We therefore have only one real solution. Graphically, the curve of the corresponding quadratic function $f(x) = ax^2 + bx + c$, touches the x-axis but does not cross it.

3. If $b^2 - 4ac < 0$, the equation has two non-real (complex) roots. Graphically, the curve of the corresponding quadratic function, $f(x) = ax^2 + bx + c$, does not touch or cross the x-axis. The curve is either entirely above the x-axis or entirely below the x-axis.

Note also that a quadratic equation is factorable if $b^2 - 4ac$ is a perfect square. (see also p. 41.)

Lesson 28 Exercises

A Solve by the quadratic formula:
1. $x^2 - 11x + 18 = 0$;   **2.** $2x^2 + 5x - 12 = 0$;   **3.** $x^2 - 4x - 1 = 0$; **4.** $x^2 - 4x + 8 = 0$.
5. $x^2 + 12x + 10 = 0$;   **6.** $x^2 + 6 - 4x = 0$ **7.** $x^2 = 3x + 8$

Solutions: **1.** $\{2, 9\}$;   **2.** $\{-4, \frac{3}{2}\}$;   **3.** $\{2 + \sqrt{5}, 2 - \sqrt{5}\}$;   **4.** $\{2 + 2i, 2 - 2i\}$; **5.** $\{-6 \pm \sqrt{26}\}$

6. $\{2 \pm i\sqrt{2}\}$; **7.** $\{\frac{3 \pm \sqrt{41}}{2}\}$

B Solve by any applicable method:
1. $x^2 - 6x + 4$;   **2.** $6x^2 + 11x - 10 = 0$;   **3.** $2x^2 - 3x + 6 = 0$;   **4.** $2x(x - 4) = 6$

Solutions: **1.** $\{3 + \sqrt{5}, 3 - \sqrt{5}\}$;   **2.** $\{\frac{2}{3}, -\frac{5}{2}\}$;   **3.** $\{\frac{3 + i\sqrt{39}}{4}, \frac{3 - i\sqrt{39}}{4}\}$ **4.** $\{2 + \sqrt{7}, 2 - \sqrt{7}\}$

Lesson 29

Applications of the Quadratic Equation

Example 1   The width of a rectangle is 3 units less than the length. If the area of the rectangle is 28 sq. units, determine the dimensions of this rectangle.

Solution

Let the length of the rectangle $= x$.

Then, the width of this rectangle $= (x - 3)$

Area of a rectangle is given by LW (Where L is the length and W is the width).

The required equation: $x(x - 3) = 28$.

$$x^2 - 3x - 28 = 0$$
$$(x - 7)(x + 4) = 0$$
$$x = 7 \text{ or } x = -4 \qquad \text{(Solving by factoring)}$$

We reject the negative value, -4, since the dimension of a rectangle cannot be negative.

When the length is 7 units, the width is 7 - 3 = 4 units.

(**Check**: If the length is 7 and the width is 4, the area is 7(4) = 28 sq. units; and also the width, 4, is 3 less than the length, 7. Therefore, the conditions in the original word problem have been satisfied.)

The length is 7 units and the width is 4 units.

Example 2   The perimeter of a rectangle is 34 units, and its area is 30 sq. units. Find the dimensions of this rectangle.

Preliminaries:

Perimeter of a rectangle $= 2L + 2W$ (where L is the length, and W is the width)

Area of a rectangle $= LW$

Solution

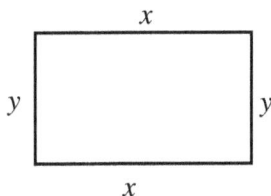

Let the dimensions be x and y, where the larger of x and y is the length and the smaller of x and y is the width.

$$2x + 2y = 34 \text{ <-----------Perimeter equation} \qquad (1)$$
$$xy = 30 \text{ <-----------Area equation} \qquad (2)$$

We will solve the above system of equations simultaneously by the substitution method.

Step 1 : Solve for y from equation (1).

  Then, $y = (17 - x)$ (3)   (obtained by subtracting $2x$ from both sides of the equation followed by dividing the equation by 2)

Step 2: Substitute for y from (3) in (2).

$$x(17 - x) = 30$$
$$17x - x^2 = 30$$
$$17x - x^2 - 30 = 0$$
$$x^2 - 17x + 30 = 0 \qquad \text{(Multiplying the equation by -1 and rearranging terms)}$$

Step 3: Solve this quadratic equation by any method.
We solve by factoring since the factors are easily recognizable.
$(x - 2)(x - 15) = 0$
$x = 2$ or $x = 15$

Step 4: When $x = 2$,
$2y = 30$; and from which $y = 15$.
when $x = 15$,
$15y = 30$; and from which $y = 2$.
From Step 4, we obtain the dimensions 15 and 2.

The dimensions are 15 units and 2 units.   (The length is 15 units and the width is 2 units)

Lesson 29 Exercises

A The perimeter of a rectangle is 36 units, and its area is 77 sq. units.
Find the dimensions of this rectangle.

Answers: Dimensions are 11 and 7.

B 1.   A number is 4 more than another number. If their product is 165, find them.

2. The sum of two numbers is 20, and their product is 96. Find these numbers.

3. The length of a rectangle is 5 units more than the width. If the area of this rectangle is 126 sq, units, find the length and width of this rectangle.

Ans: **1.** 15 and 11; and also -11 and -15;   **2.** 8 and 12;   **3.** Length = 14 units, width = 9 units.

CHAPTER 9
FUNCTIONS

Lesson 30: **Sets, Relations, Functions, Comparison of Relations and Functions**

Lesson 31: **Functional Notation; Defined Functions; Excluded Values, Domain and Range**

Lesson 32 Extra: **Algebra of Functions**

Lesson 30

Sets, Relations, Functions, Comparison of Relations and Functions

Ordered Pair

An **ordered pair** of numbers is an arrangement of two numbers in a specified order. In an x-y rectangular coordinate system of axes, the first element (or component) is the x-value and the second element is the y-value.

Example 1   (a)   (1, 2) <--- $(x = 1, y = 2)$
               (b)   (2, 3) <---- $(x = 2, y = 3)$
               (c)   (5, -1) <---- $(x = 5, y = -1)$

Note that each ordered pair represents a point in an x-y coordinate system of axes.

Set of numbers

A **set of numbers** is a well-defined collection of numbers. The numbers are called the elements or members of the set.

Example 2: If we denote the set of the numbers 2, 5 and 6 by A, then we may write $A = \{2, 5, 6\}$

Example 3: The set B of the ordered pairs $(1, 2), (2, 3),$ and $(5, -1)$ is given by
        $B = \{(1, 2), (2, 3), (5, -1)\}$.

Relation

A **relation** is a set of ordered pairs. (A collection of ordered pairs of numbers)

Example 4: The set $E = \{(2, 3), (2, 5), (4, 6)\}$ is a relation, <---There are three ordered pairs.

Example 5: The set $C = \{(6, 2), (7, 4), (11, 5)\}$ is a relation.

Definition of a Function

A function may be defined in a number of ways, namely,

(a) in terms of ordered pairs; (b) in terms of a rule involving two variables;
(c) in terms of a rule for inputs and outputs; (d) in terms of correspondence of two sets; (e) as a graph

Definition 1:   In terms of ordered pairs
A **function** is a relation in which no two distinct ordered pairs have the same first component; or a function is a set of ordered pairs in which for any two different ordered pairs, the first elements are different. The set in Example 5, above, is a function but the set in Example 4 is not a function, because the first two ordered pairs have the same first element, namely 2.

The set of all the first elements of the ordered pairs is called the **domain** of the function; and the set of all the second elements is called the **range** of the function.

Example 6: In the function $C = \{(6, 2), (7, 4), (11, 5)\}$. The domain, $D = \{6, 7, 11\}$ (first elements)
                                        The range, $R = \{2, 4, 5\}$.   (second elements)

Other definitions of a function

Definition 2: If x and y are two variables. then we say that y is a function of x if there is a rule which gives just one corresponding value of y for **each** value of x. The variable x is called the independent variable, and a variable y is called the dependent variable. The rule may be specified in the form of a set, in the form of a graph, in the form of a table, or in the form of an equation or formula.

The **domain** of a function is the set of numbers that can be assigned to x (the independent variable).
The **range** of a function is the set of all the corresponding numbers y (the dependent variable) associated by the function (rule) with the numbers, x, in the domain .

We symbolize that f is a function of x by $f(x)$, where x is called the independent variable, y is called the dependent variable. **Note:** $f(x)$s is read as f of x.

The following are examples of how the rules for functions may be specified:

(a) In the form of an equation or a formula: $y = 2x$.
(b) In the form of a set: $\{(2,3),(1,4),(7,5)\}$.
(c) In the form of a table for x and y: See Table 1.
(d) In the form of a graph. See Figure

Table 1:: y = 2x

$x =$	0	1	2	3	4
$y =$	0	2	4	6	8

Figure: Graph of $y = 2x$

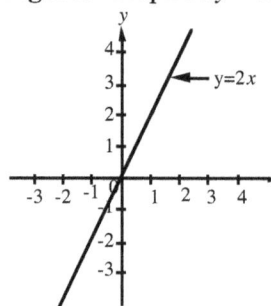

Definition 3 In terms of the correspondence of two sets (Fig. 1)

A function is a correspondence between a first set, say set A and a second set, say Set B such that **each** element of set A corresponds to exactly one element of set B. The set of all the elements of set A is a called the **domain** of the function, and the set of all the corresponding elements of set B is called the **range** of the function.

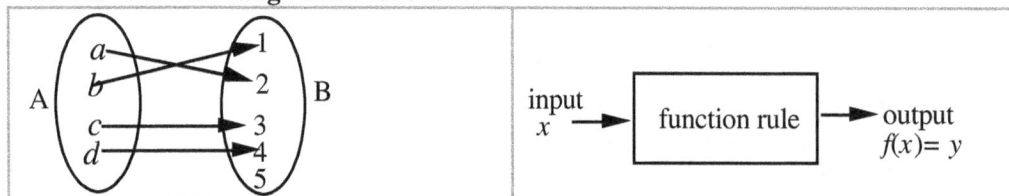

Fig 1   Fig 2

Definition 4   In terms of a rule for inputs and outputs (Fig. 2)

A function is a rule which assigns to each input number exactly one output number. The set of all input numbers that the rule is applicable to is called the **domain** of the function; and the set of all the corresponding output numbers is called the **range** of the function.
A variable representing an input number is called the independent variable, and a variable representing an output number is called the dependent variable.

Given a graph   (Vertical line test)

A given graph is that of a function if every possible vertical line drawn to intersect the graph intersects (cuts) the graph exactly once (i.e., at one point only).

Comparison of a Function and a Relation

Similarities: Each is a set of ordered pairs.

Differences: In a relation, two or more ordered pairs may have the same first component; but in a function, no two distinct ordered pairs may have the same first component.

Example: The set $D = \{(1,6),(3,4),(3,5),(4,6)\}$ is only a relation and **not** a function. because the second and third ordered pairs have the same first component, which is 3.

Example: The set $E = \{(1,2),(2,3),(4,5),(7,5)\}$ is a function (even though the second components of the third and fourth ordered pairs are the same).

A function is a relation, but a relation is not necessarily a function.

Determining if a given graph is a relation or a function

We will use the so-called **vertical line test.**

Procedure

Step 1: Draw as many vertical lines as possible (This can be done visually.) to intersect the graph.

Step 2: If any of the possible lines intersects (cuts) the graph at more than one point, then the given graph is not a function but a relation. However, if each of the possible vertical lines intersects the graph only once (at one point only), then the graph represents a function. Figure.. is a graph which is a relation but not a function.

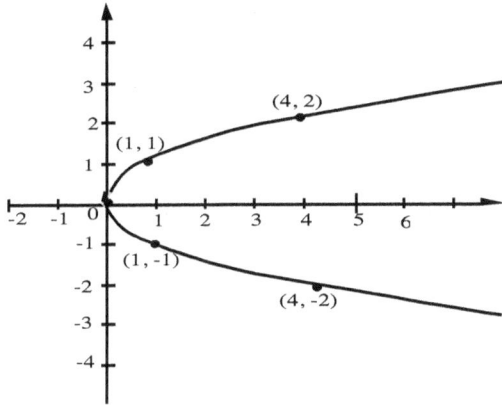

Figure: Graph of $y = \pm\sqrt{x}$ or $x = y^2$
This graph is a relation but not a function

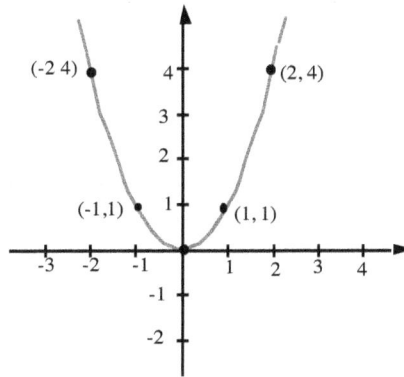

Figure: Graph of $y = x^2$.
This graph is that of a function.

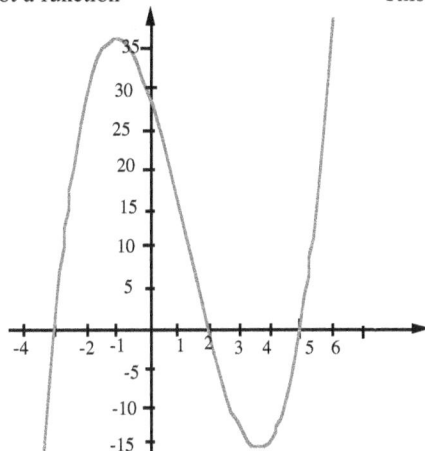

Figure: Graph of $y = (x-5)(x-2)(x+3)$. This graph is that of a function.

Lesson 30 Exercises

Determine which of the following are graphs of functions.

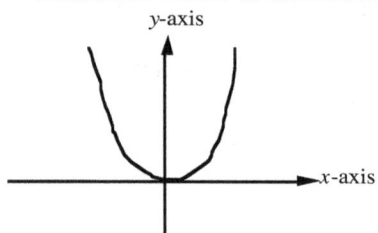

Figure (a)

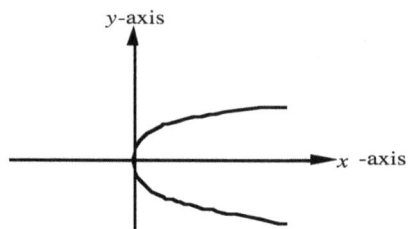

Figure (b)

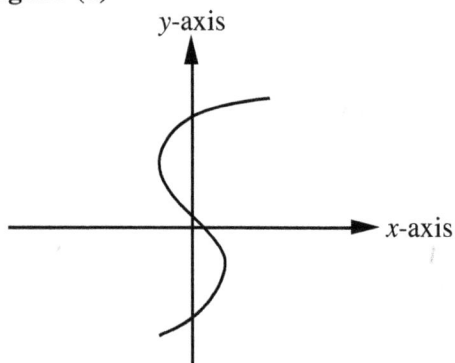

Figure (c)

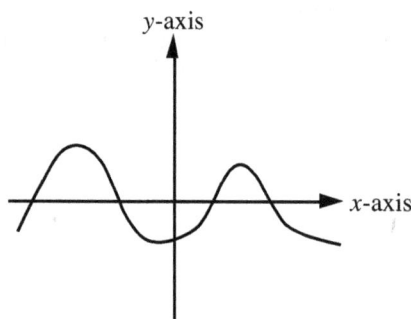

Figure (d)

Answers: (a) A function; (b) Not a function; (c) Not a function; (d) A function

Lesson 31

Functional Notation; Defined Functions; Excluded Values; Domain and Range

Functional Notation

Let a function $f(x)$ be specified by the rule $f(x) = x^2 + 3$.   (1)

In equation (1), $f(x)$ is read "f of x" or f is a function of x.

To evaluate a function for a particular value of x, we substitute that value of x in the rule that defines $f(x)$. Note that $f(x)$ does not mean f times x but that $f(x)$ is written as a symbol.

Example 1   Given that $f(x) = x^2 + 3$, find (a) $f(-1)$.; (b) $f(x_0 + h) - f(x_0)$

Solution: (a) $f(x) = x^2 + 3$

$$f(-1) = (-1)^2 + 3 \qquad \text{(replacing } x \text{ in the given equation by -1)}$$
$$= 1 + 3$$
$$f(-1) = 4$$

(b) (Replace x in the given equation by $(x_0 + h)$ and x_0, accordingly in $f(x) = x^2 + 3$

$$f(x_0 + h) \; - \; f(x_0) = [(x_0 + h)^2 + 3] - (x_0^2 + 3)$$
$$= [x_0^2 + 2hx_0 + h^2 + 3] - x_0^2 - 3$$
$$= x_0^2 + 2hx_0 + h^2 + 3 - x_0^2 - 3$$
$$= 2hx_0 + h^2$$

Example 2   Find $f(2)$, given that $f(x) = x + 7$

Solution

$$f(x) = x + 7$$
$$f(2) = 2 + 7$$
$$= 9$$

Example 3   If $f(x) = 2 - \dfrac{1}{x - 4}$, find (a) $f(-3)$;   (b) $f(x_0 + h)$;   (b) $f(-x)$.

Solution   (a) $f(x) = 2 - \dfrac{1}{x - 4}$

$$f(-3) = 2 - \frac{1}{(-3) - 4} \qquad \text{(replacing } x \text{ in the given equation by -3)}$$
$$= 2 - \frac{1}{-7}$$
$$= 2 + \frac{1}{7}$$
$$= 2\tfrac{1}{7}$$

(c)   (Replace x in the given equation by $x_0 + h$

$$f(x) = 2 - \frac{1}{x-4} \quad \text{<----given equation}$$

$$f(x_0 + h) = 2 - \frac{1}{(x_0+h)-4} \quad \text{<---replacing } x \text{ by } x_0 + h$$

$$= 2 - \frac{1}{x_0 + h - 4}$$

$$= \frac{2(x_0+h-4)-1}{x_0+h-4}$$

$$= \frac{2x_0 + 2h - 8 - 1}{x_0 + h - 4}$$

$$= \frac{2x_0 + 2h - 9}{x_0 + h - 4}$$

(c)   (Replace x in the given equation by $-x$)

$$f(x) = 2 - \frac{1}{x-4} \quad \text{<----given equation}$$

$$f(-x) = 2 - \frac{1}{(-x)-4} \quad \text{<-----}(replacing\ x\ by - x)$$

$$= 2 - \frac{1}{-x-4}$$

$$= \frac{2(-x-4)-1}{-x-4}$$

$$= \frac{-2x-8-1}{-x-4}$$

$$= \frac{-2x-9}{-x-4}$$

$$= \frac{-(2x+9)}{-(x+4)} \quad (factoring\ out -1)$$

$$= \frac{2x+9}{x+4}$$

Note above that in (b) the final result contains x. This is so, because we replaced x by $-x$. In the case of (a), we replaced x by the integer -3 , and the final result was purely a numerical value.

Furthermore, in Example 3, $f(-a) = \dfrac{2a+9}{a+4}$

Defined Real-Valued Function

Meaning of a defined function of *x*

A real-valued function $f(x)$ is said to be defined for a variable x if the following conditions are satisfied:

      1. The x and $f(x)$ must be real (i.e., x and $f(x)$ should not be the square root or an even root of a negative number). Thus, a value such as $\sqrt{-4}$ or $2i$ is not allowed.

      For example, in $f(x) = \sqrt{x - 4}$, we have to make sure that $x \geq 4$, since otherwise, we obtain imaginary numbers.

      2. The value of x when substituted in the functional equation should yield specific real numbers. (i.e., the value of x when substituted in the functional equation should **not** make the denominator become zero.)

Condition (2) implies that the function should not become undefined when the value of x is substituted in the functional equation. When the function involves a denominator, we have to make sure that the denominator is not allowed to be zero. A particular example of this function occurs when the given function is the ratio of two polynomial functions. (We call such functions rational functions.)

In this book, it is agreed that a function is real-valued unless otherwise specified.

Examples of rational functions are: (*a*) $f(x) = \dfrac{x^2 + 4}{x - 1}$;   (*b*) $f(x) = \dfrac{1}{(x - 3)(x + 4)}$

Excluded Values, Domain and Range

The **excluded values** are (usually) the values which when substituted in the functional equation make the function either undefined or imaginary.

The **domain** (say , D) of a function $f(x)$ consists of those real values of the independent variable say, x, for which $f(x)$ is real and defined.

The **range** (say, R) of the function consists of the corresponding values which $f(x)$ assumes for x in the domain of the function .

Specification of Domain and Range

The domain and the range of a given function may be specified in several ways, namely in the form of a set, in the form of a table, in the form of an equation, in the form of an inequality, or in the form of a graph.

(In some of the examples that follow, we will use a graphing calculator or a computer grapher to generate graphs which will be useful in determining the range of some of the functions. Determining the range of some of the functions analytically may require advanced methods (which we do not cover in this book) , and therefore, we use graphing as an aid.)

Example 1: In the form of a set

Find the domain and range of the function specified by $\{(1, 2), (3, 2), (4, 4), (5, 5)\}$.

Solution: The domain consists of the first components of the ordered pairs.
      The domain, $D = \{1, 3, 4, 5\}$.
      The range consists of the second components of the ordered pairs.
      The range, $R = \{2, 4, 5\}$.

Example 2: In the form of a table of values

Find the domain and range of the function specified by the table of values below.

x	y
1	2
3	2
4	4
5	5

Solution: The domain consists of (the x-values) $1, 3, 4,$ and 5.

The range consists of the (y-values) $2, 4$ and 5.

Note that this form is the set form written in a different format.

Example 3: In the form of an equation:

The domain consists of those real values of x for which the function is defined.
The corresponding values of $f(x)$ form the range of the function.

Example 4: In the form of a graph

Find the domain and range of the function specified by the graph below.

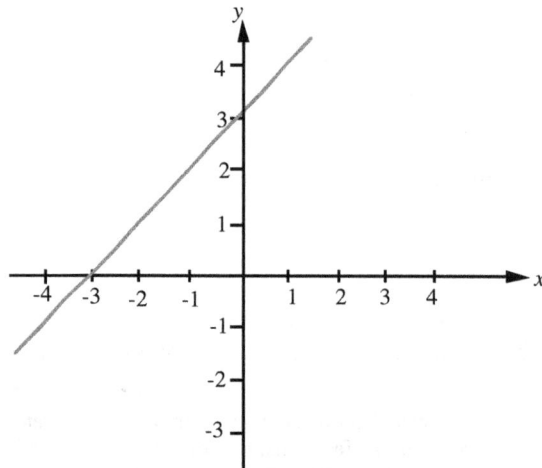

Figure: The graph of $y = x + 3$

Solution: From the graph, the function is defined for all real values of x.

The domain consists of all real x-values. The range consists of all real y-values.

Implicit and Explicit Specification of the Domain of a Function

The domain of a function may be specified either implicitly or explicitly.

     Consider the function $f(x) = x^2$
As it stands, the domain is implicitly specified. The above function is that of a polynomial and as such the domain consists of all real numbers. Here, we assume the largest possible domain.

     Now, consider $f(x) = x^2$ $0 \le x \le 5$
In this case, the inequality written to the right of the function specifies explicitly and restricts the domain. The domain of this function is such that x is between 0 and 5, including 0 and 5.
 If the inequality to the right had not been indicated, the domain would have consisted of all real x-values.

The restrictions on the domains are very important and useful in sketching the graphs of functions.

 Other functions with explicitly specified domains are

       (a) $y = \sin x$ $0 \le x \le 2\pi$

       (b) $y = \cos x$ $-2\pi \le x \le 2\pi$

Determining the Excluded Values, Domain and Range of a Function   148

Case 1: Polynomial functions

Example 1 (a) For what values of x is the following function not defined? (b) what is the domain?

(c) what is the range? $f(x) = x^2 - 3x + 1$

Solution: All polynomial functions are defined for all real values of the independent variable.

(a) Since the given function is a polynomial function, it is defined for all real values of x. We may note that the right-hand side of the equation does not involve denominators or square roots (or even roots) of the independent variable. There are **no** excluded values.

(b) Generally, determining the range of a polynomial function may require advanced methods. However, since the given function is a quadratic function, we may apply the range inequality

formula, $y \geq \dfrac{4ac - b^2}{4a}$

From the function, $a = 1, b = -3$, and $c = 1$.

Substituting these values, $y \geq \dfrac{4(1)(1) - (-3)^2}{4(1)} = -\dfrac{5}{4}$ (i.e., $y \geq -\dfrac{5}{4}$)

The range consists of all real numbers greater than or equal to $-\dfrac{5}{4}$.

(Note that any horizontal line drawn below the point $(\frac{3}{2}, -\frac{5}{4})$ will **not** intersect the curve, but any horizontal line through or above this point will intersect the curve)

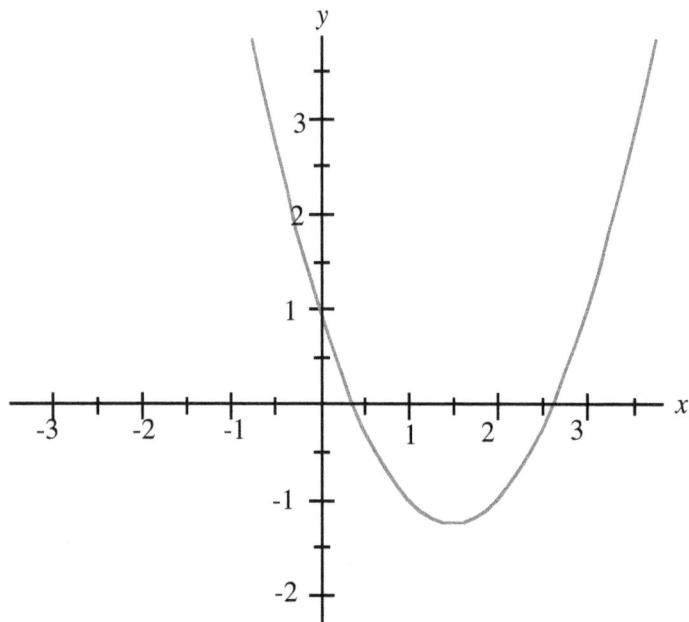

Figure: Graph of $f(x) = x^2 - 3x + 1$

Case 2: Rational functions

Note: A rational function is a function which is the ratio of two polynomial functions.

Example 2 (a) For what values of x is the following function not defined? (b) what is the domain?
(c) what is the range?

$$f(x) = \frac{3x - 2}{x - 1}$$

Solution   Step 1: Setting the denominator to zero,
$$x - 1 = 0$$
Step 2: Solving for x, $x = 1$.

(a) The function is not defined when $x = 1$. (The excluded value of x is 1.)

(b) The domain is all real values of x, except 1. same as $\{x \mid x$ is a real number and $x \neq 1\}$

(c) The range (from graph) if found by being guided by the horizontal asymptote, $y = 3$.
   The range is given by the set $\{y \mid y < 3$ or $y > 3\}$ or simply $\{y \mid y \neq 3\}$
   (Note that a horizontal line drawn through $(0, 3)$ will **not** intersect the curve)

Checking for $x = 1$, $f(1) = \frac{3(1) - 2}{1 - 1} = \frac{3 - 2}{0} = \frac{1}{0}$ which is undefined .(The right-hand side is division by zero).

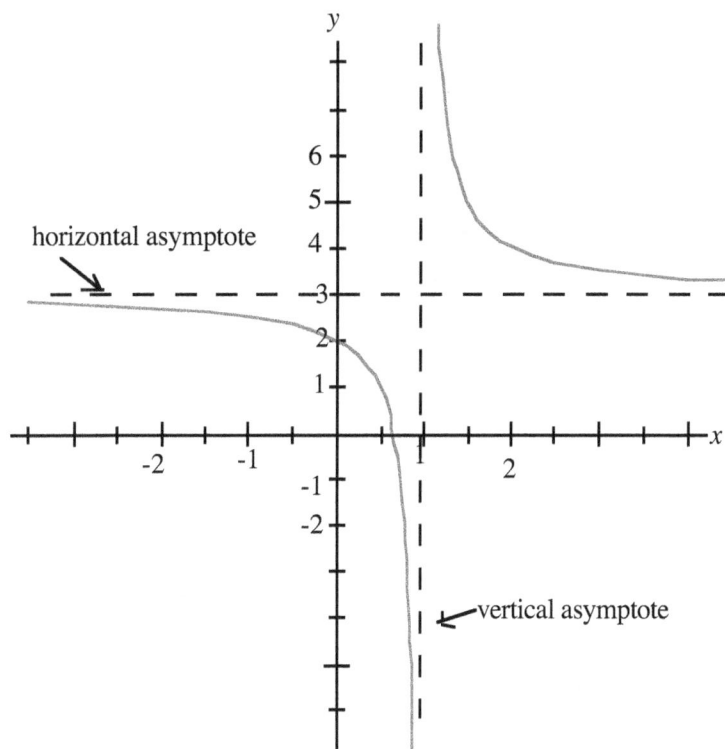

Figure: Graph of $f(x) = \frac{3x - 2}{x - 1}$

Lesson 31: Functional Notation; Defined Functions; Excluded Values, Domain and Range

Example 3 (a) For what values of x is the given function not defined? (b) what is the domain?
   (c) what is the range?

$$f(x) = \frac{2(x-1)}{(x-2)(x+4)}$$

Solution Step 1: Setting the denominator to zero,
$$(x-2)(x+4) = 0$$

   Step 2: Solving for x, $x = 2$, or $x = -4$.

(a) The function is not defined when $x = 2$ and -4. (The excluded values of x are 2 and -4.)

(b) The domain is all real values of x, except 2 and -4.
   Set-builder notation: $\{x \mid x$ is a real number and $x \neq -4$, $x \neq 2\}$

(c) The range (from graph) is all real y.
   Set-builder notation: $\{y \mid y$ is a real number$\}$
(Note that any horizontal line drawn through the y-axis will intersect the curve)

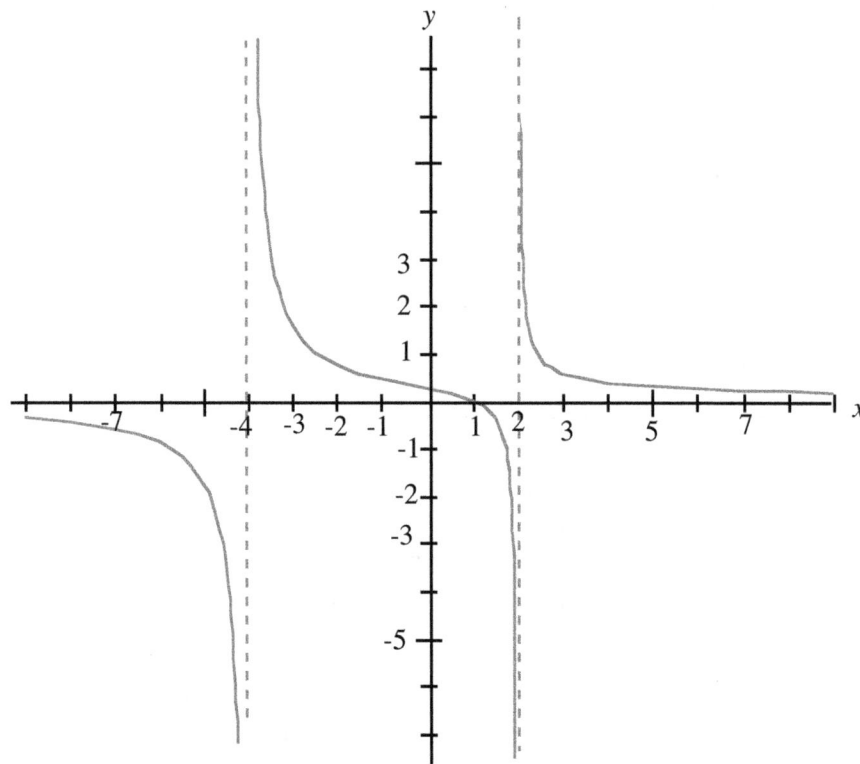

Figure: Graph of $f(x) = \dfrac{2(x-1)}{(x-2)(x+4)}$

Lesson 31: Functional Notation; Defined Functions; Excluded Values, Domain and Range

Example 4 (a) For what values of x is the given function not defined? (b) what is the domain?
(c) what is the range?

$$f(x) = \frac{1}{x}$$

Solution Setting the denominator to zero,
$$x = 0$$

(a) The function is not defined when $x = 0$. (The excluded value of x is 0.)

(b) The domain is all real values of x, except 0.
  Set-builder notation: $\{x \mid x$ is a real number and $x \neq 0\}$

(c) The range (from graph) is given by the set $\{y \mid y < 0$ or $y > 0\}$ or simply $\{y \mid y \neq 0\}$
  (the horizontal asymptote is $y = 0$)
  (Note that any horizontal line drawn through the y-axis, except through the point $(0, 0)$, will intersect the curve.

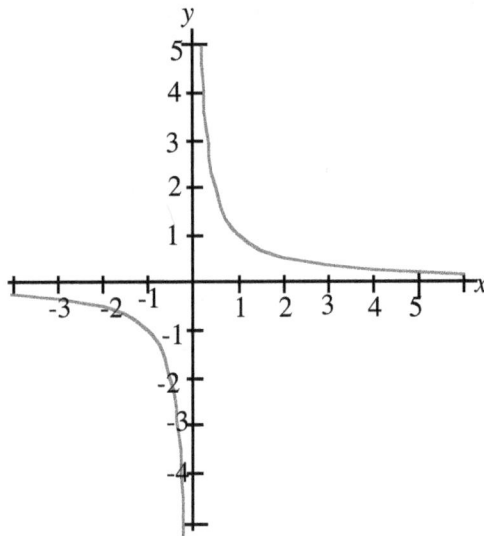

Figure: Graph of $f(x) = \frac{1}{x}$

Lesson 31: Functional Notation; Defined Functions; Excluded Values, Domain and Range

Example 5 (a) For what values of x is the given function not defined? (b) what is the domain?
(c) what is the range?

$$f(x) = \frac{x^3 + x^2 + 2}{x^2 - 16}$$

Solution: Setting the denominator to zero and solving,
$$x^2 - 16 = 0$$
$$(x + 4)(x - 4) = 0$$
$$x = -4 \text{ or } x = 4$$

The function is not defined when $x = -4$ or 4. (The excluded values of x are -4 and 4.)
(a) The domain is all real x except -4 and 4.
   $\{x \mid x \text{ is a real number and } x \neq -4 , x \neq 4 \}$

(b) The range (from graph) is all real values of y. same as $\{y \mid y \text{ is a real number}\}$
(Note that any horizontal line drawn through the y-axis will intersect the curve)

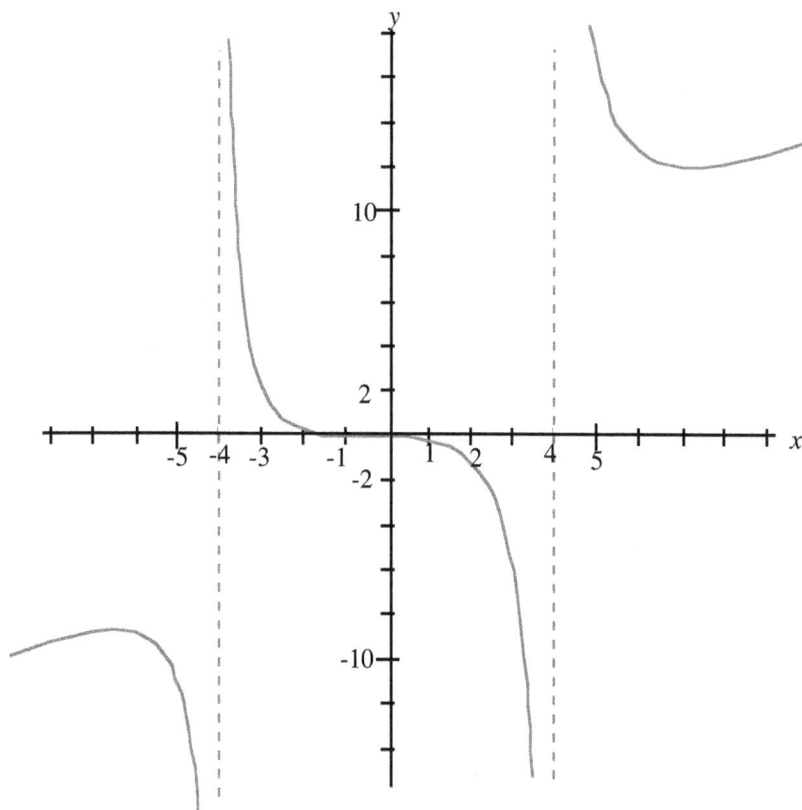

Figure: Graph of $\dfrac{x^3 + x^2 + 2}{x^2 - 16}$

Lesson 31: Functional Notation; Defined Functions; Excluded Values, Domain and Range

Example 6 (a) For what values of x is the given function not defined? (b) what is the domain?
(c) what is the range?

$$f(x) = \frac{8}{x^2 - 4}$$

Solution Step 1: Setting the denominator to zero,

$$x^2 - 4 = 0$$
$$(x + 2)(x - 2) = 0.$$

Step 2: Solving for x, $x = 2$, or $x = -2$.

(a) The function is not defined when $x = 2$ and -2. (The excluded values of x are 2 and -2.)

(b) The domain is all real values of x except 2 and -2.

Same as $\{x \mid x$ is a real number and $x \neq -2$, $x \neq 2\}$

(c) The range (from graph) is given by the set $\{y \mid y \leq -2$ or $y > 0\}$
(Note that any horizontal line drawn through or below $y = -2$ or above $y = 0$ will intersect the curve)

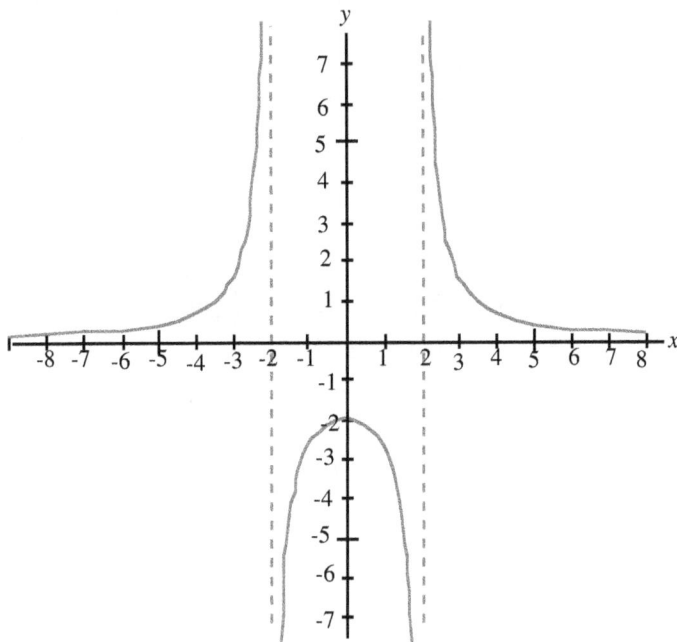

Figure: Graph of $f(x) = \frac{8}{x^2 - 4}$

Example 7 (a) For what values of x is the given function not defined? (b) what is the domain?
(c) what is the range?

$$f(x) = \frac{8}{x^2 + 4}$$

Solution Step 1: Setting the denominator to zero, and solving, we obtain non-real values. Since we are dealing with real-valued functions, we conclude that

Step 2: (a) There are **no** excluded values.
$x^2 + 4$ is positive for all real values of x and never zero, since the square of any nonzero real number is always positive.

(b) The function is defined for all real values of x.

(c) The range (from graph) is given by the set $\{y \mid 0 < y \le 2\}$

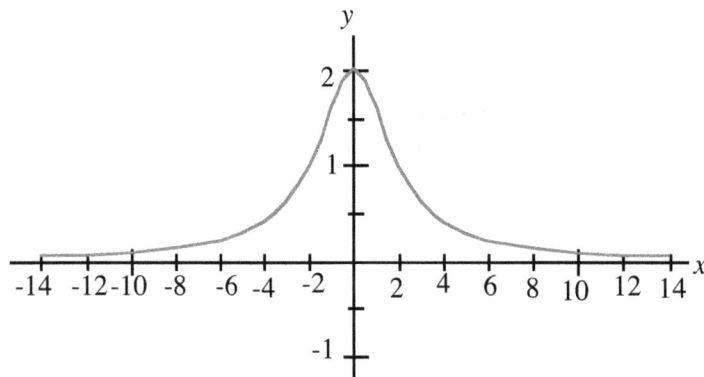

Figure: Graph of $f(x) = \frac{8}{x^2 + 4}$

Case 3: Functions containing even roots (e.g., square root) of polynomials

Example 8 (a) For what values of x is the function $f(x) = \sqrt{x - 5}$ not real?

(b) What is the domain?   What is the range?

Solution (a) For real roots, the radicand, $(x - 5)$ must be positive or zero.
Symbolically, $x - 5 \ge 0$
$x \ge 5$ (solving for x).

The function is not real when $x < 5$ (because the square root would be that of a negative number. The square root of a negative number is imaginary. Any x-value less than 5 is excluded.

Checking for some specific values of x:

1. When $x = 5$, $f(5) = \sqrt{x - 5} = \sqrt{5 - 5} = 0$, which is real.   (i.e., for $x = 5$).

2. When $x = 6$, $f(6) = \sqrt{6 - 5} = 1$, which is real.   (i.e., for $x > 5$).

3. When $x = 4$, $f(4) = \sqrt{4 - 5} = \sqrt{-1}$, which is imaginary (non-real (i.e., for $x < 5$).

(b) The domain consists of all real values of x such that $x \ge 5$.

(c) The range for $y = \sqrt{x - 5}$ is from $y = 0$ to $y = \infty$.
(The range is obtained by considering $x \ge 5$ ($x = 5$ and $x > 5$) in $y = \sqrt{x - 5}$, or from graph (next page, Fig. 2)

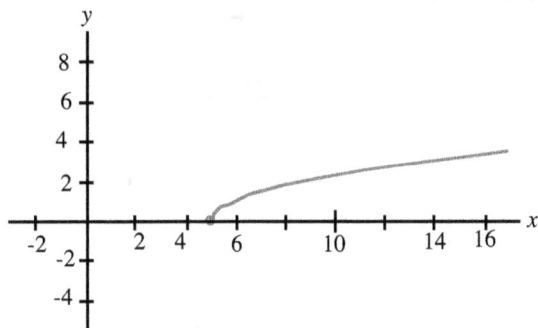

Figure 2: Graph of $f(x) = \sqrt{x-5}$

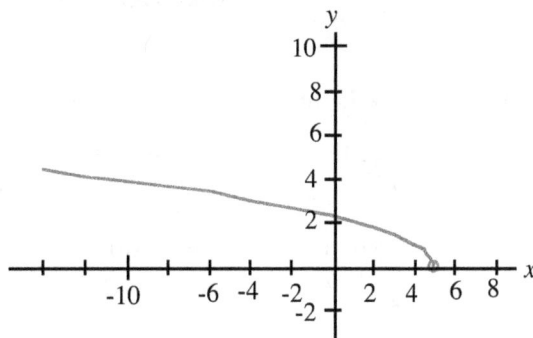

Figure 3: Graph of $f(x) = \sqrt{5-x}$

Example 9 (a) For what values of x is $f(x) = \sqrt{5-x}$ not real?

(b) What is the domain? What is the range?

Solution

(a) $5 - x \geq 0$ (for real values of $f(x)$, the radicand must be positive or zero).

$x \leq 5$ (solving for x)

The function is not real if $x > 5$. (Any x-value greater than 5 is excluded).

(b) The domain is all real x such that $x \leq 5$.

(c) The range is from $y = 0$ to $y = \infty$ (obtained by considering $x \leq 5$ ($x = 5$ and $x < 5$) in $y = \sqrt{5-x}$., or (from graph)

Example 10 Find (a) the domain and (b) the range of $f(x) = \dfrac{\sqrt{x+6}}{x-1}$

Solution Here, two conditions must be satisfied.

Condition 1: $x + 6 \geq 0$ and from which $x \geq -6$ (For real root, the radicand must be positive or zero)

Condition 2: $x - 1 \neq 0$ and from which $x \neq 1$.

(a) The domain consists of all real numbers greater than or equal to -6, except 1, that is $x \geq -6, x \neq 1$.

(b) The range (from graph) is given by the set $\{y \mid y$ is a real number$\}$. Note that when $x = -6$, $y = 0$

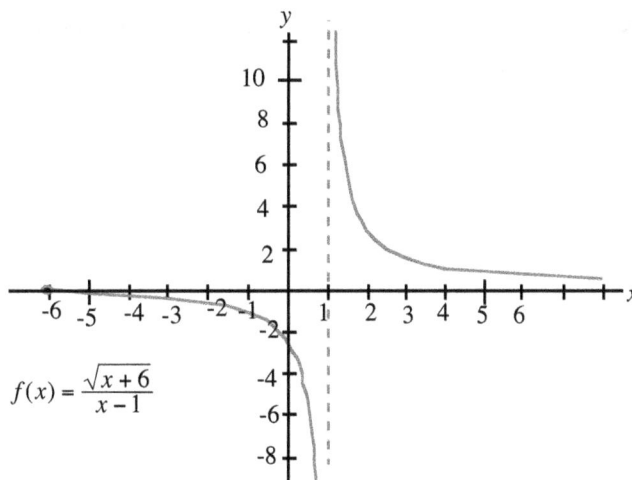

Fig.4: $f(x) = \dfrac{\sqrt{x+6}}{x-1}$

Lesson 31 Exercises

A

1. If $f(x) = x^2 - 5x + 2$, find (a) $f(-2)$; (b) $f(-1)$, (c) $f(-x)$, (d) $f(a+h)$, (e) $\dfrac{f(a+h) - f(a)}{h}$;

2. If $f(x) = 4 - \dfrac{1}{x-2}$, find (a) $f(-3)$; (b) $f(-x)$; (c) $f(a)$.

3. If $f(x) = x^3 - x^2 - x - 1$, find $f(-1)$.

Answers: **1.** (a) 16; (b) 8; (c) $x^2 + 5x + 2$; (d) $a^2 - 5a + 2ah - 5h + h^2 + 2$; (e) $2a + h - 5$

2. (a) $4\frac{1}{5}$; (b) $\dfrac{4x+9}{x+2}$; (c) $\dfrac{4a-9}{a-2}$; **(3)** -2

B

1. What is meant by the domain of a function?

2. What conditions do you look out for in determining the domain of a function?

3. What is meant by the range of a function?

Find the domain and range in each of the following:

4. $A = \{(2,4),(3,9),(4,16),(5,25)\}$; **5.** $B = (2,1),\ (5,1),\ (6,1),\ (7,3)$

6. $y = x + 2$; **7.** $y = x^2 + 3x + 7$; **8.** $y = -x - 4$; **9.** $y = x^3 + x^2 - 5$

Answers: **4.** Domain : $\{2,3,4,5\}$;, range: $\{4,9,16,25\}$

5. Domain : $\{2,5,6,7\}$;, range: $\{1,3\}$ Note: we do not repeat the "1".

6. Domain : All real values of x: range: all real values of y.

7. Domain : All real x; range: all real y such that $y \geq 4.75$.

8. Domain : All real values of x: range: all real values of y.

9. Domain : All real values of x.

C Determine the excluded values (if any) for the following:

1. $y = \dfrac{2}{x-3}$; **2.** $f(x) = \dfrac{2x-4}{x^2-1}$; **3.** $f(x) = \dfrac{3x}{x^2+1}$; **4.** $f(x) = \dfrac{1}{x}$; **5.** $f(x) = \dfrac{(x-1)}{(x-1)(x+2)}$

In Problems 6 & 7 below, for what values of x is the function not real?

6. $y = \sqrt{x-3}$; **7.** $y = \sqrt{4-x}$ **8.** Find the domain of $f(x) = x - \dfrac{1}{x}$

Extra: Why is it important to note the excluded values in dealing with functions and solving equations?

Answers: **1.** 3; **2.** $\{-1, 1\}$; **3.** None; **4.** 0; **5.** $\{-2, 1\}$; **6.** $x < 3$; **7.** $x > 4$; **8.** All real x except 0.

Lesson 32
Algebra of Functions

In the past, we covered the addition, subtraction, multiplication and division of algebraic expressions. Below, we similarly cover the same basic operations on functions.

Let f and g be functions of x such that x is in the domains of both f and g. Then

1. Sum: $(f + g)(x) = f(x) + g(x)$ (Domain: All real numbers common to the domains of f and g)

2. Difference: $(f - g)(x) = f(x) - g(x)$ Domain: All real numbers common to the domains f and g)

3. Product: $(f \bullet g)(x)$ or $(fg)(x) = f(x) \bullet g(x)$ or $f(x)g(x)$ (Domain: All real numbers common to the domains f and g

4. Quotient $\left(\dfrac{f}{g}\right)(x) = \dfrac{f(x)}{g(x)}$ (Domain: All real numbers common to the domains f and g and $g(x) \neq 0$.

Examples

If $f(x) = x^2 + 2$ and $g(x) = x - 2$; find (a) $(f + g)(x)$; (b) $(f - g)(x)$; (c) $(f \bullet g)(x)$;

(d) $\left(\dfrac{f}{g}\right)(x)$; (e) $(f + g)(3)$; (f) $\left(\dfrac{f}{g}\right)(3)$; (g) $\left(\dfrac{f}{g}\right)(2)$.

Solution

(a) $(f + g)(x) = f(x) + g(x)$
$$= (x^2 + 2) + (x - 2)$$
$$= x^2 + 2 + x - 2$$
$$= x^2 + x$$

(b) $(f - g)(x) = f(x) - g(x)$
$$= (x^2 + 2) - (x - 2)$$
$$= x^2 + 2 - x + 2$$
$$= x^2 - x + 4$$

(c) $(f \bullet g)(x) = f(x) \bullet g(x)$
$$= (x^2 + 2)(x - 2)$$
$$= x^3 - 2x^2 + 2x - 4$$

(d) $\left(\dfrac{f}{g}\right)(x) = \dfrac{f(x)}{g(x)}$
$$= \dfrac{x^2 + 2}{x - 2} \text{ or } x + 2 + \dfrac{6}{x - 2} \text{ (long division)}$$

(e) From (a), $(f + g)(x) = f(x) + g(x)$
$$= x^2 + x$$
$$(f + g)(3) = (3)^2 + 3 = 12$$
Method 2: $(f + g)(3) = f(3) + g(3)$
$$f(3) = (3)^2 + 2 = 11$$
$$g(3) = 3 - 2 = 1$$
$$(f + g)(3) = 11 + 1 = 12$$

(f) From (d), $\left(\dfrac{f}{g}\right)(x) = \dfrac{f(x)}{g(x)}$
$$= \dfrac{x^2 + 2}{x - 2}$$
$$\left(\dfrac{f}{g}\right)(3) = \dfrac{(3)^2 + 2}{(3) - 2}$$
$$= \dfrac{9 + 2}{3 - 2}$$
$$= 11.$$
Method 2: $f(3) = (3)^2 + 2 = 11$
$$g(3) = 3 - 2 = 1$$
$$\left(\dfrac{f}{g}\right)(3) = \dfrac{f(3)}{g(3)} = \dfrac{11}{1} = 11$$

(g) From (d), $\left(\dfrac{f}{g}\right)(x) = \dfrac{x^2 + 2}{x - 2}$
$$\left(\dfrac{f}{g}\right)(2) = \dfrac{(2)^2 + 2}{(2) - 2}$$
$$= \dfrac{6}{0} \text{ is undefined.}$$
Therefore, $\left(\dfrac{f}{g}\right)(2)$ is undefined. Note: $g(2) = 0$.

On the next page, we cover the domains for algebraic of functions.

Domains for Algebra of Functions

In the first step, we determine the common domain (intersection of domains) of the functions. In the second step, we check for the domain of the resulting function. (sum. difference, product or quotient). We add any excluded values to those from Step 1 and if there is an overlap in the domains, we determine the intersection of the domains from Step 1 and Step 2. Pay attention to radical functions such as $f(x) = \sqrt{x-2}$, determine the domain and find the intersection for the overall domain.

Example

Given $f(x) = \frac{3}{x}$ and $g(x) = \frac{3x-4}{x+2}$, find the domains of the following:

(a) $(f + g)(x)$; (b) $(f - g)(x)$; (c) $(f \bullet g)(x)$; (d) $\left(\frac{f}{g}\right)(x)$;

Solution:

Domain of $f : \{x \mid x \text{ is a real number and } x \neq 0\}$

Domain of $g : \{x \mid x \text{ is a real number and } x \neq -2\}$.

For (a), (b), and (c):

domain of $f + g =$ domain of $f - g =$ domain of $f \bullet g = \{x \mid x \text{ is a real number and } x \neq 0,\ x \neq -2\}$

(d) $\left(\frac{f}{g}\right)(x) = \frac{f(x)}{g(x)}$

$= \frac{3}{x} \div \frac{3x-4}{x+2}$ $(x \neq 0,\ x \neq -2)$

$= \frac{3}{x} \bullet \frac{x+2}{3x-4}$

$(x \neq \frac{4}{3}.$ See $= \frac{3(x+2)}{x(3x-4)}$ below)

Note: $\frac{f(x)}{g(x)} = \dfrac{\frac{3}{x}}{\frac{3x-4}{x+2}}$

Setting $x(3x - 4) = 0$

$x = 0$ or $3x - 4 = 0$ and from which $x = \frac{4}{3}$

The excluded values are $-2, 0$ and $\frac{4}{3}$.

Domain of $\left(\frac{f}{g}\right)(x)$: $\{x \mid x \text{ is a real number and } x \neq 0,\ x \neq -2 \text{ and } x \neq \frac{4}{3}\}$

Extra

Example: The intersection of the domains of $f(x) = \sqrt{x-2}$ and $g(x) = \frac{x-4}{x+3}$ is

$\{x \mid x \text{ is a real number and } x \geq 2\}$ or $[2, \infty)$ (from $\{x \mid x \geq 2\} \cap \{x \mid x \neq -3\}$)

It seems the radical function is "King".

Lesson 32 Extra Exercises

If $f(x) = x^2 + 2$ *and* $g(x) = x - 2$; find (a) $(f + g)(x)$; (b) $(f - g)(x)$; (c) $(f \bullet g)(x)$; (d) $\left(\frac{f}{g}\right)(x)$; (e) $(f + g)(3)$; (f) $\left(\frac{f}{g}\right)(3)$; (g) $\left(\frac{f}{g}\right)(2)$; (h) $(f - g)(4)$.

Ans: (a) $x^2 + x$; (b) $x^2 - x + 4$; (c) $= x^3 - 2x^2 + 2x - 4$; (d) $\frac{x^2+2}{x-2}$ or $x + 2 + \frac{6}{x-2}$;
(e) 12; (f) 11 (g) undefined; (h) 16.

CHAPTER 10

Sketching Parabolas

Lesson 33: **By General method**

Lesson 34: **By location of vertex, axis of symmetry and intercepts.**

The quadratic functions $y = f(x) = ax^2 + bx + c$ $(a \neq 0)$ and their graphs

We will cover two methods: a general method and the method of the location of the vertex, axis of symmetry and the x- and y-intercepts.

Lesson 33

General method of sketching the graph of the quadratic function

Example 1 Sketch the graph of $y = x^2 - 5$ for $x = 0, \pm 1, \pm 2, \pm 3$.

ordered pairs

Step 1: When $x = 0,$ $y = (0)^2 - 5$
$y = -5$ $\}\longrightarrow$ $(0,-5)$

When $x = 1,$ $y = (1)^2 - 5$
$y = 1 - 5$
$y = -4$ $\}\rightarrow$ $(1,-4)$

When $x = 2,$ $y = (2)^2 - 5$
$y = 4 - 5$
$y = -1$ $\}\longrightarrow$ $(2,-1)$

When $x = 3,$ $y = (3)^2 - 5$
$y = 9 - 5$
$y = 4$ $\}\longrightarrow$ $(3, 4)$

When $x = -1,$ $y = (-1)^2 - 5$
$y = 1 - 5$
$y = -4$ $\}\longrightarrow$ $(-1,-4)$

When $x = -2,$ $y = (-2)^2 - 5$
$y = 4 - 5$
$y = -1$ $\}\longrightarrow$ $(-2,-1)$

When $x = -3,$ $y = (-3)^2 - 5$
$y = 9 - 5$
$y = 4$ $\}\longrightarrow$ $(-3, 4)$

Step 2: Plot the points $(0,-5), (1,-4), (2,-1), (3,4), (-1,-4), (-2,-1)$ and $(-3,4)$ on a rectangular coordinate system of axes (on graph paper) and connect the points by a smooth curve noting that the parabola is U-shaped and that since the coefficient of the x^2-term is positive, the parabola opens upwards (**Figure 1**).

Note above that the y-values are the same for $x = 1$ and $x = -1$; the same for $x = 2$ and $x = -2$ and the same for $x = 3$ and $x = -3$.

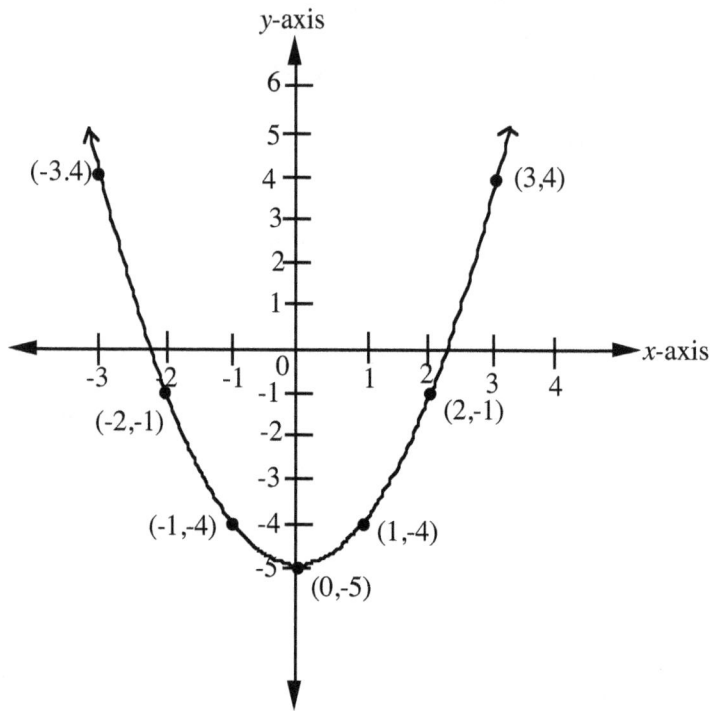

Figure 1: Graph of $y = x^2 - 5$

Example 2 Sketch the graph of $y = -x^2 + 5$ for $x = 0, \pm 1, \pm 2, \pm 3$.

Step 1: See Example 1 above and perform similar calculations to obtain
    $(0,5), (1,4), (2,1), (3,-4), (-1,4), (-2,1), (-3,-4)$.

Step 2: Plot the points obtained from Step 1 on a rectangular coordinate system of axes (on graph paper) and connect the points by a smooth curve noting that the parabola is U-shaped and that since the coefficient of the x^2-term is negative (the coefficient of $-x^2$ is -1), the parabola opens downwards (inverted U-shape). See Figure 2 below.

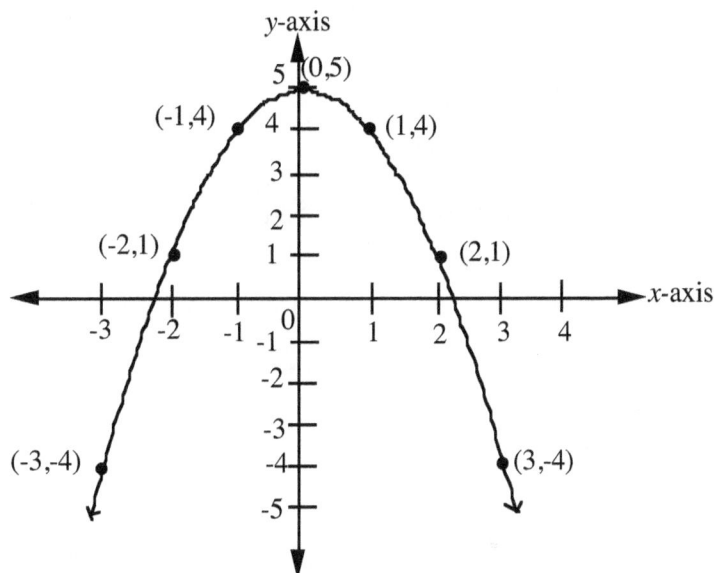

Figure 2: Graph of $y = -x^2 + 5$

Lesson 33 Exercises

Sketch the graphs of the following:
1. $y = x^2 - 4$ for $x = 0, \pm 1, \pm 2, \pm 3$;   **2.** $y = -x^2 + 1$ for $x = 0, \pm 1, \pm 2, \pm 3$.

1. Parabola opens upwards;   **2.** parabola opens downwards

Lesson 34

By location of vertex, axis of symmetry and intercepts

An alternative method of sketching the graphs of quadratic functions is by the location of the vertex, axis of symmetry and the x- and y-intercepts.

Example 1 Given the equation of the parabola $y = f(x) = x^2 - 6x + 8$, perform the following:
1. Find the x- and y-intercepts.
2. Find the axis of symmetry.
3. Find the coordinates of the vertex (or turning point)
4. Find the domain and range.
5. Sketch the graph of $y = f(x) = x^2 - 6x + 8$ (The graph of a quadratic function is called a parabola.)

Solution

1. To find the x-intercept, let $y = 0$ in $y = x^2 - 6x + 8$.

  Then, $0 = x^2 - 6x + 8$ or
  $x^2 - 6x + 8 = 0$ <-------This is a quadratic equation. Solve by any method
  $(x - 2)(x - 4) = 0$ <------We solve by factoring since the factors are easily recognizable.
  Solving, $x = 2$, or $x = 4$
  The x-intercepts are 2 and 4 . (The x-intercepts are at the points (2,0), and (4,0))

  To find the y-intercept, let $x = 0$ in $y = x^2 - 6x + 8$
  Then, $y = (0)^2 - 6(0) + 8$
  $y = 0 - 0 + 8$
  $y = 8$

  The y-intercept $= 8$.   (The y-intercept is at the point (0,8).)

2. $a = 1$ (the coefficient of the x^2-term), $b = -6$ (the coefficient of the x-term).

  Axis of symmetry is the line $x = -\dfrac{b}{2a}$ <------formula.

  $$= -\frac{-6}{2(1)}$$
  $$= 3$$

The axis of symmetry is the line $x = 3$. (see also p. 163)
The x-coordinate of the vertex $= 3$,

The y-coordinate of the vertex $= y = f(-\dfrac{b}{2a}) = f(3) = 3^2 - 6(3) + 8$

$$= 9 - 18 + 8$$
$$y = -1.$$

3. The vertex is at $(3,-1)$. This point is the turning point on this curve. It is also a minimum point, since a is positive. This minimum point is also the lowest point on this curve.

4. The domain is all real x. (The given function is a polynomial.)
  The range is specified by using the y-coordinate of the vertex obtained in Step 2.
  The range is given by the inequality $y \geq -1$.
  The range consists of all real numbers greater than or equal to -1.

(Note: If we did not know the x-coordinate of the vertex, we could obtain the range by

Substituting $a = 1$. $b = -6$, $c = 8$, in $y \geq \dfrac{4ac - b^2}{4a}$ (Note the " $\geq$") to obtain

$y \geq \dfrac{4(1)(8) - (-6)^2}{4(1)}$ $\geq \dfrac{32 - 36}{4}$ $\geq \dfrac{-4}{4} \geq -1$; That is $y \geq -1$)

Lesson 34: By location of vertex, axis of symmetry and intercepts.

5. To sketch the graph: On graph paper and using a rectangular coordinate system of axes,   163

Step 1: Locate the x- and y-intercepts from (**1**) above, by plotting the points (2,0), (4,0) and (0,8).
Step 2: Using a broken line draw the line $x = 3$, the axis of symmetry (from (**2**) above)
Step 3: Locate the vertex by plotting the point (3,-1). (from (**3**) above).
Step 4: Locate the point B using symmetry as follows: From the point A (the y-intercept), count
        horizontally say h units ($h = 3$ in this problem) to the axis of symmetry (the line $x = 3$). Next, c
        ontinue and count horizontally h units (h = 3) from the axis of symmetry and stop at the point B
      (this is another point on the curve). You may also locate the point B by the intersection of the lines
        $y = 8$ and $x = 6$.
Step 5: Connect the above points by a smooth U-shaped curve (**Figure 1**), noting that the
        parabola opens upwards (is concave up), since a is positive ($a = 1$), and also noting
        that the graph is symmetric about the axis of symmetry (the line $x = 3$, in this example).

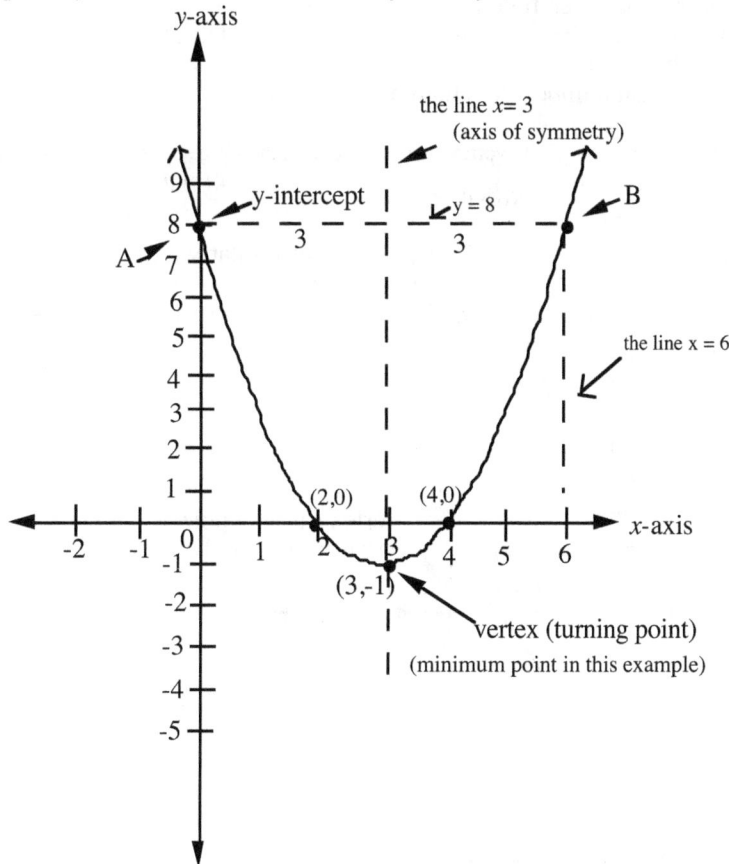

Figure 1: Graph of $y = x^2 - 6x + 8$

Note above that in sketching the graph, you may choose convenient x-values on either side of the axis of symmetry, calculate corresponding y-values to obtain additional points, especially, when there are no (real) x-intercepts.(This occurs when in solving the quadratic equation, we obtain solutions involving imaginary numbers.)
Again: When in doubt as to the location of the curve, pick x-values, calculate y, plot the points and connect them.

Example 2 Given the equation of the parabola $y = f(x) = -x^2 + 6x - 8$, perform the following:

1. Find the x- and y-intercepts.   **2**. Find the axis of symmetry.
3. Find the coordinates of the vertex (or turning point).   **4**. Find the domain and range.
5. Sketch the graph of $y = f(x) = -x^2 + 6x - 8$

Solution

1. Using the procedure in Example 1,

The x-intercepts are 2 and 4.   (The x-intercepts are at the points (2,0), and (4,0))

The y-intercept = -8.   (The y-intercept is at (0,-8))

2. a = -1 (the coefficient of the x^2-term), b = +6 (the coefficient of the x-term)

Axis of symmetry is the line $x = -\dfrac{b}{2a} = -\dfrac{+6}{2(-1)} = 3$.

The axis of symmetry is the line $x = 3$. The x-coordinate of the vertex = 3,

and the y-coordinate of the vertex = $y = f(-\dfrac{b}{2a}) = f(3) = -3^2 + 6(3) - 8 = 1$

3. The vertex is at the point (3, 1). Since a is negative (a = -1), the vertex is a maximum point.

4. The domain is all real x. (The given function is a polynomial)

The range is specified by using the y-coordinate of the vertex obtained in Step 2.

The range is given by the inequality $y \le 1$.   (Note the " $\le$ ")

The range consists of all real numbers less than or equal to 1.

(Note: If we did not know the x-coordinate of the vertex, we could obtain the range by substituting. Substituting

a = -1. b = +6, c = -8 in $y \le \dfrac{4ac - b^2}{4a}$ (Note the " $\le$ ") to obtain $y \le \dfrac{4(-1)(-8) - (+6)^2}{4(-1)}$; $y \le 1$.)

5. To sketch the graph, plot the points from (**1**), (**2**), (**3**) above; similarly locate the point B (as was
done in Example 1) and connect the points by a smooth U-shaped curve, noting that the parabola
opens downwards (**Figure 2**), since a is negative (a = -1). Also note the symmetry as in Example 1.

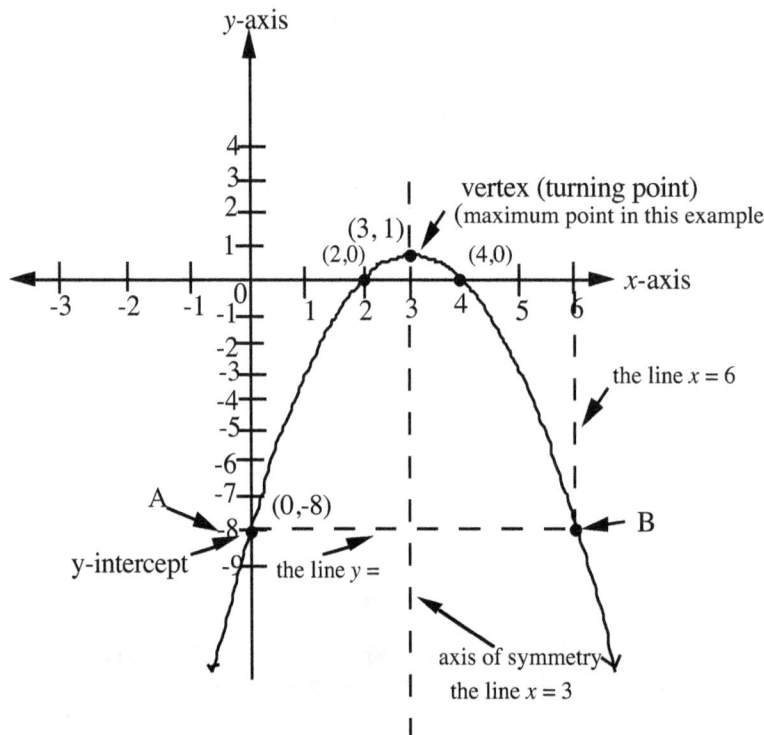

Figure 2: Graph of $y = -x^2 + 6x - 8$

Lesson 34: By location of vertex, axis of symmetry and intercepts.

165

EXTRA: Axis of Symmetry , Vertex and Minimum Value

Example 1

Given $f(x) = x^2 - 12x + 8$ (A)

Let $y = x^2 - 12x + 8$ (B)

$= x^2 - 12x + (-6)^2 - (-6)^2 + 8$

$= x^2 - 12x + (-6)^2 - 36 + 8$

$y = (x - 6)^2 - 28$ <---- knowing this

$y + 28 = (x - 6)^2$ <----- or this

Vertex is at $(6, -28)$

Axis of symmetry is the line $x = 6$

Minimum value is -28.

Alternatively

From (B), $a = 1$, $b = -12$

Axis of symmetry is given by $x = -\frac{b}{2a}$

$\qquad = -\frac{-12}{2(1)}$

$\qquad = \frac{12}{2}$

$\qquad x = 6$

$f(6) = 6^2 - 12(6) + 8$ (subst.. in (A))

$\qquad = 36 - 72 + 8$

$\qquad = -36 + 8$

$\qquad y = -28$

when $x = 6$. $y = -28$

Vertex is at $(6, -28)$

Axis of symmetry is the line $x = 6$

Minimum value is -28.

Example 2

Given $f(x) = x^2 - 6x + 8$ (A)

Let $y = x^2 - 6x + 8$ (B)

$= x^2 - 6x + (-3)^2 - (-3)^2 + 8$

$= x^2 - 6x + (-3)^2 - 9 + 8$

$y = (x - 3)^2 - 1$ <---- knowing this

$y + 1 = (x - 3)^2$ <----- or this

Vertex is at $(3, -1)$

Axis of symmetry is the line $x = 3$

Minimum value is -1.

Alternatively

From (B), $a = 1$, $b = -6$

Axis of symmetry is given by $x = -\frac{b}{2a}$

$\qquad = -\frac{-6}{2(1)}$

$\qquad = \frac{6}{2}$

$\qquad x = 3$

$f(3) = 3^2 - 6(3) + 8$ (subst.. in (A))

$\qquad = 9 - 18 + 8$

$\qquad = -9 + 8$

$\qquad y = -1$

when $x = 3$. $y = -1$

Vertex is at $(3, -1)$

Axis of symmetry is the line $x = 3$

Minimum value is -1.

Lesson 34 Exercises

Sketch the graphs of the following by the location of the vertex, axis of symmetry and the intercepts.:

1. $y = x^2 - 4$;   **2.** $y = -x^2 + 1$.   **3.** $y = x^2 - 6x + 8$

Answers: See above

CHAPTER 11

Graphing Linear Inequalities in two Variables

Lesson 35: **Graphing and determining solution plane by point-testing**
Lesson 36: **Graphing and determining solution plane by the "Seesaw" view Method**
Lesson 37: **Graphing Special Cases of Linear Inequalities**

Geometrically, the solution set of a linear inequality in two variables is a section of the x-y plane. The plane may be open or closed. If the plane is closed, the symbol used would be " ≥ " or " ≤ "; but if plane is open, the symbol would be " > " or "<". The solution set for the closed half-plane includes the boundary line, but the solution set for the open half-plane does not include the boundary line. We will learn how to graph the solution sets of some inequalities.

Lesson 35

Graphing and determining solution plane by point-testing

General steps in graphing linear inequalities

Step 1: Replace the inequality symbol by the equality symbol " = ".

Step 2: Using convenient points, graph the equation (now a linear equation) from Step 1. To do this, choose convenient x-values and calculate corresponding y-values. Connect the points by a straight solid line if the inequality is the symbol " ≥ " or " ≤" ; but use broken (dashed or dotted) lines if the given inequality is the symbol " > " or " < ".

Note: You may also use any other specialized methods (for graphing straight lines) such as the "slope-intercept" method or the x- and y-intercepts method. Note: The **slope-intercept method** (p. 57) is faster..

Step 3: We will determine which side of the straight line forms the solution. Choose any convenient point not on the boundary (of the straight line drawn) and substitute its x- and y- coordinates in the original inequality. If these values satisfy the inequality, then the chosen point is in the solution set of the inequality, and consequently, that side of the line (half-plane) where this point is located is the solution set. However, if the tested point does not satisfy the original inequality, then it is the other side of the line (the half-plane) which is the solution set We must note that the straight line drawn in Step 2 always divides the x-y plane (area or surface) into two halves.

Step 4: Shade the half-plane which is the solution set.

Example 1 Graph the solution set for the inequality $3x + 2y \geq 4$.
Solution

Step 1: Replace the inequality symbol by the equality symbol and solve for y.
$$3x + 2y = 4$$
$$y = -\frac{3}{2}x + 2$$

Step 2: Choose convenient x-values and calculate corresponding y-values
(To avoid fractional values of y, solve for y first, and choose x-values divisible by the denominator of the x-term)

For $x = 0, y = 2 \Leftrightarrow (0, 2)$

For $x = 2, y = -1 \Leftrightarrow (2, -1)$

Step 3: Plot the ordered pairs (0, 2) and (2, -1). See Figure .
(Note: You may use the **slope intercept method** (p. 57).

Step 4: Connect the two points by a **solid straight line** since the inequality symbol is " ≥ "
    (That is, the half-plane is closed)

Step 5: Now, determine which half of the half-plane to shade (or is the solution set). Let us choose the origin $(0, 0)$ to test the inequality. Substituting $x = 0, y = 0$ in the original inequality.

$$3(0) + 2(0) \overset{?}{\geq} 4$$

$$0 \overset{?}{\geq} 4 \qquad \textbf{No} \quad \text{(False inequality)}$$

Thus, the origin and the half-plane in which the origin is located **is not** in the solution set. Therefore, the solution set is **the other** half-plane. The solution contains, for example, the point $(3, 4)$.

Step 5: Shade this half-plane (the solution half-plane) as in Figure 43.

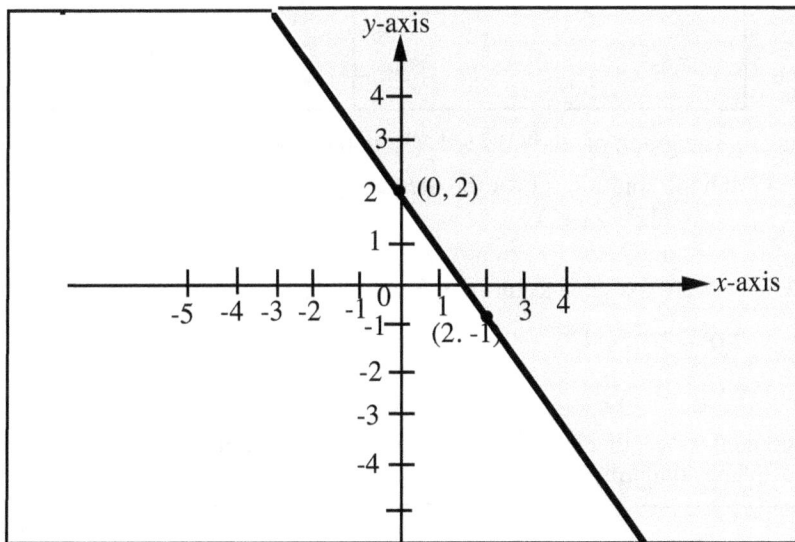

Figure Graph of $3x + 2y \geq 4$ (Note the solid boundary line)

The use of the origin (0 ,0) in determining the solution half-plane

Using the origin, $x = 0, y = 0$ makes the checking relatively easy and fast because the product of zero and a number is zero; and also when zero is added to a number, the number remains unchanged. However, if the boundary line passes through the origin $(0, 0)$, then choose some other convenient point for testing. (Do not use $(0, 0)$ for testing if there are inequalities such as $y \geq ax, \ y \leq ax, \ x \geq 0$ and $y \geq 0$.)

Example 2 Graph the solution set for the inequality: $3x + 2y > 4$

Solution

Steps 1, 2, 4, and 5 are the same as those in the last example (Example 1) except that in Step 3, we use **broken** (dashed) **lines in** connecting the two points , since the inequality symbol is " > ". and consequently, the solution half is open. See Figure 44.

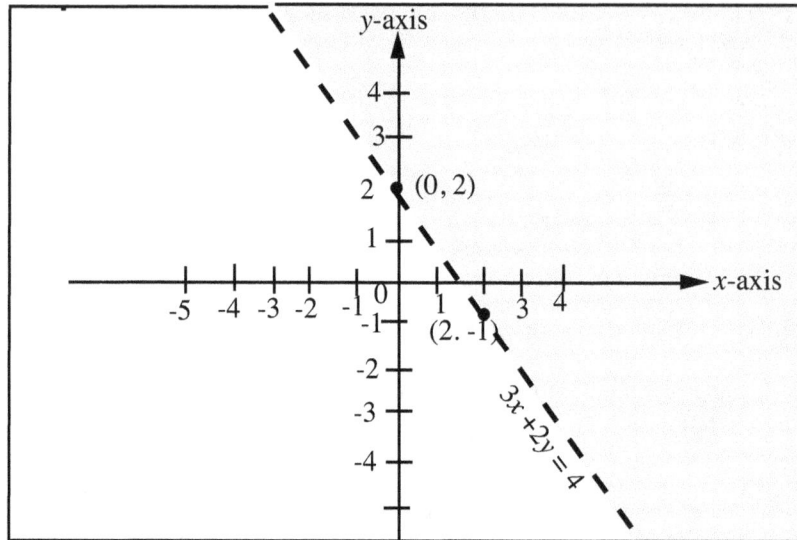

Figure   The graph of $3x + 2y > 4$ (Note the broken boundary line)

Example 3 Graph the solution set for the inequality
$$3x + 2y \le 4$$

Solution

Steps 1, 2, 3, and 4, are the same as those in Example 1, except that in Step 5. we have

$$3(0) + 2(0) \overset{?}{\le} 4$$

$$0 \overset{?}{\le} 4 \qquad \textbf{Yes} \quad \text{(True inequality)}$$

The true statement shows that the origin $(0, 0)$ **is in** the solution set. We therefore shade the solution half-plane containing $(0, 0)$.. See Figure

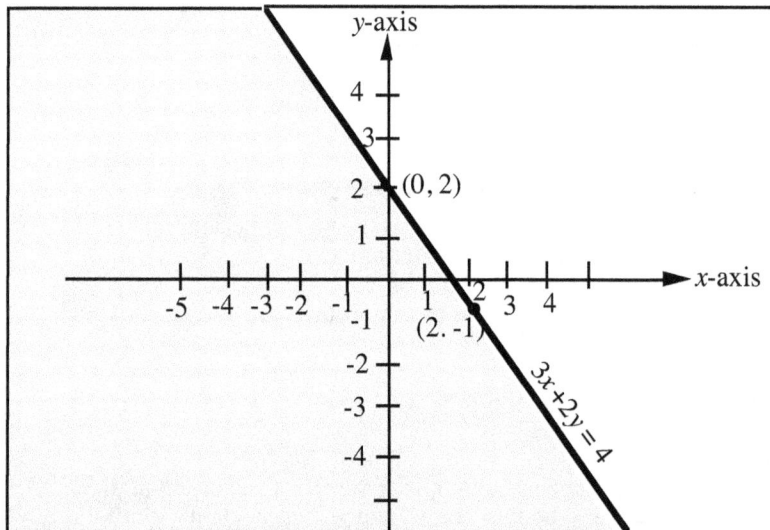

Figure: The graph of $3x + 2y \le 4$

Example 4   From Example 3, the solution set for the inequality $3x + 2y < 4$ is graphed in Figure 46. Note the broken line (compared to the solid line in Example 3).

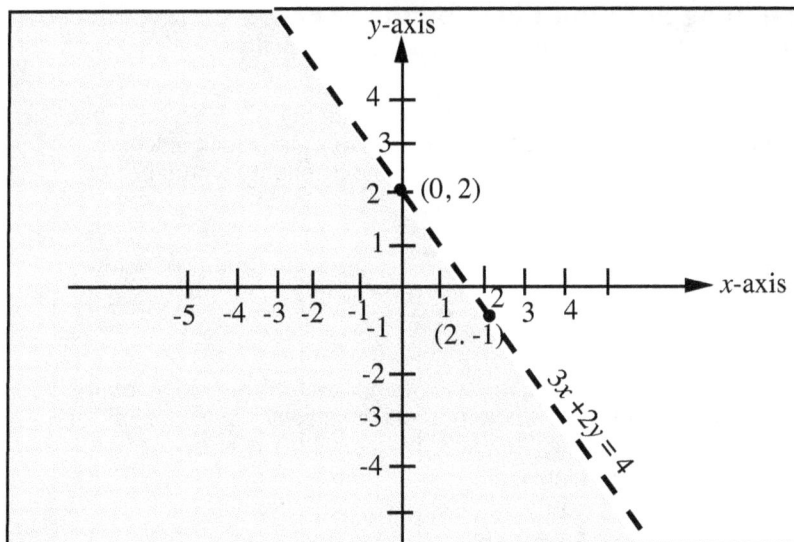

Figure: The graph of $3x + 2y < 4$

Lesson 35 Exercises

1. Graph the solution set for the inequality $3x + 2y \geq 4$

2. Graph the solution set for the inequality: $3x + 2y > 4$

3 Graph the solution set for the inequality $3x + 2y < 4$ is

4. Graph the solution set for the inequality $3x + 2y \leq 4$

Solution: See above

Lesson 36

Graphing: Determining Solution Plane by the "Seesaw" View Method
(Another method of determining the solution half-plane)

This method depends on two main principles. First, we solve the inequality for y. Secondly we should be able to point out which side of the boundary line is " above", and which side is "below" the boundary line. (When the inequality is solved for y, the inequality would be of the form $y \geq ax + b$, $y \leq ax + b$, $y < ax + b$, or $y > ax + b$)

The symbols " $>$ and $\geq$" will imply " above "; and the symbols " $\leq$ and $<$ " will imply "below ". To determine the " above" and the " below " of the boundary line (the straight line drawn) we use the following mnemonic device:

Assume the line, ℓ, and the y-axis form a seesaw system, with the y-axis forming the pivot at where the axis meets the line. We will call the side where one sits or stands " the side above the line and we will call the other side " the side below the line ". In Figures (a), (b), (c), (d), we have shaded the half-planes above the lines for clarity.

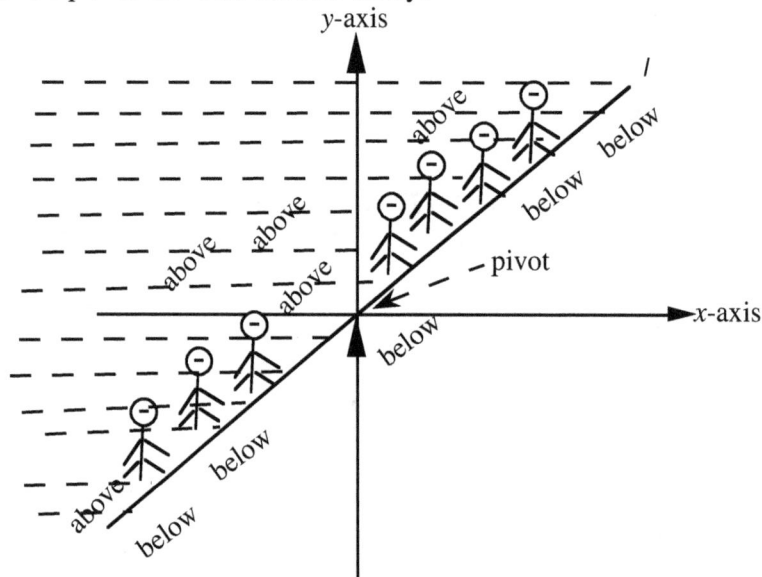

Figure: Seesaw-View Method

Note: The seesaw-view method is an original method devised by the author.

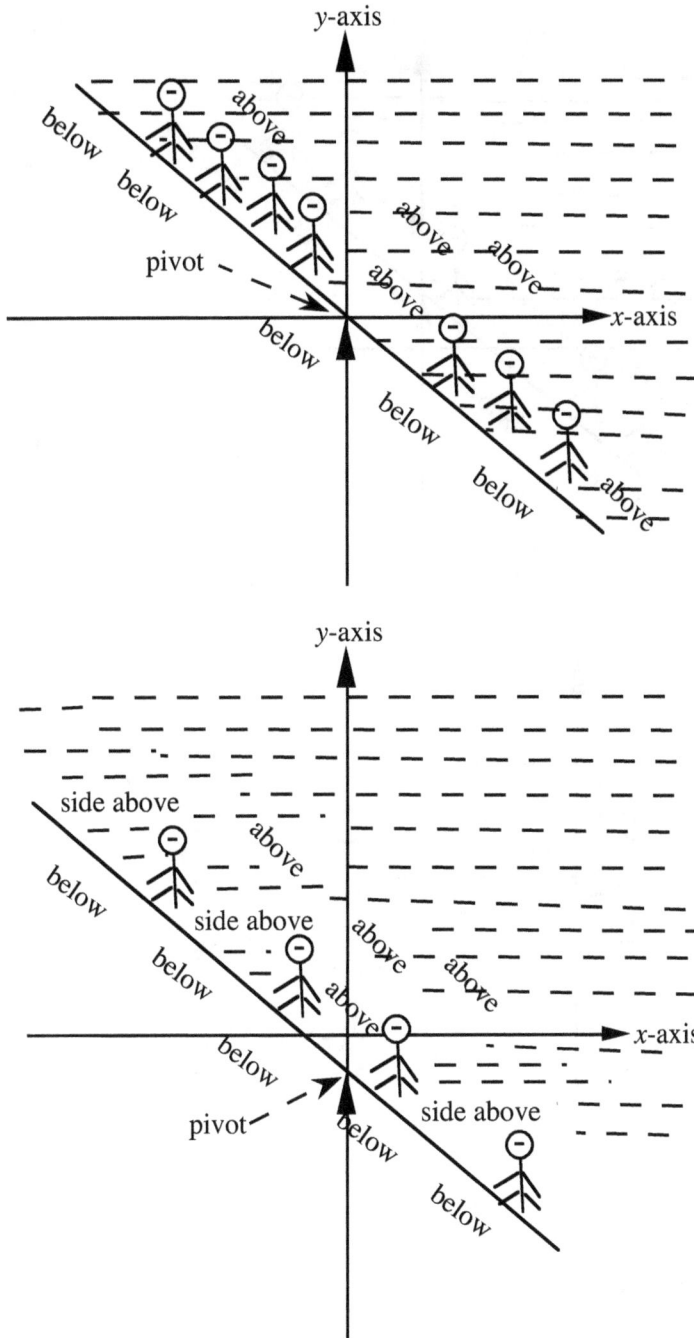

Figure: Seesaw-View Method

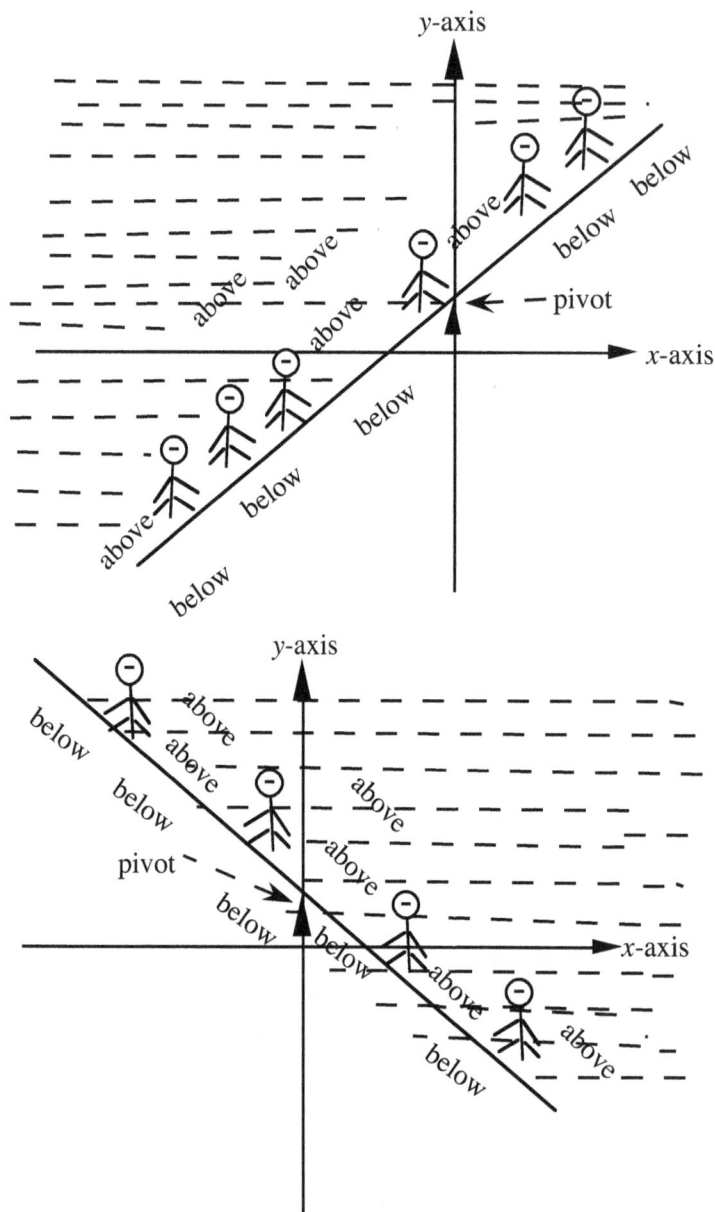

Figure: Seesaw-View Method

We will redo the previous examples using the " seesaw view" method, noting that::
the symbols " **>** and **≥** " (greater than symbols) will imply " **above the line**" and the symbols "**<** and **≤** (less than symbols) will imply "**below the line** ". The inequality must be placed in the form $y \geq$ (or $>, \leq, <$) $ax + b$. (That is, solve the inequality for y first before using this method.)

Example 5 (Example 1 redone)

 Graph the solution set for the inequality

$$3x + 2y \geq 4$$

Solution (The solution is a closed half-plane including the boundary because of the " $\geq$ ".)

Step 1: Solve the inequality for y.

$$3x + 2y \geq 4$$

$$y \geq -\tfrac{3}{2}x + 2$$

Step 2: Replace the inequality symbol by the equality symbol and find two convenient ordered pairs to graph the straight line.

$$y = -\tfrac{3}{2}x + 2$$

For $x = 0, y = 2 \Leftrightarrow (0, \ 2)$

For $x = 2, y = -1 \Leftrightarrow (2, -1)$

Step 2: Plot the ordered pairs (0, 2) and (2, -1). See Fig. .
      (Note: You may use the **slope intercept method** (p. 57).

Step 3: Connect the two points by a **solid straight line** since the inequality symbol is " $\geq$ "
      (That is, the half-plane is closed)

Step 4: Since the sense of the inequality with y solved for is " $>$ " , the **greater than** symbol, we shade half-plane **above** the line . See Figure.
      The solution set is a closed half-plane including .

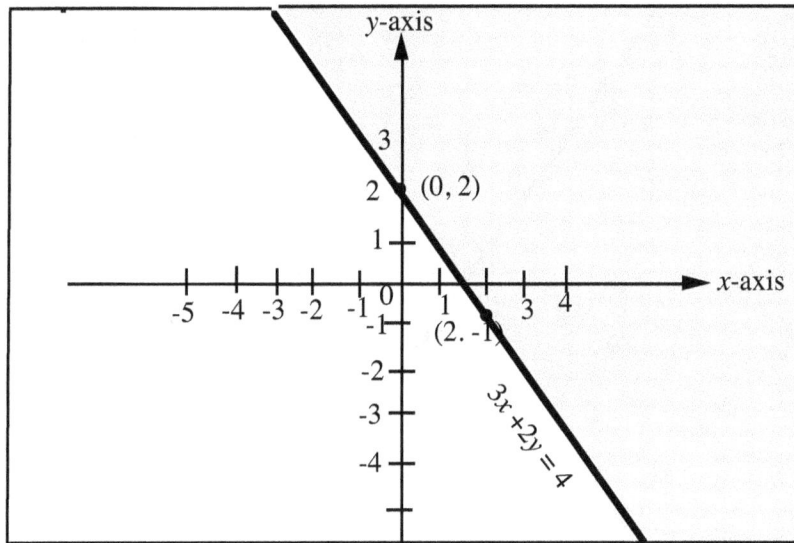

Figure: Graph of $3x + 2y \geq 4$

Lesson 36: Graphing and determining solution plane by the "Seesaw" view Method

Example 6 (Example 3 redone)

Graph the solution set for the inequality

$$3x + 2y \le 4$$

Step 1: Solve the inequality e for y.

$$3x + 2y \le 4$$

$$y \le -\tfrac{3}{2}x + 2$$

Step 2: Replace the inequality symbol by the equality symbol and find two convenient ordered pairs to graph the straight line.

$$y = -\tfrac{3}{2}x + 2$$

For $x = 0, y = 2 \Leftrightarrow (0,\ 2)$

For $x = 2, y = -1 \Leftrightarrow (2, -1)$

Step 2: Plot the ordered pairs (0, 2) and (2, -1). See Figure..

Step 3: Connect the two points by a **solid straight line** since inequality symbol is " $\ge$ "
(That is, the half-plane is closed)

Step 4:   Since the sense of the inequality with y solved for is" $<$ " , the **less than** symbol , we
shade half-plane **below** the line $y = -\tfrac{3}{2}x + 2$. See Figure
The solution set is a closed half-plane including the boundary.

We will now use the " seesaw view method " to graph more inequalities without explanation

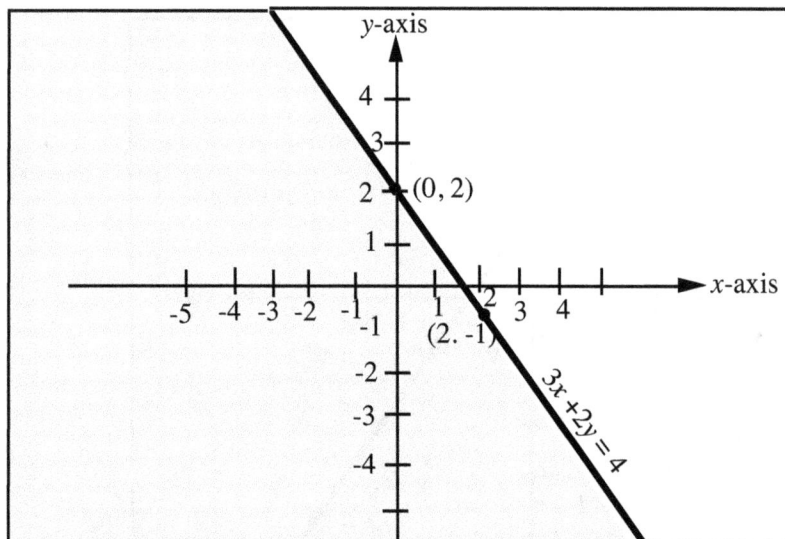

Figure Graph of $3x + 2y \le 4$

Lesson 36: Graphing and determining solution plane by the "Seesaw" view Method

Example 7 Graph the solution set for the inequality $y + 2x \geq 6$.

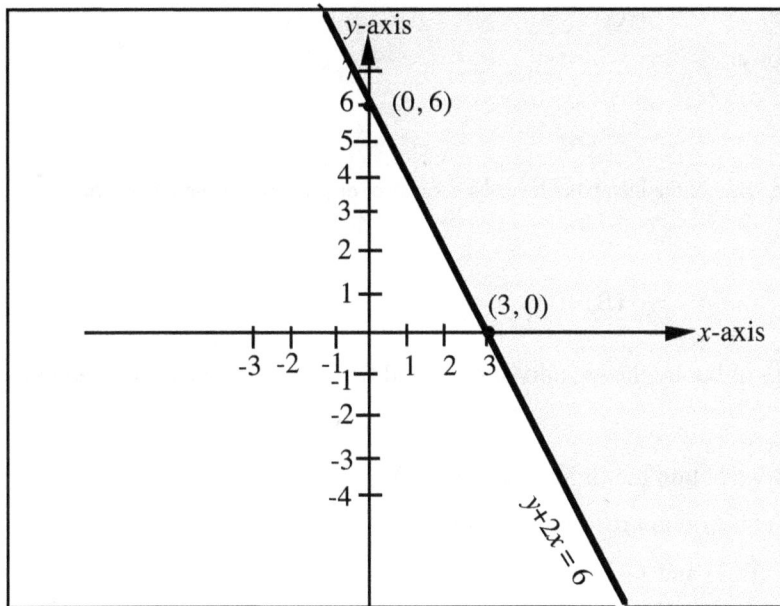

Figure : The graph of $y + 2x \geq 6$.

Example 8: Graph the solution set for the inequality $y + 2x \leq 6$

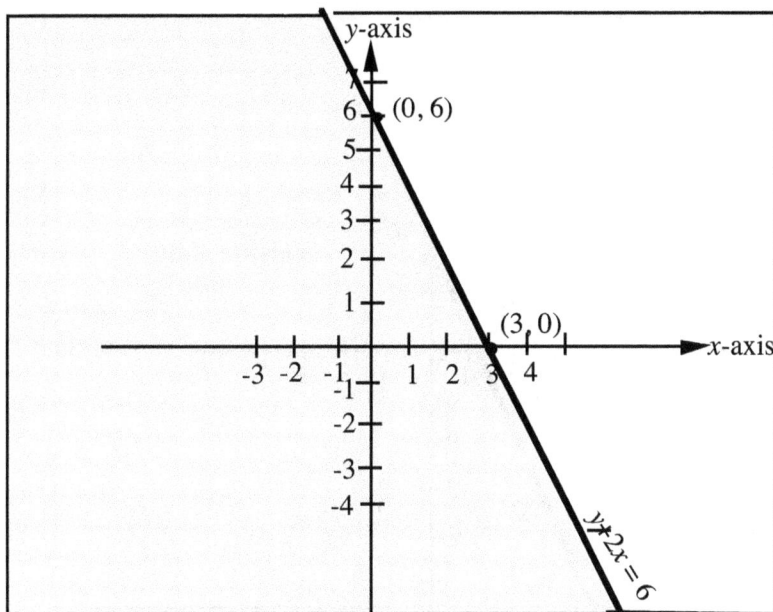

Figure The solution set for the inequality $y + 2x \leq 6$

Example 9 Graph the solution set for the inequality

$$4x - 2y \leq -5 \qquad (A)$$

Step 1: Solve the inequality for y.

$4x - 2y \leq -5$

$\quad -2y \leq -5 - 4x$

$\qquad y \geq \frac{-5}{-2} - \frac{4x}{-2}$ (The sense of the inequality reverses because of division by a negative number)

$\qquad y \geq \frac{5}{2} + 2x$

$\qquad \boxed{y \geq 2x + \frac{5}{2}} \qquad (B)$

Step 2: Replace the inequality symbol by the equality symbol and find two convenient ordered pairs to graph the line.

$\quad y = 2x + \frac{5}{2}$

For $x = 0$, $y = \frac{5}{2}$, and we obtain the ordered pair $(0, \frac{5}{2})$

For $y = 0$, $x = -\frac{5}{4}$, and we obtain the ordered pair $(-\frac{5}{4}, 0)$

Step 2: Plot the ordered pairs $(0, \frac{5}{2})$ and $(-\frac{5}{4}, 0)$ See Fig..

Step 3: Connect the two points by a **solid straight line** (boundary included).

Step 4: Since the sense of the inequality with y solved for is the " **greater than** "symbol inequality B) we shade half-plane **above** the line $y = 2x + \frac{5}{2}$ (See Figure) . For a quick check, test with $(0, 0)$

Note: In Step 4, do not use implicitly inequality (A) but use inequality (B).

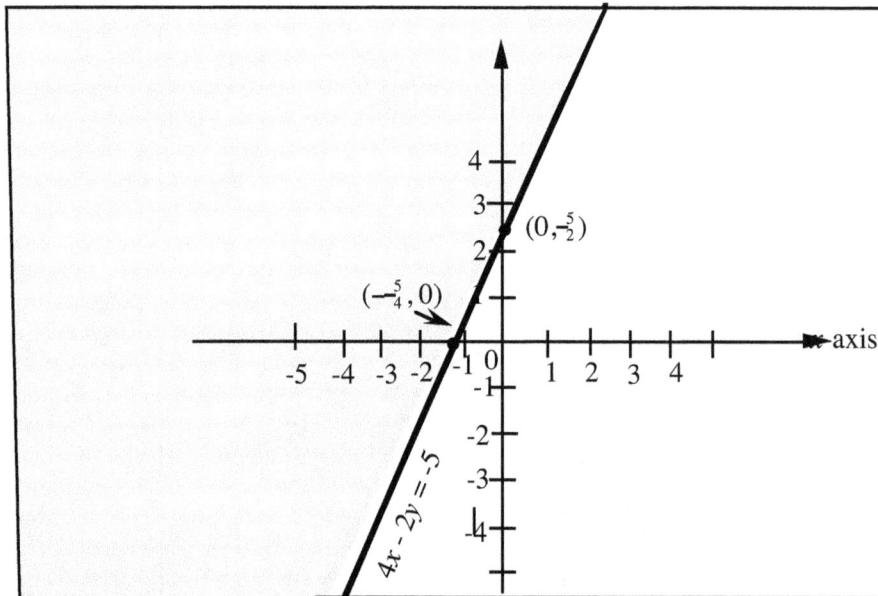

Figure : Graph for $4x - 2y \leq -5$

Lesson 36 Exercises

Use the "see-saw view" method to graph the following:

1. Graph the solution set for the inequality $3x + 2y \geq 4$

2. Graph the solution set for the inequality: $3x + 2y > 4$

3 Graph the solution set for the inequality $3x + 2y < 4$ is

4. Graph the solution set for the inequality $3x + 2y \leq 4$

5. Graph the solution set for the inequality $y + 2x \geq 6$.

6. Graph the solution set for the inequality $4x - 2y \leq -5$

Solution: See above

Lesson 37

Graphing Special Cases of Linear Inequalities and Distinction between a Point, a Set of Points, a Line, and a Half-plane

We must distinguish between, for example, the set of points $x \geq 4$ on a number line and the half-plane $x \geq 4$ just as we distinguish between the point $x = 4$ on a number line and the line $x = 4$ in an x-y plane.

The graph of the point $x = 4$:

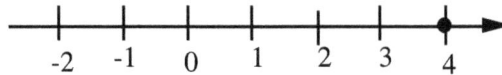

The graph of the set of points $x \geq 4$:

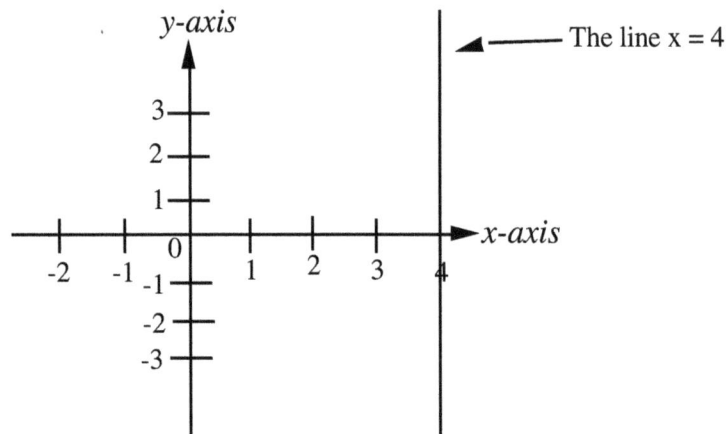

The line x = 4

Figure : The graph of the line $x = 4$

Example 8 Graph the linear inequality $x \geq 4$.

Step 1: Replace " > " by " = ".
   Then $x = 4$

Step 2: Draw the vertical line $x = 4$ using a solid line (Fig. 50)
   To determine the half-plane which is the solution set, either choose a point and test or use the sense of the inequality.

   Point test: Substitute $x = 0, y = 0$, (the origin) in $x \geq 4$, then $0 \geq 4$, which is false. Therefore, the origin is not in the solution. The solution set is the other side of the line $x = 4$

   Sense test: Since the sense of the inequality symbol is " >", the solution half-plane is the half-plane to the right of the line $x = 4$ (noting that $x > 4$). The solution includes the boundary line.

Step 4: Shade the half-plane to the right of the line $x = 4$.

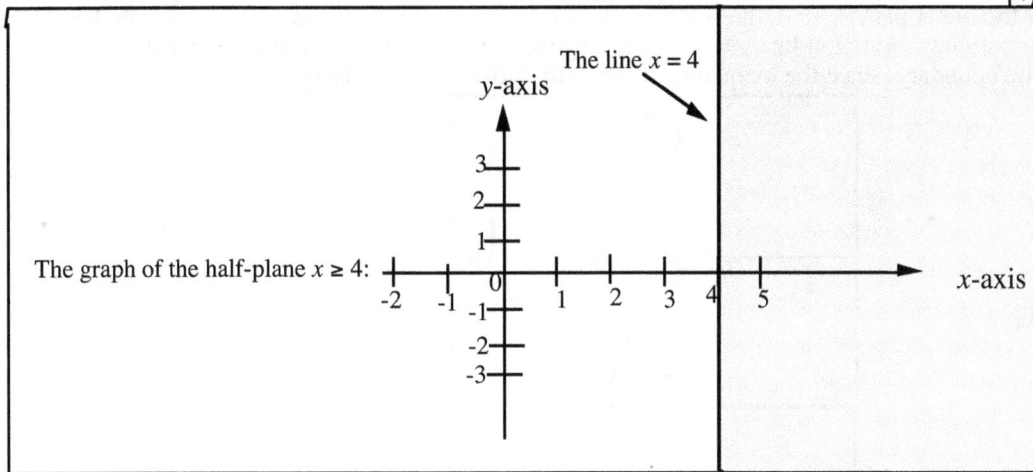

We present, below, similar special cases without the procedures:

(*a*) For the linear plane $x \le 4$, the solution plane is the half-plane to the left of the line $x = 4$ (noting that $x < 4$), including the boundary. (Figure)

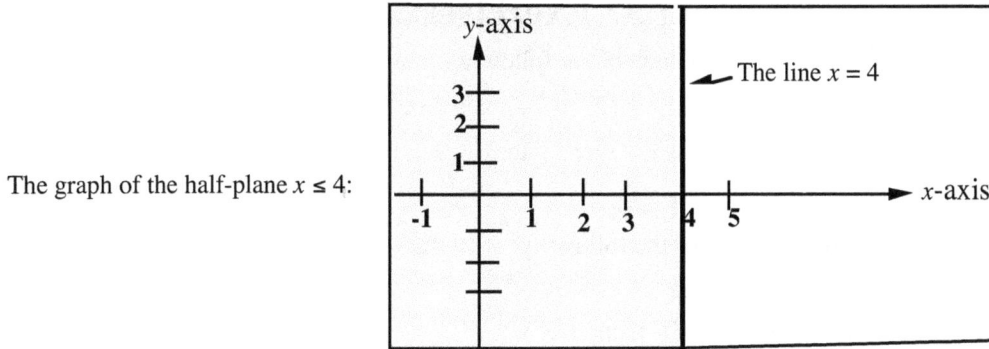

(b) For the linear plane $y > 2$, the solution plane (Figure ..) is the half-plane above (for the **greater** than symbol) the horizontal line $y = 2$, excluding the boundary; and we use a broken (dotted) line to draw the boundary, since the inequality symbol does **not include** the equality symbol.

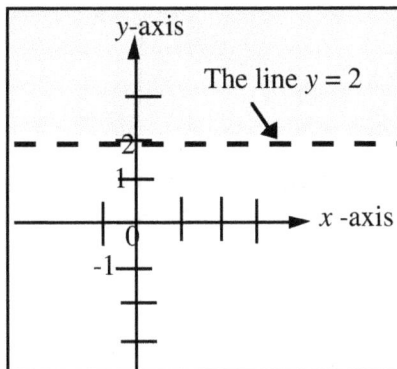

Figure: half-lane $y > 2$,

(c) For the linear plane $y \leq -4$, the solution plane (Fig. 53) is the half-plane below (for the **less** than symbol) the horizontal line $y = -4$, including the boundary and we use a solid line to draw the boundary, since the inequality symbol **includes** the equality symbol.

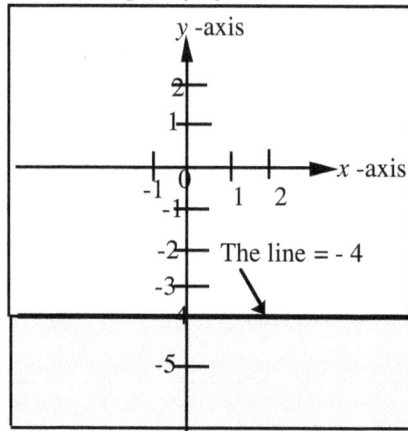

Figure: The graph of the linear plane $y \leq -4$

Lesson 37 Exercises

Questions 1-16: Graph the solution-plane of each of the following:

1. $y + 4x < 2$; **2.** $x + 5y + 5 \geq 0$; **3.** $-4y + x \leq 12$; **4.** $-x + y \geq 0$;

5. $3x - 2y < 6$; **6.** $y \geq |x|$; **7.** $y \leq |x|$; **8.** $y \leq -|x|$; **9.** $y < x$ **10.** $y \geq x$ **11.** $y \geq 4$; **12.** $y \geq -2$;

13. $y \leq 4$; **14.** $x \leq -2$; **15.** $|y| \geq 2$; **16.** $|y| - 2 \leq -|x|$;

Questions 17-20: Draw the graphs for each of the following:

 17. The point $x = 3$;   **18.** The line $x = 3$;

 19. The set of points $x \geq 3$;   **20.** The half-plane $x \geq 3$.

Answers:

1. $y + 4x < 2$

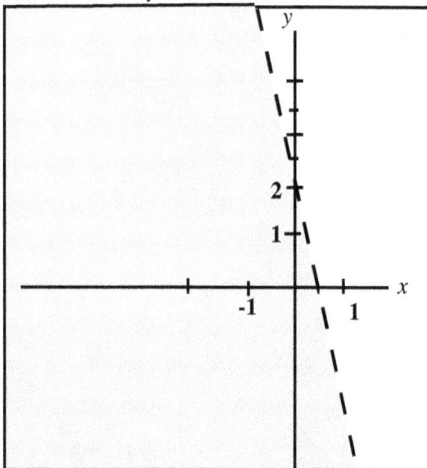

2. $x + 5y + 5 \geq 0$

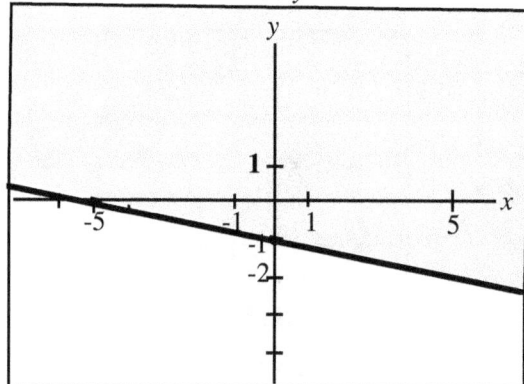

3.

$-4y + x \le 12$

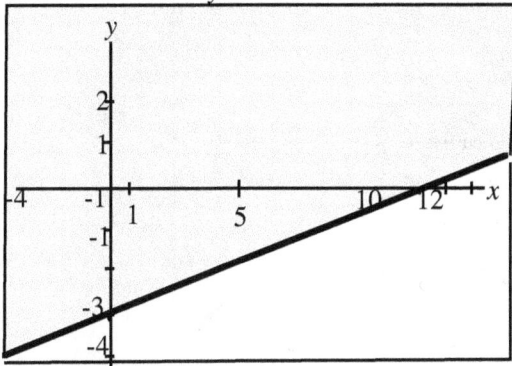

4.

$-x + y \ge 0$

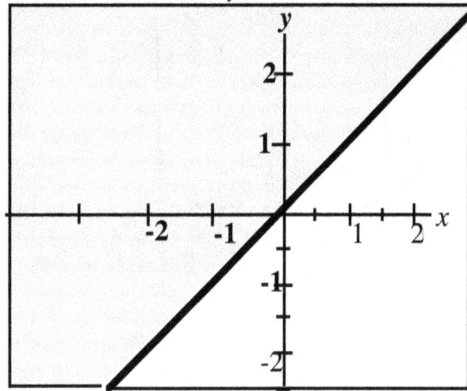

5.

$3x - 2y < 6$

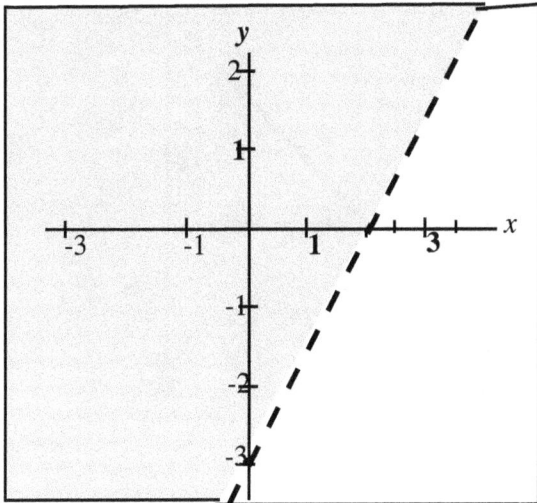

6.

$y \ge |x|$

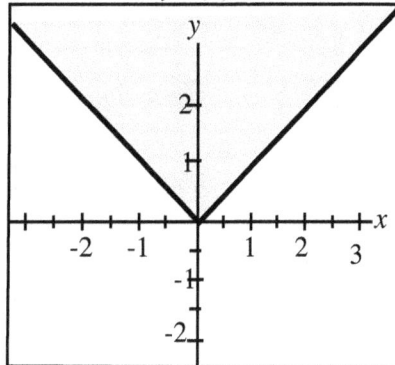

7.

$y \le |x|$

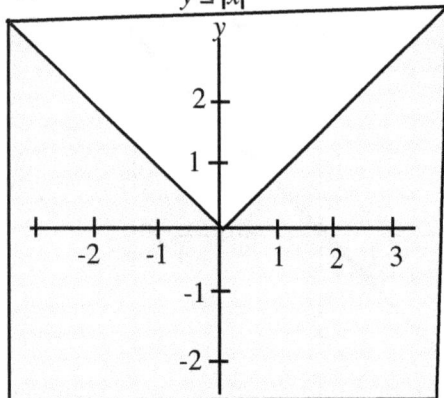

8.

$y \le -|x|$

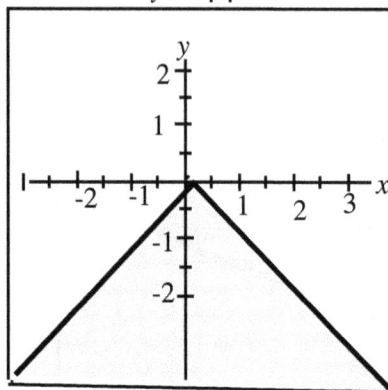

9.

$y < x$

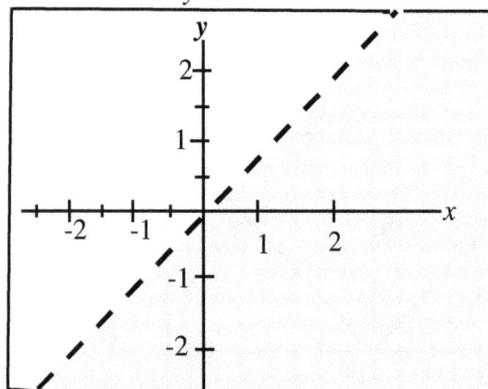

10.

$y \ge x$

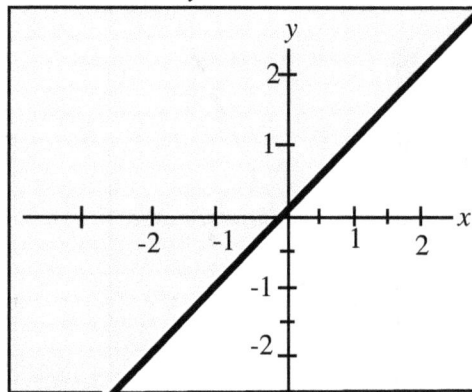

11.

$y \ge 4$

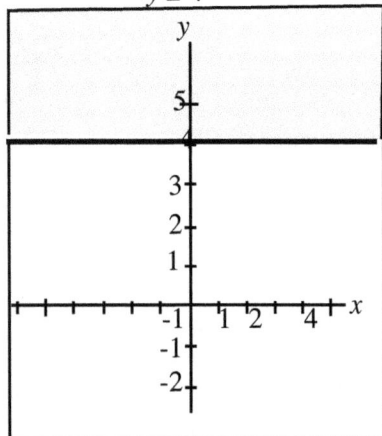

12.

$y \ge -2$

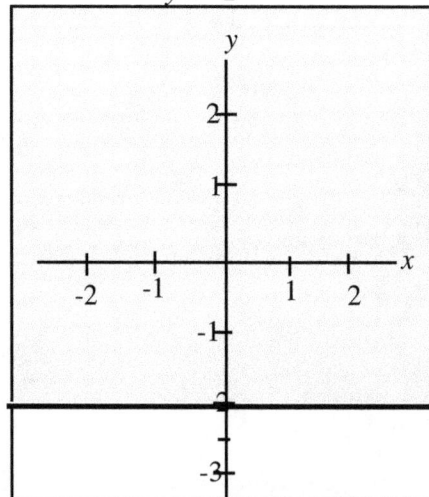

13. $y \le 4$

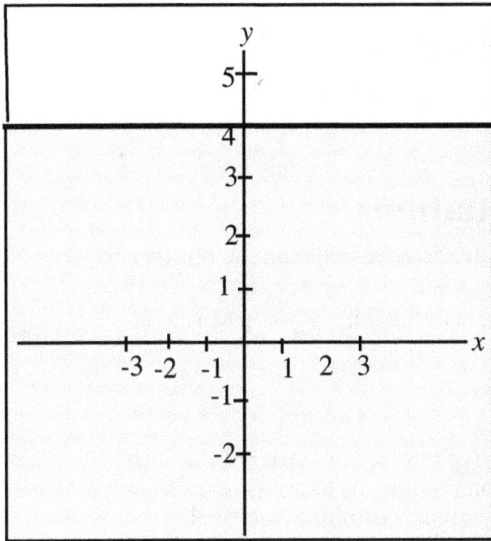

14. $x \le -2$

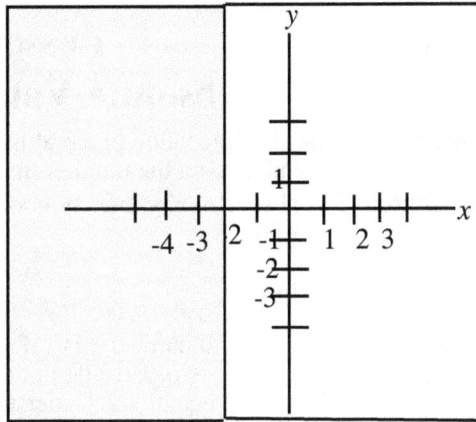

15. $|y| \ge 2$

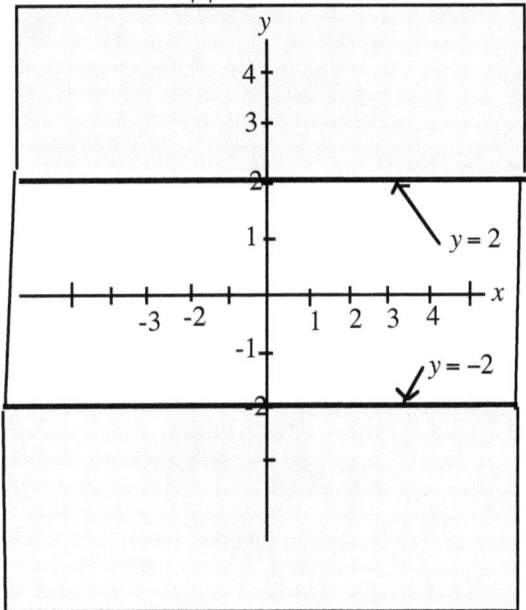

16. $|y| - 2 \le -|x|$ or $|y| + |x| \le 2$

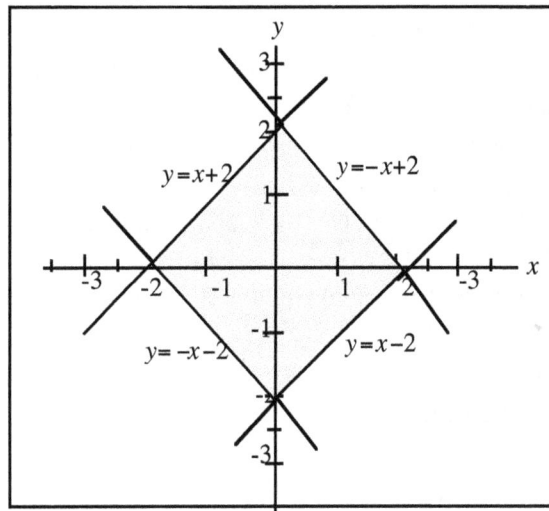

For solutions to Problems 17-20, see page 178-180 of the present Lesson and imitate.

CHAPTER 12

Lesson 38: **Absolute Value Equations**
Lesson 39: **Absolute Value Inequalities**

Lesson 38
Absolute Value Equations

Geometric definition: The absolute value of a real number is the distance between the number and zero on the number line.

Algebraic definition: The absolute value of x symbolized, $|x|$, is defined by a two-part rule as follows:

$$|x| = \begin{cases} x \text{ if } x \geq 0 \\ -x \text{ if } x < 0 \end{cases}$$

In words, the absolute value of a real number x is x if x is a positive number or zero; but it is $-x$ if x is a negative number. The above two-part rule in the definition requires us to consider two cases when solving absolute value equations. Understanding the above definition very well will save us from having to memorize the solution methods for absolute value equations.

Absolute value equations: Absolute value equations are equations in which the absolute value terms contain variables.

Example l Solve for x: $|x| = 2$

We will consider two methods.

Method 1 (Using the geometric definition)

$|x| = 2$ means the distance between x and 0 on the number line is 2.
Therefore $x = 2$ or -2 Check: $|2| = 2$; $|-2| = 2$.

Solution set is $\{-2, 2\}$

Method 2: If $x \geq 0$, then $|x| = x = 2$
If $x < 0$, then $|x| = -x = 2$, and from which $x = -2$.
Again the solution set is $\{-2, 2\}$

Graph for $\{-2, 2\}$

Example 2 Solve for x: $|x - 3| = 4$

Method 1 (Using the geometric definition)

$|x - 3| = 4$ means, on the number line, the distance between x and 3 is 4.

Step l: Locate the point 3 on the number line. (To obtain 3, set $x - 3$ to zero and solve for x.)

Step 2: From this point, count 4 units to the right and stop, the coordinate of this stopping point, 7, is one of the solutions. Similarly from the point 3, count 4 units to the left and stop, the coordinate of this stopping point is -1, and it is another solution. The solution set is $\{-1, 7\}$

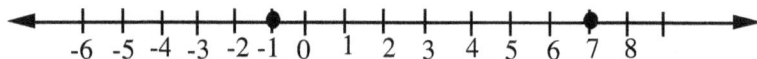

Graph for $\{-1, 7\}$

Method 2 (Using the algebraic definition)

Solution $|x - 3| = 4$ is equivalent to

$x - 3 = 4$ or $-(x - 3) = 4$

Solve each equation for x:

$x - 3 = 4$ or $-(x - 3) = 4$ (or $-x + 3 = 4$ and from which $x = -1$)

$x - 3 = 4$ or $(x - 3) = -4$ (Multiplying each side of the second equation by -1)

$x - 3 = 4$ or $x - 3 = -4$

$\underline{+3 +\ \ 3}$ $\underline{+3\ \ +3}$

$x = 7$ or $x = -1$ Check #1: $|7 - 3| = |4| = 4$; Check #2: $|-1-3| = |-4| = 4$

The solution set is $\{-1, 7\}$

Example 3 Solve for x: $4 - |3x - 4| = -7$

Solution: $4 - |3x - 4| = -7$

$\underline{-4 \hspace{3cm} -4}$

$-|3x - 4| = -11$

$|3x - 4| = 11$ (Multiplying both sides of the equation by -1)

Now, we apply the algebraic definition of the absolute value:

Then, $3x - 4 = 11$ or $-(3x - 4) = 11$

$\underline{+4 \hspace{0.5cm} +4}$ $3x - 4 = -11$ (Multiplying both sides of the equation by -1)

$3x = 15$ $\underline{+4 \hspace{0.8cm} +4}$

$x = 5$ $3x = -7$ and $x = -\frac{7}{3}$

Checking for $x = 5$:

$4 - |3(5) - 4| \overset{?}{=} -7$; $4 - |15 - 4| \overset{?}{=} -7$

$4 - |11| \overset{?}{=} -7$; $4 - 11 \overset{?}{=} -7$

$-7 = -7$ True (LHS = RHS)

Checking for $x = -\frac{7}{3}$: $4 - |3 \cdot (-\frac{7}{3}) - 4| \overset{?}{=} -7$; $4 - |-7 - 4| \overset{?}{=} -7$;

$4 - |-11| \overset{?}{=} -7$; $4 - 11 \overset{?}{=} -7$ (note: $|-11| = 11$)

$-7 = -7$ True

The solution set is $\{-\frac{7}{3}, 5\}$.

Example 4: Solve for x: $|x - 2| = |x + 6|$. (A) (See Appendix B, p.306 for another approach)

(We apply the geometric definition which says that the absolute value of a real number is its distance from zero on the number line)

Applying the geometric definition, either the two expressions of the equation are the same number, or are opposites. For examples, $|3| = |3|$, $|-3| = |-3|$ (same number) or $|-3| = |3|$, $|3| = |-3|$ opposites.

Case 1: If the two expressions represent the same number, then $x - 2 = x + 6$	**Case 2**: If the two expressions represent (opposite numbers, then $x - 2 = -(x + 6)$
$x - 2 = x + 6$	$x - 2 = -x - 6$, and from which
Solving, $-2 = 6$,	$x = -2$
which is a contradiction.	We check for $x = -2$. in (A)
There is therefore no solution for **Case 1**.	Then $|-2 - 2| \overset{?}{=} |-2 + 6|$; $|-4| \overset{?}{=} |4|$; $4 \overset{?}{=} 4$. Yes.
	The solution is -2.

Note above that sometimes, there may be solutions to both cases.

Lesson 38 Exercises

Solve for x: **1.** $|x - 5| = 2$; **2.** $|2x - 1| = 5$;   **3.** $|2x + 1| = 4$;   **4.** $3 - |2x - 1| = -8$

5. $|x + 4| = 3$;   **6.** $|3x - 2| = 4$;   **7.** $6 - |x - 3| = -2$;   **8.** $|x - 10| = -5$

Answers: **1.** $\{3, 7\}$; **2** $\{-2,3\}$; **3.** $\{-\frac{5}{2}, \frac{3}{2}\}$; **4.** $\{-5,6\}$; **5.** $\{-1, -7\}$; **6.** $\{-\frac{2}{3}, 2\}$; **7.** $\{-5, 11\}$; **8.** No solution

Lesson 39
Absolute Value Inequalities

We shall consider two cases

Case 1: $|ax + b| < c$. where c is positive. Example: $|x - 2| < 1$. (There is no solution if c is negative)

Case 2: $|ax + b| > c$, where c may be positive or negative. Example: $|x - 2| > 1$.

Examples   1. $|x - 2| < 1$ means the distance from x to 2 is less than 1.

2. $|x - 2| > 1$ means the distance from x to 2 is greater than 1.

Case 1: $|ax + b| < c$ is equivalent to

$\boxed{-c < ax + b < c}$, the " **and**" problem; or $-c < ax + b$ and $ax + b < c$.

Therefore, we can solve this inequality using the condensed method.
Generally, the solution is a one-interval solution. Case 1 is a conjunction problem.

Case 2: $|ax + b| > c$ is equivalent to

either $\boxed{ax + b > c}$ or $\boxed{-(ax + b) > c}$

Here, we solve an " or" problem. Generally, we have a two-interval solution.
Note that the form $-(ax + b) > c$ which is equivalent to $ax + b < -c$. (obtained by multiplying by -1 and reversing the sense) is easier to recall, since it follows immediately from the absolute value definition.
Case 2 is a disjunction problem.

Example 1: Solve for x: $|x - 2| < 3$.
Solution $|x - 2| < 3$ is equivalent to
$-3 < x - 2 < 3$
$\underline{+2 \quad + 2 \quad +2}$
$-1 < x \quad < 5$ (Solving for x by adding +2 to all three sides of the inequality)

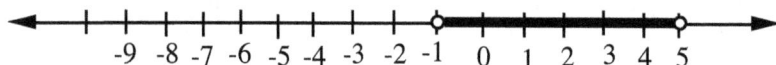
Graph for $\{x \mid -1 < x < 5\}$

Example 2 Solve for x: $|x - 2| > 3$

Solution $|x - 2| > 3$ is equivalent to
$(x - 2) > 3$ or $-(x - 2) > 3$
$x - 2 > 3$ or $\quad x - 2 < -3$ or $-x + 2 > 3$
$\underline{+ 2 \ +2} \qquad \underline{+2 \quad +2} \qquad \underline{-2 \ - 2}$
$\quad x > 5 \quad$ or $\quad x < -1 \qquad \quad -x > 1$
$\qquad \qquad \qquad \qquad \qquad \qquad \quad x < -1$

Solution set is $\{x \mid x < -1 \text{ or } x > 5\}$

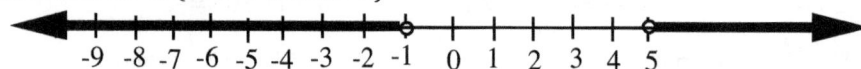
Graph for $\{x \mid x < -1 \text{ or } x > 5\}$

Example 3   Solve for x: $|x + 3| \geq 4$

Solution $|x + 3| \geq 4$ is equivalent to
$(x + 3) \geq 4$ or $-(x + 3) \geq 4$
$x + 3 \geq 4$ or $x + 3 \leq -4$
$\underline{-3\ -3}$ $\underline{-3\ \ \ -3}$
$x \geq 1$ or $x \leq -7$

Solution set is $\{x \mid x \leq -7 \text{ or } x \geq 1\}$

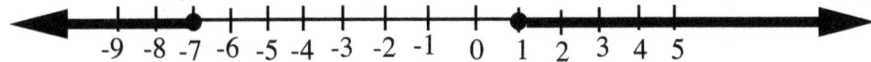

Graph for $\{x \mid x \leq -7 \text{ or } x \geq 1\}$

Lesson 39 Exercises

A   Solve for x: **1.** $|x - 4| < 2$;   **2.** $|x - 4| > 2$;   **3.** $|x + 5| < -4$

Answers: **1.** $\{x \mid 2 < x < 6\}$;   **2.** $\{x \mid x < 2 \text{ or } x > 6\}$;   **3.** No solution

B   Solve for: **1.** $|x - 5| < 2$;   **2.** $|x - 5| > 2$;   **3.** $|x - 5| \geq 2$;   **4.** $|x - 5| = 2$

5. $|2x + 3| \geq 3$

Answers: **1.** $\{x \mid 3 < x < 7\}$;   **2.** $\{x \mid x < 3 \text{ or } x > 7\}$;   **3.** $\{x \mid x \leq 3 \text{ or } x \geq 7\}$;   **4.** $\{3, 7\}$;
5. $\{x \mid x \leq -3 \text{ or } x \geq 0\}$

CHAPTER 13
Lesson 40
Solving Fractional or Rational Equations

Example 1 Solve for x: $\dfrac{4}{x-1} = 5$

Solution

Step 1: Determine the excluded values. These are the x-values which make the denominators zero. Set the denominator to zero and solve for the variable.

$$x - 1 = 0$$
$$\underline{+1 \quad +1}$$
$$x = 1$$

The excluded value is 1.

Step 2: Multiply both LHS and RHS of the equation by $x - 1$

$$\frac{4}{x-1} = 5$$

$$\frac{(x-1)\,4}{(x-1)} = \frac{5(x-1)}{1}$$

$$4 = 5x - 5$$
$$\underline{+5 \qquad +5}$$
$$9 = 5x$$
$$\frac{9}{5} = x$$

Since $\dfrac{9}{5}$ is not an excluded value, it may be accepted as a solution; however, for good practice, we check it in the original equation.

Step 3: $\qquad \dfrac{4}{\frac{9}{5} - 1} \overset{?}{=} 5$

$$\frac{4}{1\frac{4}{5} - 1} \overset{?}{=} 5$$

$$\frac{4}{\frac{4}{5}} \overset{?}{=} 5$$

$$\frac{4}{1} \times \frac{5}{4} \overset{?}{=} 5$$

$$5 = 5 \qquad \text{True} \ (\text{RHS} = \text{LHS})$$

$\therefore$ the solution is $\dfrac{9}{5}$.

Example 2 Solve for x: $\dfrac{4}{x-3} = \dfrac{3}{x-4}$

Step 1:   Determine the excluded values.
Set $x - 3 = 0$ and solve for x to obtain $x = 3$
Set $x - 4 = 0$ and solve for x to obtain $x = 4$
The excluded values are 3 and 4. Thus in solving the above equation, any value of x obtained may be checked for acceptance as a solution **except** 3 and 4.

Step 2: Multiply both sides of the equation by the LCM of the denominators.
(The LCD $= (x - 3)(x - 4)$

$$\dfrac{4}{(x-3)} \cdot \dfrac{(x-3)(x-4)}{1} = \dfrac{3}{(x-4)} \cdot \dfrac{(x-3)(x-4)}{1}$$

$$4(x-4) = 3(x-3)$$

$$4x - 16 = 3x - 9$$
$$\underline{-3x \qquad\quad -3x}$$
$$x - 16 = -9$$
$$\underline{+16 \quad +16}$$
$$x = 7$$

This value, 7, is not one of the excluded values (from Step 1) and so we may accept it as a solution but for good practice, we check it in the original equation.

$$\dfrac{4}{7-3} \overset{?}{=} \dfrac{3}{7-4}$$
$$\dfrac{4}{4} \overset{?}{=} \dfrac{3}{3} \quad \textbf{True} \;\;(\text{Also both the LHS and RHS of the equations are defined.})$$
The solution is 7.

Example 3   Solve for y.

$$y + \dfrac{2y}{y-2} = \dfrac{4}{y-2}$$

Step 1: Set $y - 2 = 0$ and solve for y.
Then, $y - 2 = 0$, and $y = 2$

The excluded value is 2.
Step 2: Multiply every term of the equation by $y - 2$ (the LCD of the fractions).

$$y(y-2) + \dfrac{2y(y-2)}{y-2} = \dfrac{4(y-2)}{y-2}$$
$$y(y-2) + \dfrac{2y(y-2)}{y-2} = \dfrac{4(y-2)}{y-2}$$
$$y^2 - 2y + 2y = 4$$
$$y^2 = 4$$
$$y = \pm\sqrt{4}$$
$$y = \pm 2$$
i.e. $y = +2$, or $y = -2$

Since +2 is an excluded value (from Step 1), it is rejected; and only -2 may be accepted as a solution.

$$\text{Check:} \quad -2 + \frac{2\,(-2)}{-2\,-2} \overset{?}{=} \frac{4}{-2\,-2}$$

$$-2 + \frac{(-4)}{-4} = \frac{4}{-4}$$

$$-2 + 1 = -1$$

$$-\,1 = -1 \quad \text{True}$$

Therefore, the solution. is -2.

Note above that +2 is an extraneous solution.

Example 4 Solve for x.

$$\frac{3x}{x + 5} - \frac{x + 1}{x + 3} = 2$$

Step 1. Find the excluded values.

Set $x + 5 = 0$ and solve for x to obtain $x = -5$.
-5 is an excluded value.

Similarly, set $x + 3 = 0$, and solve for x to obtain $x = -3$

The excluded values are -5 and -3. Note above that we could obtain the excluded values by inspection (mentally).

Step 2: Multiply each term of the equation by the LCM of the denominators (to undo the denominators). The LCM is $(x + 5)(x + 3)$.

$$(x + 5)(x + 3)\,\frac{3x}{x + 5} - (x + 5)(x + 3)\frac{(x + 1)}{x + 3} = 2(x + 5)\,(x + 3)$$

$$\cancel{(x + 5)}(\,x + 3) \cdot \frac{3x}{\cancel{x + 5}} - (x + 5)(\,\cancel{x + 3}\,)\,\frac{(x + 1)}{\cancel{x + 3}} = 2(x + 5)\,(\,x + 3)$$

Step 3: Simplify and solve for x.

$$(x + 3\,)(3x) - (x + 5\,)(x + 1) = 2(x + 5)(x + 3)$$

$$3x^2 + 9x - (x^2 + 6x + 5) = 2(x^2 + 8x + 15)$$

$$3x^2 + 9x - x^2 - 6x - 5 = 2x^2 + 16x + 30 \qquad \text{(simplifying the left-hand side)}$$

$$2x^2 + 3x - 5 = 2x^2 + 16x + 30$$

$$2x^2 - 2x^2 + 3x - 16x - 5 - 30 = 0 \qquad \text{(eliminating the right-hand-side)}$$

$$-13x - 35 = 0$$

$$\frac{-35x}{-13} = \frac{+35}{-13}$$

$$x = -\frac{35}{13}$$

After checking (not shown), the solution is $-\frac{35}{13}$.

Note in Example **4** above, that in Step 3, the equation contained x^2-terms; but when we proceeded to solve the equation, the x^2-terms dropped out. Sometimes, this happens; in some cases these terms do not drop out, as in Example 3 above.

Note:

The excluded values for $\dfrac{8}{x^2 - 4} = 1$ are -2 and +2 (obtained by setting $x^2 - 4 = 0$ and solving for x);

however, note that for $\dfrac{8}{x^2 + 4} = 1$, $x^2 + 4$ is positive for all real values of x and never zero, since

the square of any nonzero real number is always positive. Therefore, there are no real excluded

values for $\dfrac{8}{x^2 + 4} = 1$. In fact the solution to $\dfrac{8}{x^2 + 4} = 1$ is the set $\{-2, 2\}$.

Lesson 40 Exercises

A Solve for x:

1. $\dfrac{9}{x + 4} = 1$;   **2.** $\dfrac{2}{x - 3} = \dfrac{5}{x - 4}$;   **3.** $\dfrac{x^2}{x + 2} = \dfrac{4}{x + 2}$ **4.** $\dfrac{7}{x - 5} + x = \dfrac{3}{x - 5}$

Answers: **1.** {5};   **2.**$\{\dfrac{7}{3}\}$ **3.** {2} Note: $x \neq -2$;   **4.** {1,4}

B Solve for x and state the excluded values:

1. $\dfrac{4}{x} + \dfrac{1}{2} = \dfrac{2}{5}$; **2.** $\dfrac{2}{3} - \dfrac{5}{x} = \dfrac{3}{4}$; **3.** $\dfrac{2x}{2x - 5} = 2$; **4.** $\dfrac{1}{x - 3} + \dfrac{1}{x + 3} = \dfrac{9}{x^2 - 9}$;   **5.** $\dfrac{1}{x - 2} = \dfrac{x^2}{x - 2} + 1$

Answers: **1.** -40 $(x \neq 0)$; **2.** -60 $(x \neq 0)$; **3.** 5 $(x \neq \dfrac{5}{2})$; **4.** $\dfrac{9}{2}$ $(x \neq -3 \text{ or } 3)$; **5.** $-\dfrac{1}{2} \pm \dfrac{\sqrt{13}}{2}$ $(x \neq 2)$

CHAPTER 14
Lesson 41
Solving Radical Equations

We define **a radical equation** as an equation in which one of the variables occurs under one or more radical signs.

Examples   **1.** $\sqrt{x} = 6$

2. $\sqrt{x - 3} = 4$

3. $\sqrt{2x + 4} = \sqrt{x + 2}$

Principle of powers: If $a = b$

Then, $a^n = b^n$. However, this principle does not always produce equivalent equations, and therefore, if we use this principle to solve an equation, we must check the solutions in the original equation.

Procedure   The technique here involves the elimination of the radical signs by raising the radical expressions to integral (integer) powers: both sides of an equation are to be raised to the same integral power. We will then solve the resulting non-radical equation for the variable. If there are more than one radical, we may sometimes have to "undo" the radical signs in a number of steps. It will also be necessary that we test any solutions in the original equation for **extraneous solutions**. Before raising each side of an equation to an integral power, the radical whose radical sign we want to eliminate must be on one side of the equation by itself.

Example 1   Solve for x: $\sqrt{x} = 6$.

Solution

Step 1:   Square both sides of the equation to undo the square root symbol (i.e., raise both sides of the equation to the second power).
Then, $(\sqrt{x})^2 = (6)^2$
$x = 36$.
Step 2: Check the solution in the original equation.
$\sqrt{36} \overset{?}{=} 6$
$6 = 6$ True (left-hand side equals right-hand side, i.e., LHS = RHS).
∴ the solution is 36.

The above problem was so simple that we could have asked: If the square root of a number is 6, what is the number? Of course, the number is 36.

Example 2 Solve for x: $\quad \sqrt{x-3} = 4$

Solution

Step 1: Square both sides of the equation to undo the square root symbol.

Then, $(\sqrt{x-3})^2 = (4)^2$

$$x - 3 = 16$$

Step 2: Solve for x.

$$\begin{array}{r} x - 3 = 16 \\ \underline{+3 \quad +3} \\ x = 19 \end{array}$$

Step 3: Check the solution 19 in the original equation.

$$\sqrt{19 - 3} \overset{?}{=} 4$$

$$\sqrt{16} \overset{?}{=} 4$$

$$4 = 4 \ \text{True} \quad (\text{LHS} = \text{RHS})$$

$\therefore$ the solution is 19.

Note above that squaring the square root of a number eliminates the square root symbol.

Example 3 Solve for x:

$$\sqrt{2x + 4} = \sqrt{x + 2}$$

Solution

Step 1: Square both sides of the equation.

$$(\sqrt{2x + 4})^2 = (\sqrt{x + 2})^2$$

$$2x + 4 = x + 2$$

Step 2: Solve for x.

$$\begin{array}{r} 2x + 4 = x + 2 \\ \underline{-x \qquad -x} \\ x + 4 = 2 \\ \underline{-4 \quad -4} \\ x = -2 \end{array}$$

Step 3: Check for $x = -2$ in the original equation

$$\sqrt{2(-2) + 4} \overset{?}{=} \sqrt{(-2) + 2}$$

$$\sqrt{-4 + 4} \overset{?}{=} \sqrt{0}$$

$$\sqrt{0} \overset{?}{=} \sqrt{0}$$

$$0 = 0 \qquad \text{True} \quad (\text{RHS} = \text{LHS})$$

$\therefore$ the solution is -2.

Example 4 Solve for x: $\qquad \sqrt{2x-5} = x-4$
Solution

Step 1: Square both sides of the equation.
$$(\sqrt{2x-5})^2 = (x-4)^2$$
$$2x - 5 = x^2 - 8x + 16$$
$$0 = x^2 - 10x + 21$$
or $x^2 - 10x + 21 = 0$

Step 2: Solve the quadratic equation by any method.
We will solve by factoring, since the factors are easily recognizable.
$$x^2 - 10x + 21 = 0$$
$$(x-3)(x-7) = 0$$
$$x - 3 = 0 \text{ or } x - 7 = 0$$
$$x = 3 \text{ or } x = 7 \quad \text{(Setting each factor equal to zero and solving for } x)$$

Step 3: Checking for $x = 3$:
$$\sqrt{2(3)-5} \overset{?}{=} 3 - 4$$
$$\sqrt{6-5} \overset{?}{=} -1$$
$$\sqrt{1} \overset{?}{=} -1$$
$$1 = -1 \text{ False} \quad \text{(LHS \textbf{not} equal to RHS)}$$
Therefore 3 is not a solution. It is an extraneous root.

Checking for $x = 7$:
$$\sqrt{2(7)-5} \overset{?}{=} 7 - 4$$
$$\sqrt{14-5} \overset{?}{=} 3$$
$$\sqrt{9} \overset{?}{=} 3$$
$$3 = 3 \quad \text{True} \quad \text{(RHS = LHS)}$$
Therefore, 7 is a solution.
Since 3 is an extraneous root, the only solution is 7.

Lesson 41 Exercises

Solve for x:
1. $\sqrt{x} = 7$; $\qquad$ **2.** $\sqrt{x-4} = 5$ $\quad$ **3.** $\sqrt{3x+5} = \sqrt{x+9}$

Answers: 1. $x = 49$; $\quad$ **2.** $x = 29$; $\qquad$ **3.** $x = 2$;

Solve and check: $\qquad$ **1.** $\sqrt{x-5} + 6 = 2$; $\qquad$ **2.** $2\sqrt{x} = 6$; $\qquad$ **3.** $\sqrt{x^2 - 8x} = 3$;

4. $3\sqrt{x+2} = 4\sqrt{x-5}$; $\qquad\qquad$ **5.** $\sqrt{x+5} - \sqrt{x} = 3$; $\qquad$ **6.** $\sqrt[3]{x+1} = 2$

Answers: **1.** No solution (21 is extraneous) ; $\quad$ **2.** { 9 } ; $\quad$ **3.** { -1, 9 } ; $\quad$ **4.** { 14 } ;

5. No solution. ($\frac{4}{9}$ is extraneous) ; $\quad$ **6.** {7}

CHAPTER 15
Complex Numbers

Lesson 42: **Definition, Powers of i, Square Root of Negative Numbers**
Lesson 43: **Addition and Subtraction of Complex Numbers**
Lesson 44: **Multiplication and Division of Complex Numbers**

Lesson 42

Definition, Powers of i, Square Root of Negative Numbers

Sometimes, in attempting to solve certain polynomial equations, we arrive at situations in which we have to find the square roots of negative numbers.

Example: If $x^2 = -1$, then $x = \pm\sqrt{-1}$
Since, for the set of real numbers, there is no provision for the square root of a negative number, we introduce a number "i" which we call the imaginary unit, with the following definition:

Definition: $i = \sqrt{-1}$ or $i^2 = -1$ (We will use both forms of the definition; memorize them)

For example, $(\sqrt{-1})(\sqrt{-1}) = (i)(i) = i^2 = -1$

Powers of i   (Cyclical property of i or i^2) : All powers of i are either equal to ± 1 or $\pm i$.
Examples

(a) $i^2 = -1$
(b) $i^3 = (i^2)i$
     $\quad = -1(i)$
     $\quad = -i$

(c) $i^4 = (i^2)(i^2)$
     $\quad = (-1)(-1)$
     $\quad = 1$
(d) $i^{10} = (i^4)(i^4)(i^2)$
     $\quad = (1)(1)(-1)$
     $\quad = -1$

Note above: Even powers of i are equal to ± 1. Odd powers of i are equal to $\pm i$.

Note that the imaginary unit "i" is only a tool in mathematics, and that it is not more imaginary (in the literal sense) than the real number 3. The introduction of the imaginary unit allows us to find the square roots of negative numbers.

Example   Find the square root: (a) -4 ;   (b) -25 ;   (c) Simplify: $\sqrt{-15}$.

Solution   (a) $\sqrt{-4} = (\sqrt{-1})(\sqrt{4})$
                      $\quad = i(2)$

    Therefore $\sqrt{-4} = 2i$

(b) $\sqrt{-25} = \sqrt{-1}(\sqrt{25})$
       $\quad = i(5)$

$\sqrt{-25} = 5i$

(c) $\sqrt{-15} = (\sqrt{-1})(\sqrt{15})$
     $\quad = i\sqrt{15}$ or $\sqrt{15}\,i$

Note: In (c) we prefer the first form of the answer; because, sometimes, if we are not careful in writing "i", the "i" may look as if it is under the radical sign . However, if no radical is involved we will leave the answers as in (a) and (b) above. In some old textbooks, you may find "i" written after the radical.

Generally, $\sqrt{-b} = (\sqrt{-1})(\sqrt{b})$ $(b \geq 0)$

         $= i\sqrt{b}$

The introduction of the imaginary unit helps us to expand the real number system to a more general system called the complex number system.

If we denote a complex number by z, then $z = a + bi$, where a and b are real numbers; a is called the real part and b is called the imaginary part. If $b = 0$, we have a pure real number (e.g., $z = 3 + 0i = 3$). If $a = 0$, we have a pure imaginary number ($z = 0 + 2i = 2i$). Thus, the product bi is called a pure imaginary number.

Distinction Between Square Roots of Negative Numbers and Roots of Equations

Roots of numbers (Principal Roots): $\sqrt{-b} = i(\sqrt{b})$ $(b \geq 0)$

Example (a) $\sqrt{-4} = (\sqrt{-1})(\sqrt{4})$

           $= 2i.$

     (b) $\sqrt{-25} = \sqrt{-1}(\sqrt{25})$

             $= 5i.$

Roots of equations: Here, we must note that we have more than one root. (two roots.)

Example 1 Consider the solution to the equation $x^2 = -16$.

Solution $x^2 = -16$

            $x = \pm\sqrt{-16}$

               $= \pm 4i$, which means $x = +4i$ or $x = -4i$.

Example 2   Solve for x:: $x^2 = -4$.

Solution

     $x = \pm\sqrt{-4}$

     $x = \pm 2i$ (or $x = +2i$ or $x = -2i$).

Lesson 42 Exercises

Find the square root or simplify:

1. $\sqrt{-9}$;   **2** $\sqrt{-49}$;   **3.** $\sqrt{-36}$;   **4.** $\sqrt{-18}$;   **5.** $\sqrt{\frac{4}{-9}}$;   **6.** i^{40};   **7.** i^{93} ;   **8.** i^{100}

9. $\sqrt{-28}$;   **10.** $\sqrt{-32}$

Answers: **1.** $3i$;   **2.** $7i$;   **3.** $6i$;   **4.** $3i\sqrt{2}$;   **5.** $\frac{2}{3}i$ **6.** 1 ;   **7.** i ;   **8.** 1 ;   **9.** $2i\sqrt{7}$;   **10.** $4i\sqrt{2}$

Lesson 43

Addition and Subtraction of Complex Numbers

In adding complex numbers, we will add the real parts, and then add the imaginary parts (in much the same way as we add like terms in polynomial addition).

Perform the indicated operations, leaving the answers in the form $a + bi$.

Example 1   Simplify: $(-3 + 5i) + (-2 + 7i)$

Step 1: Remove the parentheses.

$$(-3 + 5i) + (-2 + 7i)$$

$$= -3 + 5i - 2 + 7i$$

Step 2: Add the real parts, and add the imaginary parts.

$$-3 + 5i - 2 + 7i$$
$$= -5 + 12i \,.$$

Scrapwork:
For the real parts: $-3 - 2 = -5$
For the imaginary parts: $5 + 7 = 12$

Example 2   Simplify: $(6 - 5i) - (-3 + 2i)$

Solution   Remove the parentheses and add.

$$(6 - 5i) - (-3 + 2i)$$
$$= 6 - 5i + 3 - 2i$$

$$= 9 - 7i$$

Example 3   Simplify: $(5 + 2i) + (-3 - 6i)$

Solution

Remove the parentheses and add.
$$(5 + 2i) + (-3 - 6i)$$
$$= 5 + 2i - 3 - 6i$$
$$= 5 - 3 + 2i - 6i \longleftarrow \text{-------you may skip this step.}$$
$$= 2 - 4i$$

or adding vertically:

$$\begin{array}{r} 5 + 2i \\ \underline{-3 - 6i} \\ 2 - 4i \end{array}$$ (Adding).

Example 4   Simplify : $(3 + 4i) - (2 - 5i)$ (subtraction)

Solution

Remove the parentheses and add.
$$(3 + 4i) - (2 - 5i)$$
$$= 3 + 4i - 2 + 5i$$
$$= 3 - 2 + 4i + 5i \text{ <-------you may skip this step.}$$
$$= 1 + 9i$$

Example 5 Simplify : $(6 - 3i) + (2 + 3i)$

Solution
$$(6 - 3i) + (2 + 3i)$$
$$= 6 - 3i + 2 + 3i$$
$$= 8 + 0i$$
$$= 8$$

Lesson 43 Exercises

Simplify: **1.** $4 + 3i - 6 + 5i$;   **2.** $(7 - 8i) + (2 + 9i)$ **3.** $(2 - 3i) - (7 - 4i)$;

4. Subtract $(1 - i)$ from $4 - 2i$;   **5.** $4 + 2i - 3(4 - 6i) + i$;   **6.** $-i^5 + i^2$; **7.** $4 - i + i^2$

Answers: **1.** $-2 + 8i$; **2.** $9 + i$; **3.** $-5 + i$;   **4.** $3 - i$;   **5.** $-8 + 21i$;   **6.** $-1 - i$;   **7.** $3 - i$

Lesson 44
Multiplication and Division of Complex Numbers

Multiplication of Complex Numbers

The approach here is multiply, replace i^2 by -1, (or higher powers of i by ± 1 or $\pm i$) add the real parts and add the imaginary parts.

Example 1 Multiply   -4 - 2i and -5 + i

Solution

Step 1:   Multiply as you multiply binomials
$(-4 - 2i)(-5 + i)$
$= -4(-5 + i) + (-2i)(-5 + i)$ <--- You may skip this step.
$= 20 - 4i + 10i - 2i^2$

Step 2:   (Replace i^2 by -1, since $i^2 = -1$ by definition)
$= 20 - 4i + 10i - 2(-1)$
$= 20 + 6i + 2$

Step 3:   (Add the like terms: add the real parts; and add the imaginary parts)
$= 22 + 6i$

Example 2   Multiply   3 + 2i and 4 + 5i
Solution
Procedure: Multiply, replace i^2 by -1, and simplify.

$(3 + 2i)(4 + 5i)$
$= 12 + 15i + 8i + 10i^2$
$= 12 + 23i + 10(-1)$
$= 12 + 23i - 10$
$= 2 + 23i$ (Adding)

Example 3   Simplify : 4(6 - 3i)

Solution

$4(6 - 3i)$
$= 24 - 12i$

Example 4   Simplify: (4 + 2i)(3 - 6i)

Solution   The above implies multiplication.
Multiply, replace i^2 by -1, and add the like terms.
$(4 + 2i)(3 - 6i)$
$= 12 - 24i + 6i - 12i^2$
$= 12 - 24i + 6i - 12(-1)$
$= 12 - 24i + 6i + 12$
$= 24 - 18i$ (Note: 12 +12 =24 and -24i + 6i = -18i)

Example 5 Simplify: $(7 + 2i)(7 - 2i)$

Solution

$(7 + 2i)(7 - 2i)$
$= 49 - 14i + 14i - 4i^2$
$= 49 - 4(-1)$
$= 49 + 4$
$= 53$

Multiplication of the square roots of negative numbers

Example 1 Find the product of $(\sqrt{-8})$ and $(\sqrt{-2})$

Solution

Step 1: Change to complex number forms.

$\sqrt{-8} = \sqrt{-1}(\sqrt{8})$
$\quad = i\sqrt{8}$

Similarly, $\sqrt{-2} = i\sqrt{2}$

Step 2: Multiply the complex forms now.

$(\sqrt{-8})(\sqrt{-2}) = i\sqrt{8} \ i\sqrt{2}$
$\qquad\qquad = i^2\sqrt{16} \qquad\qquad (\sqrt{16} = 4)$
$\qquad\qquad = i^2 \ (4)$
$\qquad\qquad = (-1)(4) \qquad\qquad (i^2 = -1)$
$\qquad\qquad = -4$

In the above problem (Example 1), you may skip Step 1 and show only Step 2.

Example 2 Find the product $(\sqrt{-49})(\sqrt{-25})$

Solution

Step 1: Change to complex number forms.

Then, $\sqrt{-49} = i\sqrt{49} = i(7) = 7i$

<---You may do Step 1 mentally and show only Step 2.

$\sqrt{-25} = i\sqrt{25} = i(5) = 5i$

Step 2: Multiply the complex forms now.

Then, $(\sqrt{-49})(\sqrt{-25}) = (7i)(5i)$
$\qquad\qquad\qquad = (7)(5)i^2$
$\qquad\qquad\qquad = 35(-1) \qquad (i^2 = -1)$
$\qquad\qquad\qquad = -35$

We must note in the last two examples that it was necessary first to express the square roots of
the negative numbers in terms of i before proceeding to multiply. Failure to do this may result
in error such as the following:

Wrong procedure---> $\qquad (\sqrt{-8}(\sqrt{-2}) = \sqrt{(-2)(-8)}$
$\qquad\qquad\qquad\qquad = \sqrt{16}$
$\qquad\qquad\qquad\qquad = 4$, **which is a wrong answer**.

Generally, it is true that $(\sqrt{a})(\sqrt{b}) = \sqrt{ab}$ if a and b are positive but it is not true if a and b are negative.

Complex conjugates

Definition : The **conjugate** of a given binomial is another binomial that differs from the given binomial only in the sign of one of the terms. The conjugate of $a + b$ is $a - b$; and the conjugate of $a - b$ is $a + b$.

The **conjugate** of $a + bi$ is $a - bi$. The conjugate of $a - bi$ is $a + bi$. A complex number and its conjugate differ only in the sign of the imaginary part. To find the conjugate of a complex number, change the sign of the imaginary part and keep the sign of the real part unchanged.

Examples:

The conjugate of $2 + 3i$ is $2 - 3i$. The conjugate of $4 - 2i$ is $4 + 2i$. The conjugate of $7 - 5i$ is $7 + 5$. The conjugate of $-3 + 6i$ is $-3 - 6i$. The conjugate of $-2 - 3i$ is $-2 + 3i$. The conjugate of $4i$ is $- 4i$. The conjugate of 7 is 7

The product of a complex number and its conjugate is a real number.

The product of $-2 - 3i$ and $-2 + 3i$. $= 4 - 6i + 6i - 9i^2 = 4 - 9(-1) = 4 + 9 = \mathbf{13}$, a real number.

The product of $a + bi$ and $a - bi = a^2 + b^2$

We can use the conjugate of a complex number to rationalize the denominator of a fraction or to divide by a complex number.

Division of Complex Numbers

Example 1   Simplify : $\dfrac{2 + 3i}{4 - 5i}$ or divide $2 + 3i$ by $4 - 5i$

To simplify the above complex number is meant we are to write it in the form $z = a + bi$.
The operation here is similar to that of the rationalization of denominators of radical expressions.
To simplify the above expression, we **multiply both the denominator and the numerator by the conjugate of the denominator.**

Solution   The conjugate of $4 - 5i$ is **$4 + 5i$.**   (See also the note below)
         Multiply both the denominator and the numerator by $4 + 5i$.

Then, we obtain: $\dfrac{2 + 3i}{4 - 5i}$

$$= \frac{(2 + 3i)}{(4 - 5i)} \cdot \frac{(4 + 5i)}{(4 + 5i)}$$

$$= \frac{8 + 10i + 12i + 15i^2}{16 + 20i - 20i - 25i^2}$$

$$= \frac{8 + 22i + 15(-1)}{16 + 0 - 25(-1)}$$

$$= \frac{8 + 22i - 15}{16 + 25}$$

$$= \frac{8 + 22i - 15}{41}$$

$$= \frac{-7 + 22i}{41}$$

$$= -\frac{7}{41} + \frac{22}{41}i$$

We may observe above that the **denominator** does **not** contain the imaginary unit "i", even though the numerator contains the imaginary unit.

Note above that we could have multiplied by $-4 - 5i$. (Try it.) It is therefore not critical in this problem which terms should differ in sign, (so far as the rationalization is concerned) provided that either the real parts differ in sign or the imaginary parts differ in sign, but **not** both. We may produce a minus sign in the denominator which we can take care of as usual.

Example 2   Simplify: $\dfrac{5 + 4i}{-3 + 2i}$

Solution Multiply both the denominator and the numerator by -3 - 2i   (The conjugate of -3 + 2i).

$$\dfrac{5 + 4i}{-3 + 2i}$$

$$= \dfrac{(5 + 4i)}{(-3 + 2i)} \cdot \dfrac{(\textbf{-3 - 2i})}{(\textbf{-3 - 2i})}$$

$$= \dfrac{-15 - 10i - 12i - 8i^2}{9 + 6i - 6i - 4i^2}$$

$$= \dfrac{-15 - 10i - 12i - 8(-1)}{9 + 0 - 4(-1)}$$

$$= \dfrac{-15 - 22i + 8}{9 + 4}$$

$$= \dfrac{-7 - 22i}{13}$$

$$= -\dfrac{7}{13} - \dfrac{22}{13}i \qquad (-15 + 8 = -7)$$

Example 3 Simplify: $\dfrac{4 - 2i}{i}$

Solution $\dfrac{4 - 2i}{i}$

$$= \dfrac{(4 - 2i)}{i} \dfrac{(\textbf{-i})}{(\textbf{-i})} \quad \text{(The conjugate of } i \text{ is } -i \text{ ; but you could also use } i)$$

$$= \dfrac{(4 - 2i)(-i)}{(i)(-i)}$$

$$= \dfrac{-4i + 2i^2}{-i^2}$$

$$= \dfrac{-4i + 2(-1)}{-(-1)} \qquad (i^2 = -1)$$

$$= \dfrac{-4i - 2}{+1}$$

$$= -2 - 4i$$

Lesson 44 Exercises

A **1.** Multiply: 4 + 3i and 2 + 5i ;   **2.** Multiply 2 - 5i and -2 + 6i ; **3.** Simplify: (6 -7i)(2 + 3i) ;
4. Simplify (1 - i)(1 + i) ;   **5.** $(2 + 3i)^2$; **6.** $(4i^2)(-8i)(i^2)$;   **7.** (5 + 3i)(5 - 3i)

Answers: **1.** -7 + 26i ; **2.** 26 + 22i ; **3.** 33 + 4i ;   **4.** 2 ; **5.** -5 + 12i ; **6.** -32i ;   **7.** 34.

B Divide: **1.** $\dfrac{8 - 6i}{2}$; **2.** $\dfrac{6 + 8i}{2i}$;   **3.** $\dfrac{\sqrt{16}}{\sqrt{-16}}$;   **4.** $\dfrac{3}{4 - \sqrt{-9}}$; **5.** $\dfrac{4 - 2i}{3 + 5i}$

Answers: **1.** 4 - 3i ; **2.** 4 - 3i ; **3.** -i ;   **4.** $\dfrac{12}{25} + \dfrac{9}{25}i$;   **5.** $\dfrac{1}{17} - \dfrac{13}{17}i$

C Find the product of each of the following:

1. $\left(\sqrt{-4}\right)\left(\sqrt{-1}\right)$; **2.** $\left(\sqrt{-16}\right)\left(\sqrt{-4}\right)$; **3.** $\left(\sqrt{-25}\right)\left(\sqrt{49}\right)$; **4.** $\left(\sqrt{-8}\right)\left(\sqrt{-2}\right)$

Answers: **1.** -2; **2.** -8; **3.** $35i$; **4.** -4.

D Simplify: **1.** $\dfrac{2+3i}{5-2i}$; **2.** $\dfrac{4}{3-4i}$; **3.** $\dfrac{5+2i}{-i}$ **4.** Divide $4+3i$ by $-2+3i$;

Simplify: **5.** $\dfrac{2-6i}{4+3i}$; **6.** $\dfrac{i-2}{2-i}$; **7.** $(3-2i)(4+2i)(i^2)$; **8.** $(6-4i)(6+4i)$

Answers: 1. $\dfrac{4}{29}+\dfrac{19}{29}i$; **2.** $\dfrac{12}{25}+\dfrac{16}{25}i$; 3. $-2+5i$; **4.** $\dfrac{1}{13}-\dfrac{18}{13}i$; **5.** $-\dfrac{2}{5}-\dfrac{6}{5}i$; **6.** -1 ; **7.** $-16+2i$; **8.** 52

CHAPTER 16
Logarithms

Lesson 45 **Definitions, Basic Properties and Applications**
Lesson 46 **Evaluation of Logarithms; Common and Natural Logarithms**

Lesson 45
Definition, Basic Properties and Applications

Definition: The logarithm of a number to a base is the power (or exponent) to which the base must be raised in order to produce that number. A logarithm is therefore an exponent.

For any positive numbers x and b $(b \neq 1)$, and y any real number,

$$\boxed{\log_b x = y \text{ if and only if } b^y = x}$$

(Read " log of x to the base b is y if and only if b to the power y is x.
 or log to the base b of x is y if and only if b to the power y is x.)

Note: $\log_b x = y$ is equivalent to $b^y = x$.

Examples: **1.** $\log_{10} 100 = 2$ (because $10^2 = 100$)

            2. $\log_2 16 = 4$ (because $2^4 = 16$)

            3. $\log_{10} \dfrac{1}{100} = -2$ (because $10^{-2} = \dfrac{1}{10^2} = \dfrac{1}{100}$)

If you memorize and understand the definition of the logarithm very well, you will , in some evaluations be able to find the log mentally. For example, if asked to evaluate $\log_2 16$, ask yourself this question: To what power should 2 (the base) be raised in order to produce 16? ; or what is the exponent that must be placed on 2 in order to produce 16?.

Basic Properties of Logarithms

There are three basic rules which are properties of logarithms. Each rule is associated with a fundamental rule of exponents.

Property 1 For any positive numbers M, N, and b ($b \neq 1$),

$$\log_b (MN) = \log_b M + \log_b N$$

In words, the log of a product equals the sum of the logs of the factors.

Examples 1. $\log_{10} [3(5)] = \log_{10} 3 + \log_{10} 5$.

           2. $\log_{10} [423(785)] = \log_{10} 423 + \log_{10} 785$.

Property 2 $\log_b \dfrac{M}{N} = \log_b M - \log_b N$ ($b \neq 1$)

In words, the log of a fraction equals the log of the numerator minus the log of the denominator.

Example: $\log_5 \dfrac{42}{23} = \log_5 42 - \log_5 23$

Property 3   For any positive numbers M and b and any real number p.

$$\log_b M^p = p \log_b M \qquad (b \neq 1)$$

In words, the log of the pth power of a number equals p times the log of the number.

Example: $\log_2 8^3 = 3 \log_2 8$

Some other useful properties of logarithms

Property 4 $\qquad\qquad \log_b b = 1 \qquad$ (equivalently, $b^1 = b$)

Property 5 $\qquad\qquad \log_b 1 = 0 \qquad$ (equivalently, $b^0 = 1$)

Property 6 $\qquad\qquad b^{\log_b M} = M$

Property 7 $\qquad\quad$ If $M = N$ then

$\qquad\qquad\qquad \log_b M = \log_b N \qquad$ (We may call this step " taking logs of both sides of the equation")

Property 8 $\qquad\quad$ If $\log_b M = \log_b N$, then

$\qquad\qquad\qquad\qquad M = N \quad$ (Taking antilogs or " undoing the logs " of both sides of the equation")

A very useful property of equality of exponents when the bases are equal:

$$\boxed{\text{If } b^x = b^y \text{ then } x = y}$$

Applications of the above properties:

Express $\log C$ in terms of the logs of t, 5, and 7; and $\log S$ in terms of the logs of A, π, R and 5.

$$\textbf{1.} \quad C = \frac{t^2 \sqrt[3]{7}}{5}; \qquad\qquad \textbf{2.} \quad S = \sqrt{\frac{5A}{\pi R}}$$

Solution

1. $\log c = \log t^2 + \log (7)^{1/3} - \log 5$ (Applying properties **7, 1, & 2** and changing radical to exponential form)

$\qquad = 2 \log t + \frac{1}{3} \log 7 - \log 5 \qquad$ (Applying property **3**)

2. $\qquad S = \sqrt{\frac{5A}{\pi R}}$

$\qquad\qquad S = \left(\frac{5A}{\pi R}\right)^{1/2} \qquad\qquad\qquad$ (Changing the radical to exponential form)

$\qquad \log S = \log \left(\frac{5A}{\pi R}\right)^{1/2} \qquad\qquad$ (Taking logs)

$\qquad \log S = \frac{1}{2} \left(\log \frac{5A}{\pi R}\right) \qquad\qquad$ (Applying property **3**)

$\qquad\qquad = \frac{1}{2} \left(\log 5 + \log A - \log \pi - \log R\right) \qquad$ (Applying properties **1, & 2**)

Note that $\dfrac{\log_b M}{\log_b N}$ is **not** equal to $\log_b \dfrac{M}{N}$ $\qquad$ ($\log_b \dfrac{M}{N} = \log_b M - \log_b N$)

In the first expression you find the logs first before dividing; in the second expression, you divide first before finding the log.

Antilogarithms

Definition : If $y = \log_a x$, then the **antilogarithm** (antilog) of y to the base a is x.

The antilog of a given number is, therefore, the number whose logarithm to a given base is the given number. Note that " taking logs" and "taking antilogs" are inverse operations of each other. Each operation reverses the action of the other. Example: $\text{antilog}_2 5 = 2^5 = 32$, since $\log_2 32 = 5$.

Lesson 45 Exercises

Find the following:: **1.** $\log_{10} 100$; **2..** $\log_2 16$; **3.** $\log_{10} \dfrac{1}{100}$

4-5: Express $\log C$ in terms of the logs of t, 5, and 7; and $\log S$ in terms of the logs of A, π, R and 5.

4. $C = \dfrac{t^2 \sqrt[3]{7}}{5}$; **5.** $S = \sqrt{\dfrac{5A}{\pi R}}$

Answers: 1. 2; **2.** 4; **3.** -2; **4.** $\log c = 2\log t + \dfrac{1}{3}\log 7 - \log 5$;

5. $\log S = \dfrac{1}{2}(\log 5 + \log A - \log \pi - \log R)$

Lesson 46

Evaluation of Logarithms; Common and Natural Logs; Writing as a Single log term

Example 1 Evaluate $\log_2 16$.   (1)

Method 1

Step 1: Use the equivalent definition of logarithms to change (1) to exponential form.

Let $\log_2 16 = y$

Then $2^y = 16$

$2^y = 2^4$ $(16 = 2^4)$

Step 2: Since bases on both sides of the equation are equal, the exponents must be equal.

$\therefore y = 4$

and $\log_2 16 = 4$

Method 2

$\log_2 16 = \log_2 2^4$

$= 4 \log_2 2$ $(\log_2 2 = 1)$

$= 4(1)$

$= 4$

Again, we obtain the same result as by Method 1.

Method 3   Ask yourself this question and answer it : To what power should 2 (the base) be raised in order to produce 16? ; or what is the exponent that must be placed on 2 in order to produce 16?. Of course the answer is 4.

Example 2   Evaluate $\log_3 9^{1/4}$

Method 1

Step 1: Let $\log_3 9^{1/4} = y$.

Then, $3^y = 9^{1/4}$

Step 2: Express the right-hand side as a power of 3.

$3^y = (3^2)^{1/4}$ $(9 = 3^2)$

$3^y = 3^{2/4}$

$3^y = 3^{1/2}$

Step 3: Since the bases are equal, equate the exponents.

then, $y = \dfrac{1}{2}$

$\therefore \log_3 9^{1/4} = \dfrac{1}{2}$

Method 2

$\log_3 9^{1/4} = \log_3 (3^2)^{1/4}$

$= \log_3 3^{2/4}$

$= \log_3 3^{1/2}$

$= \dfrac{1}{2} \log_3 3$ $(\log_3 3 = 1)$

$= \dfrac{1}{2}$

Note: In evaluations and proofs involving logs, whenever you do not know how to proceed, it is good practice to begin by saying " let the given log = say, y, and then apply the equivalent exponential definition and continue.

Common Logarithms

Logarithms to the base 10 are called common logarithms. The base 10 is sometimes omitted. Thus, $\log_{10} x$ may be written as $\log x$. In changing a logarithm to an exponential forms, it helps to indicate the base 10. The common log of a number is the power to which 10 must be raised to produce that number. Thus, the common log of a power of 10 is the exponent on 10.

Examples **1.** $\log_{10} 100 = 2$ (because $10^2 = 100$)

2. $\log_{10} .01 = \log_{10} \dfrac{1}{100} = \log_{10} \dfrac{1}{10^2} = \log_{10} 10^{-2} = -2$ (because $10^{-2} = .01$)

3. $\log 2 = .3010$ (because $10^{.3010} = 2$)

More evaluations

Example 1: If $\log 2 = .3010$; $\log 3 = .4771$; $\log 7 = .8451$, evaluate the following:

(a) $\log \dfrac{9}{4}$;   (b) $\log 24$;   (c) $\log 7^{1/2}$.

Solution

Procedure: Express each log in terms of $\log 2$, $\log 3$, $\log 7$ or $\log 10$, and simplify.

(a) $\log \dfrac{9}{4} = \log 9 - \log 4$ $(\log \dfrac{M}{N} = \log M - \log N)$

$= \log 3^2 - \log 2^2$

$= 2\log 3 - 2\log 2$ $(\log M^p = p\log M)$

$= 2(.4771) - 2(.3010)$ (We are given that $\log 3 = .4771$; and $\log 2 = .3010$)

$= .9542 - .6020$

$= .3522$

(b) $\log 24 = \log 8(3)$

$= \log 8 + \log 3$

$= \log 2^3 + \log 3$

$= 3 \log 2 + \log 3$

$= 3(.3010) + .4771$

$= .9030 + .4771$

$= 1.3801$

(c) $\log 7^{1/2} = \dfrac{1}{2}\log 7$

$= \dfrac{1}{2}(.8451)$

$= .4226$

Example 2 Evaluate (a) $\log .00036$; (b) $\log 15$ using the given values in Example 1.
Solution

(a) $\log .00036 = \log (36 \times 10^{-5})$

$= \log (4 \times 9 \times 10^{-5})$

$= \log (2^2 \times 3^2 \times 10^{-5})$

$= \log 2^2 + \log 3^2 + \log 10^{-5}$

$= 2\log 2 + 2\log 3 - 5\log 10$

$= 2(.3010) + 2(.4771) - 5(1)$ $(\log 2 = .3010; \log 3 = .4771; \log 10 = \log_{10} 10 = 1)$

$= .6020 + .9542 - 5$

$= -3.4438$

(b) $\log 15 = \log(3 \times 5)$

$\qquad = \log(3 \times \dfrac{10}{2})$

$\qquad = \log 3 + \log \dfrac{10}{2}$ <-------------You may skip this step.

$\qquad = \log 3 + \log 10 - \log 2$

$\qquad = .4771 + 1 - .3010$ $\qquad\qquad\qquad\qquad$ $(\log 10 = \log_{10} 10 = 1)$

$\qquad = 1.1761$

Example 3 Given that $f(x) = \log_2 x$

$\qquad\qquad$ Find (a) $f(4)$; (b) $f(\dfrac{1}{16})$.

Solution

$\qquad\qquad$ (a) $f(4) = \log_2 4$ $\qquad\qquad\qquad$ $(x = 4)$

Method 1 $\qquad$ Let $y = \log_2 4$

$\qquad\qquad$ Then, $2^y = 4$

$\qquad\qquad\qquad$ $2^y = 2^2$ and

$\qquad\qquad\qquad\quad$ $y = 2$ $\qquad\qquad\qquad\qquad$ (Equating exponents)

Method 2 $\qquad$ $\log_2 4$

$\qquad\qquad = \log_2 2^2$

$\qquad\qquad = 2 \log_2 2$

$\qquad\qquad = 2(1)$ $\qquad\qquad\qquad\qquad\qquad$ $(\log_2 2 = 1)$

Method 3 $\qquad$ Ask yourself this question and answer it : To what power should 2 (the base) be raised in order to produce 4 ? ; or what is the exponent that must be placed on 2 in order to produce 4 ?. Of course, the answer is 2.

(b) $\qquad\qquad$ $f(\dfrac{1}{16}) = \log_2(\dfrac{1}{16})$

Method 1

$\qquad \log_2(\dfrac{1}{16}) = \log_2 1 - \log_2 16$ $\qquad\qquad$ $(\log \dfrac{M}{N} = \log M - \log N)$

$\qquad\qquad\qquad = 0 - \log_2 2^4$ $\qquad\qquad\qquad$ $(\log_2 1 = 0; \qquad \log_2 2 = 1)$

$\qquad\qquad\qquad = -4 \log_2 2$

$\qquad\qquad\qquad = -4(1)$

$\qquad\qquad\qquad = -4$

Method 2

$\qquad\qquad \log_2(\dfrac{1}{16}) = \log_2(\dfrac{1}{2^4})$

$\qquad\qquad\qquad\quad = \log_2 2^{-4}$

$\qquad\qquad\qquad\quad = -4 \log_2 2$ $\qquad$ $(\log M^p = p \log M)$

$\qquad\qquad\qquad\quad = -4(1)$

$\qquad\qquad\qquad\quad = -4$

Writing as a single log term

Example   Write as a single log term: $\log 4x + 2\log(x+1) - \log(x^2 - 1)$

$$\log 4x + 2\log(x+1) - \log(x^2 - 1)$$
$$= \log 4x + \log(x+1)^2 - \log(x^2 - 1)$$
$$= \log \frac{4x(x+1)^2}{x^2 - 1}$$
$$= \log \frac{4x(x+1)(x+1)}{(x+1)(x-1)} \qquad \text{(Applying } \log \frac{MN}{T} = \log M + \log N - \log T \text{ backwards)}$$
$$= \log \frac{4x(x+1)(\;\cancel{x+1}\;)}{\cancel{(x+1)}(\;x-1)}$$
$$= \log \frac{4x(x+1)}{x-1} \text{ or } \log \frac{4x^2 + 4x}{x-1}$$

Natural Logarithms

Logarithms to the base e (e $\approx$ 2.72) are called natural logarithms (or Napierian logarithms).
Example The natural log of x is written: $\ln x$ or $\log_e x$.
Conversions: $\ln x = \log_e x \approx 2.3 \log_{10} x$ **or** $\log x = \log_{10} x \approx .43 \log_e x$

Lesson 46 Exercises

A **1.** $\log_3 81$;   **2.** $\log_6 6$;   **3.** $\log_2 64$;   **4.** $\log_3 27^2$;   **5.** $\log_5 \frac{1}{5}$; **6.** $\log_2 32$;

7.. $\log_3 81$; **8.** $\log_3 243$; **9.** $\log_{10} 1000$; **10.** $\log_3 \frac{1}{27}$;   **11.** $\log 0.001$; **12.** $\log_{10} 1$

Answers: **1.** 4 ;   **2.** 1 ;   **3.** 6 ;   **4.** 6 ;   **5.** -1 ; **6.** 5 ;   **7.** 4 ; **8.** 5 ; **9.** 3 ; **10.** -3 ; **11.** -3 ; **12.** 0

B   If $\log 2 = .3010$; $\log 3 = .4771$; $\log 7 = .8451$, evaluate the following:

1. $\log 16$;   **2.** $\log 20$;   **3.** $\log \frac{21}{2}$;   **4.** $\log .0072$;   **5.** $\log 50$; **6.** $\log \sqrt{3}$

7. $\log 5^{.01}$; **8.** $\log \frac{1}{600}$;   **9.** $\log \frac{1}{2^4}$

Answers: **1.** 1.204 ;   **2.** 1.3010 ;   **3.** 1.0212 ;   **4.** - 2.1428 ;   **5.** 1.699 ;   **6.** 0.2386 ;
7. 0.007 ;   **8.** -2.7781 ;   **9.** -1.204

C   **a:** Given that $f(x) = \log_3 x$,   find **1.** $f(9)$;   **2.** $f(81)$; **3.** $f(\frac{1}{9})$;

b: If $f(x) = \log_4 x$,   find **1.** $f(256)$;   **2.** $f(\frac{1}{64})$

Answers: **a 1.** 2 ;   **2.** 4 ;   **3.** - 2; **b 1.** 4 ; **2.** -3

D Write as a single log term: **1.** $\log(5x + 10) - \log(x + 2) + \log x$;

2. $3\log(x^2 + 7x + 12) - 3\log(x + 3)$;   **3.** $\log_4 x - \log_4 y + \log_4 z$;   **4.** $2\log_4 5 + 3\log_4 2$;

5. $\log(x^2 - 25) - \log(x + 5)$;

Answers: **1.** $\log 5x$; **2.** $\log(x+4)^3$ or $3\log(x+4)$; **3.** $\log_4 \frac{xz}{y}$; **4.** $\log_4 200$; **5.** $\log(x-5)$

CHAPTER 17
Lesson 47: Exponential Equations
Lesson 48: Logarithmic Equations
Lesson 47
Exponential Equations

An **exponential equation** is an equation in which a variable occurs as an exponent.
Example 1   Solve the exponential equation:
$$15^x = 32$$
Step 1: Take logs of both sides of the equation.

$$\log 15^x = \log 32$$
$$x \log 15 = \log 32 \qquad (\log M^p = p \log M)$$
$$x = \frac{\log 32}{\log 15}$$
$$x = \frac{1.5051}{1.1761} \qquad \text{(From tables or calculator, } \log 32 = 1.5051; \log 15 = 1.1761)$$
$$x = 1.28$$

Example 2   Solve for x: $4^{x+1} = 16$

Method 1	**Method 2:**
$4^{x+1} = (2^2)^{x+1} = 2^4$	$4^{x+1} = 4^2$
$2^{2(x+1)} = 2^4$	$x + 1 = 2$ (Equating exponents,since bases are equal)
$2(x + 1) = 4$ (Equating exponents,since bases are equal)	$x = 1$
$2x + 2 = 4$	
$2x = 2$ and from which $x = 1$	**Method 3** Use the method in Example above.

Example 3   Solve for x:
$$7^{x^2 + x} = 34$$
Step 1: Take logs of both sides of the equation:

$$\log 7^{(x^2 + x)} = \log 34$$
$$(x^2 + x)\log 7 = \log 34$$
$$x^2 + x = \frac{\log 34}{\log 7}$$
$$x^2 + x = \frac{1.5315}{.8451} \qquad (\log 34 = 1.5315 \, ; \log 7 = .8451)$$
$$x^2 + x = 1.8122$$
$$x^2 + x - 1.8122 = 0$$

Step 2:   Solve for x by the quadratic formula.

$$x = \frac{-1 \pm \sqrt{1 + 7.2488}}{2} \qquad (a = 1, b = 1, c = -1.8122)$$
$$x = \frac{-1 \pm \sqrt{8.2488}}{2}$$
$$x = \frac{-1 \pm 2.8271}{2}$$
$$x = \frac{-1 + 2.8271}{2} \quad \text{or } x = \frac{-1 - 2.8271}{2}$$
$$x = -1.9360 \text{ or } .9360. \quad \text{The solution set is } \{-1.9360, .9360\}$$

Lesson 47 Exercises

Solve for x: **1.** $12^x = 42$; **2.** $3^{x+1} = 243$; **3.** $2^{x^2+1} = 1024$; **4.** $5^{x^2+x} = 42$

Answers: **1.** 1.5041 ; **2.** 4 ; **3.** Solution set is $\{-3, 3\}$; **4.** Solution set is $\{-2.1038, 1.1038\}$

Lesson 48
Logarithmic Equations

A logarithmic equation is an equation which contains the logarithm of an expression involving the variable. We must check the solutions of logarithmic equations, since the logarithmic function is not defined for negative numbers.

Example 1   Solve for x: $\log(x + 9) - \log x = 1$

Step 1: Write the logarithmic term as a single logarithmic term.

$$\text{Then we obtain } \log_{10}\left(\frac{x+9}{x}\right) = 1 \qquad\qquad (\log A - \log B = \log\frac{A}{B})$$

Step 2: $10^1 = \dfrac{x+9}{x}$ (Applying the equivalent exponential definition)

$$10 = \frac{x+9}{x}$$
$$10x = x + 9$$
$$x = 1 \qquad \text{(Solving for } x\text{)}$$

Checking: Substitute $x = 1$ in the original equation.

$$\log(1 + 9) - \log 1 \stackrel{?}{=} 1$$
$$\log 10 - \log 1 \stackrel{?}{=} 1$$
$$1 - 0 = 1 \qquad\qquad (\log_{10} 10 = 1)$$
$$1 = 1 \quad \text{True}$$

The solution is 1.

Example 2 Solve for x: $\log x + \log(x - 21) = 2$

Step 1: $\log x + \log(x - 21) = 2$

$\log[x(x - 21)] = 2$ (Writing as a single term)

$\log_{10}[x(x - 21)] = 2$ (Indicating the understood base so as to help transition to exponential form)

$10^2 = x(x - 21)$ (Applying the equivalent exponential definition)

$100 = x^2 - 21x$

Step 2:   Solve the quadratic equation by any method.

We solve by factoring, since the factors are easily recognizable.

$x^2 - 21x - 100 = 0$

$(x + 4)(x - 25) = 0$

$x = -4 \text{ or } x = 25$

Step 3: Since for $x = -4$, $\log(-4) + \log(-4 - 21)$ is not defined (The log of negative number is not defined) we reject the solution -4, and check for only 25.

Checking for $x = 25$: $\log 25 + \log(25 - 21) \stackrel{?}{=} 2$

$$\log 25 + \log 4 \stackrel{?}{=} 2$$
$$\log 25(4) \stackrel{?}{=} 2$$
$$\log_{10} 100 \stackrel{?}{=} 2 \quad (\log_{10} 100 = 2)$$
$$2 = 2 \quad \text{True} \quad (\text{LHS} = \text{RHS})$$

The solution is 25.

Example 3   Solve for x: $\log \dfrac{3}{x} - 2 = -\log 6x^2$ $(x \neq 0)$

Solution

Step 1: Collect the log terms on the left-hand side.

$$\log \dfrac{3}{x} + \log 6x^2 = 2$$

Step 2: Write the log terms as a single log term.

$$\log \left[\dfrac{3}{x} \cdot 6x^2\right] = 2$$

After checking (not shown) the solution is $\dfrac{50}{9}$.

Step 3: $\log \dfrac{18x^2}{x} = 2$

$\log_{10} 18x = 2$ (simplifying and indicating the base 10)

$10^2 = 18x$ (using the equivalent exponential definition)

$100 = 18x$

$\dfrac{100}{18} = x$, and $x = \dfrac{50}{9}$

Example 4   Solve for x: $\log_x 2 = \dfrac{1}{3}$

Step 1: Apply the equivalent exponential definition.

If $\log_x 2 = \dfrac{1}{3}$

then $x^{1/3} = 2$

Step 2: $(x^{1/3})^3 = (2)^3$

(Cubing both sides of the equation)

$x = 2^3$

$x = 8$

Checking :

Step 1: $\log_8 2 \overset{?}{=} \dfrac{1}{3}$

Let $\log_8 2 = k$

Then, $8^k = 2$

(Cubing $(2^3)^k = 2$

$2^{3k} = 2^1$

$3k = 1$

$k = \dfrac{1}{3}$

Step 2: $\log_8 2 = \dfrac{1}{3}$

$\dfrac{1}{3} = \dfrac{1}{3}$ True

$\therefore$ the solution is 8.

Equivalent check:

$8^{1/3} \overset{?}{=} 2$

$\sqrt[3]{8} \overset{?}{=} 2$

$2 = 2$ True

Example 5 Solve for x: $\log(x + 2) + \log(x - 5) = \log 8$

Step 1: $\log(x + 2) + \log(x - 5) = \log 8$

$\log(x + 2)(x - 5) = \log 8$ ($\log MN = \log M + \log N$)

$(x + 2)(x - 5) = 8$ (If $\log M = \log N$, then $M = N$, undoing the logs.)

$x^2 - 3x - 10 - 8 = 0$

Step 2: $x^2 - 3x - 18 = 0$ (Solve this quadratic equation)

$(x + 3)(x - 6) = 0$

$x + 3 = 0$ or $x - 6 = 0$

$x = -3$ or $x = 6$

Step 3: We check the values of $x = -3$ and $x = 6$ in the original logarithmic equation.

Checking for $x = -3$: $\log(-3 + 2) + \log(-3 - 5) \overset{?}{=} \log 8$, is not defined , since the logarithm of a negative number is not defined, and we reject the value -3 as a solution.

Checking for $x = 6$: $\log(6 + 2) + \log(6 - 5) \overset{?}{=} \log 8$

$\log 8 + \log 1 \overset{?}{=} \log 8$

$\log 8(1) \overset{?}{=} \log 8$ ($\log MN = \log M + \log N$)   **or** $\log 8 + 0 \overset{?}{=} \log 8$ ($\log 1 = 0$)

$\log 8 = \log 8$ True   (LHS = RHS) $\log 8 \overset{?}{=} \log 8$

Since $x = 6$ satisfies the original logarithmic equation, the solution is 6.

General discussion about solving exponential and logarithmic equations

1. Given the exponential equation, we (take logs) go through the logarithmic forms to simple algebraic equations (e.g., linear or quadratic equations).

Thus, **exponential** equation---->**logarithmic** equation (and then solve) or ---->**algebraic** equation and then solve. However, sometimes, it may not be necessary to use logarithms. Example: To solve for x, given $2^{x+3} = 8$, we may either do the following : $2^{x+3} = 2^3$; $x + 3 = 3$ (equating the exponents) and from which $x = 0$ or we may do the following: $(x + 3) \log 2 = \log 2^3$

$$x + 3 = \frac{3\log 2}{\log 2}$$

$$x + 3 = 3 \text{ and from which } x = 0.$$

2. Given the logarithmic equation, we change to the exponential equivalent form, and then to some basic algebraic equations.

Thus, **logarithmic** equation --->**exponential** equation (and then solve) or--->**algebraic** equation and then solve.

Changing from one base to another base (Change of base formula)

Example   Change $\log_b M$ to base c.

Step 1: Let $\log_b M = y$.

Then, $M = b^y$ (Equivalent exponential definition)

Step 2: Now, take logs of both sides of the equation and introduce the new base, c.

Then, $\log_c M = \log_c b^y$

Step 3: Solve for y:

$$\log_c M = y \log_c b$$

$$\frac{\log_c M}{\log_c b} = y \qquad \text{(Solving for } y\text{)}$$

Therefore, $\log_b M = \dfrac{\log_c M}{\log_c b}$ <-----This is known as the change of base formula.

(Here we changed from base b to base c.)

Lesson 48 Exercises

A Solve for x:

1. $\log(x + 4) - \log x = 1$;   2. $\log_2 x + \log_2(x - 7) = 3$;   3. $\log(x + 26) = \log(x - 2) + \log 5$

Answers: 1. $\{\frac{4}{9}\}$;   2. $\{8\}$;   3. 9

B Solve for x: 1. $7^x = 3$;   2. $2^x = 7$;   3. $5^{x-2} = 16$;   4. $\log x = 4$;

5. $\log(3x + 4) = 2$;   6. $\log_2(x - 7) + \log_2 x = 3$;   7. $\log_3(x - 2) = 4$;

8. Solve for t: $4^{2t} = \dfrac{9}{4}$

Answers: **1.** 0.5645 ;   **2.** 2.807 ;   **3.** 3.722 ;   **4.** 10,000 ;   **5.** 32 ; **6.** 8 ; **7.** 83 ;   **8.** 0.2925

CHAPTER 18
Basic Review For Geometry

Lesson 49: **Points, Lines, Angles, Congruency, Complimentary and Supplementary Angles; Parallel and Perpendicular Lines**

Lesson 50: **Statements in Geometry, Axioms, Postulates, Parallel Lines and Transversal; Corresponding Angles, Triangles, Sum of the Measures of the angles of a triangle**

Lesson 49

Points, Lines, Angles, Congruency, Complimentary and Supplementary Angles; Parallel and Perpendicular Lines

The undefined terms in geometry are a point, a line, and a plane.
By using these terms we can define other geometric terms.

Point: We use a point to indicate a position. On a piece of paper, we represent a point by a dot. A point has no size. (The "thinner" the dot, the better it represents a point; the tip of a well-sharpened pencil will be an acceptable size). We name a point by a capital letter written near the point (See the points P, A and B below).

P.   •   B
       A   •

Figure 1

Line: We may consider a line as an infinite set of points. A line may be a straight line or a curved line. A straight line may be extended indefinitely in both directions.
We may name a line by using a single lower case letter or two capital letters representing two points on the line, with a double-headed arrow over the capital letters (see Figure 2 below)

The line passing through the points A and B may be named as $\overleftrightarrow{AB}$ or ℓ

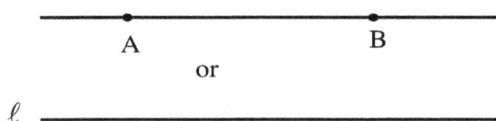

A •——————————• B

or

ℓ ——————————

Figure 2

Line segment: A line segment or segment is a part of a line between two points on the line, including the two points (end points). We denote a line segment by placing a bar over the two capital letters which represent the two end points (see Figure 3 below).

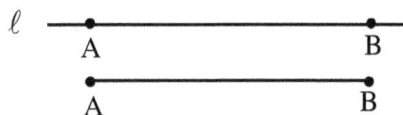

ℓ •——————————•
  A   B
  •——————————•
  A   B

•——————————•
C   D

Line segment AB is represented by $\overline{AB}$ Line segment CD is represented by $\overline{CD}$

Figure 3

Ray: A ray is a part of a line to one side of a point on the line. This point is the endpoint of the ray. The ray begins at the end point and extends indefinitely in one direction. The arrow drawn begins at the end point . We name a ray by using two capital letters: the first letter represents the end point and the other letter represents any other point on the ray (Fig.4).

P   Q

B   A

ray PQ is represented by $\overrightarrow{PQ}$ ray AB is represented by $\overrightarrow{AB}$

Figure 4

Measurement of a line segment : The **length** of a line segment is the distance between its end points.

The length of line segment $\overline{AB}$ is denoted by AB (i.e. without a bar over AB)

A   B

Figure 5

Angle: An angle (symbol: ∠) is the union of two rays having the same end point, the endpoint being called the vertex of the angle. An angle is formed when two line segments (or lines) meet at a common point (Figure 6).

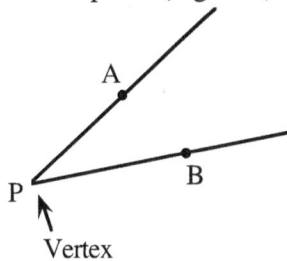

A

B

P

Vertex

Figure 6

Naming of angles We may name an angle by using capital letters or lowercase letters as follows:

1. **If** the angle is the only angle at the vertex, then we can use either a single capital letter or three capital letters. However, if there are two or more angles at the vertex, then we must use three capital letters. In using three capital letters, the middle letter is the letter at the vertex.

2. We may use a lower case letter , a Greek letter, or a number .The letter or the number is placed inside the angle (as shown in Figure 7).

Examples: ∠ 2 and ∠ DAB denote the same angle
∠ 1 and ∠ CAD denote the same angle.

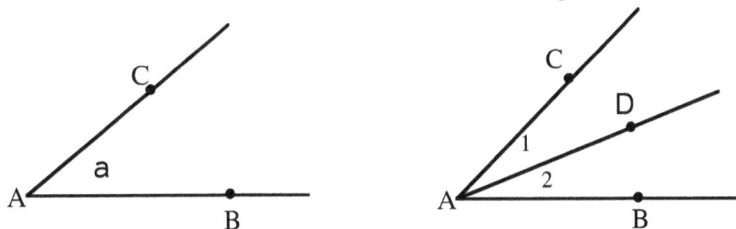

C

a

A

B

C

D

1

2

A

B

Figure 7

Congruent Figures: Congruent figures have the same shape and size.
Congruent figures can be made to coincide.
The symbol for "congruent" or "congruent to" is ≅ .

Congruency of Line Segments: As shown in Figure 8, if we can place $\overline{AB}$ and $\overline{CD}$ so that
they coincide, then, we say that $\overline{AB}$ and $\overline{CD}$ are congruent,
and we write $\overline{AB}$ ≅ $\overline{CD}$ (read: line segment $\overline{AB}$ is
congruent to line segment $\overline{CD}$)

Congruent line segments have the same length.

$$\overline{AB} \cong \overline{CD} \text{ if and only if } AB = CD$$

(that is, $\overline{AB}$ is congruent to $\overline{CD}$ if the length of $\overline{AB}$ is equal to the length of $\overline{CD}$, and conversely, the length of $\overline{AB}$ is equal to the length of $\overline{CD}$, if $\overline{AB}$ is congruent to $\overline{CD}$).

Indication of congruency of line segments: We may use the same number of slashes or strokes (as shown in Figure 8) to indicate that two line segments are congruent.

Example (a) $\overline{AB} \cong \overline{CD}$ (b) $\overline{EF} \cong \overline{GH}$

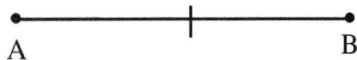

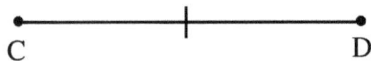

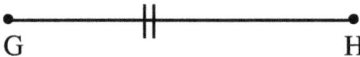

Figure 8

Congruency of angles: As shown in Figure 9, if ∠ P can be placed on ∠ Q so that the angles coincide, then we say that ∠ P is congruent to ∠ Q and we write
$$\angle P \cong \angle Q$$

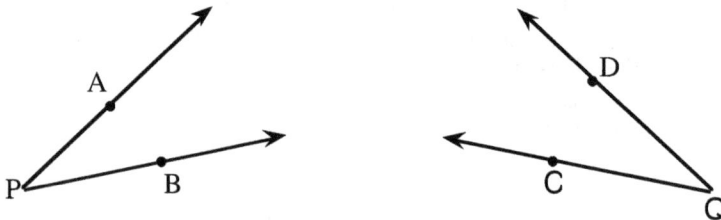

Figure 9

Note that congruent angles have equal measures.

Indication of congruency of angles: We can use arcs or a combination of arcs and slashes to indicate congruency (as shown in Figure 10)

Examples $\angle 1 \cong \angle 2$ (the angle marked 1 is congruent to the angle marked 2)
$\angle 3 \cong \angle 4$ (the angle marked 3 is congruent to the angle marked 4)
$\angle 5 \cong \angle 6$ (the angle marked 5 is congruent to angle marked 6)

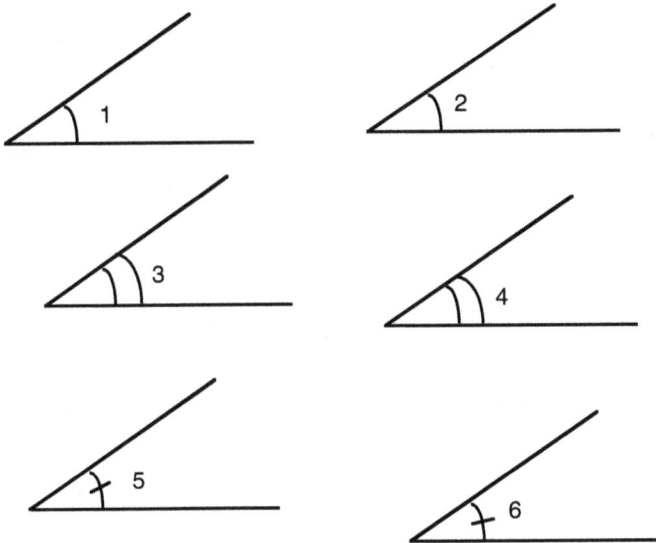

Figure 10

Units of Measuring angles : The main unit of measuring angles in geometry is the degree. There are smaller units of the degree, namely the minute, and the second.

The degree and the smaller units are as follows: 1 degree = 60 minutes
1 minute = 60 seconds

Classification of angles by the measures of the angles

Straight angle: An angle whose measure is 180°

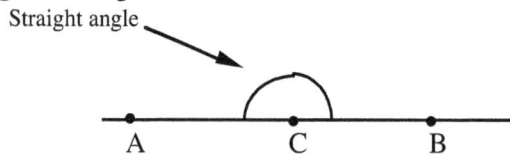

Right angle: A right angle is an angle whose measure is 90°.
In Figure 11, below, ∠ 5 is a right angle.

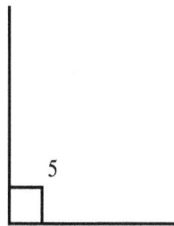

Figure 11

Acute angle: An angle whose measure is between 0° and 90°.
In Figure 12, below, ∠ 6 is an acute angle.

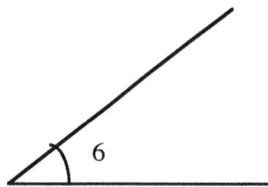

Figure 12

Obtuse angle: An angle whose measure is between 90° and 180°.
In Figure 13, below, ∠ 8 is an obtuse angle.

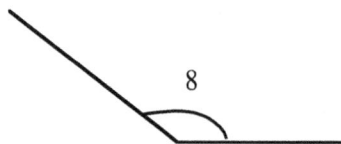

Figure 13

Reflex angle: An angle whose measure is between 180° and 360°.
In Figure 14, below, ∠ 7 is reflex angle.

Figure 14

Pairs of Angles

Complementary angles : If the sum of the measures of two angles is 90°, then the two angles are complementary.
The complement of a 40° angle is a 50° angle
In Figure 15, below, ∠ 1 and ∠ 2 are complementary angles.(i.e., m ∠ 1 + m ∠ 2 = 90°)

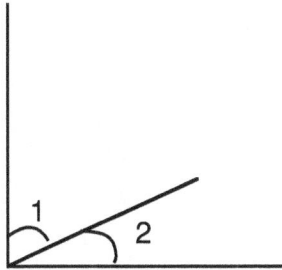

Figure 15

Supplementary angles: If the sum of the measures of two angles is 180°, then the two angles are supplementary.
The supplement of a 70° angle is a 110° angle.

In Figure 16, ∠ 3 and ∠ 4 are supplementary angles. (i.e., m ∠ 3 + m ∠ 4 = 180°)

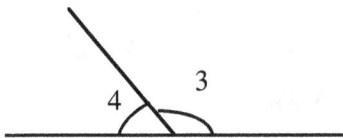

Figure 16

Example 1 In the figure below , find x

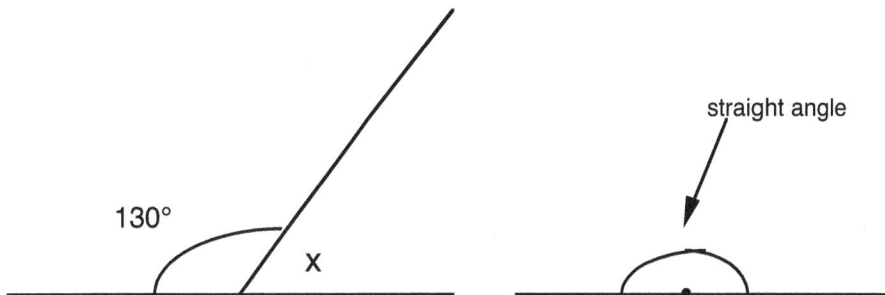

Solution: $x + 130° = 180°$ Reason : A straight angle has a measure of 180°
$x = 50°$

Vertical angles: As shown in Fig.17, below, when two lines intersect (cross each other) a pair of opposite angles such as ∠ 1 and ∠ 3 are called vertical angles. Similarly, ∠ 2 and ∠ 4 are vertical angles.

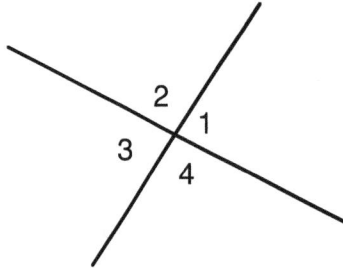

Figure 17

Theorem: When two lines intersect, the vertical angles are congruent.
In Figure 17, ∠ 1 and ∠ 3 are congruent.
∠ 2 and ∠ 4 are congruent.

Adjacent angles: Two angles which are in the same plane and which have a common vertex and a common side . In Figure 17, ∠ 1 and ∠ 2 are adjacent angles .
∠ 2 and ∠ 3 are adjacent angles; ∠ 3 and ∠ 4 are adjacent angles.

Perpendicular lines (symbol for " perpendicular to": ⊥): Perpendicular lines are two lines which meet at right angles.
In Figure 18,. line ℓ_1 is perpendicular to line ℓ_2 and we write
$$\ell_1 \perp \ell_2$$

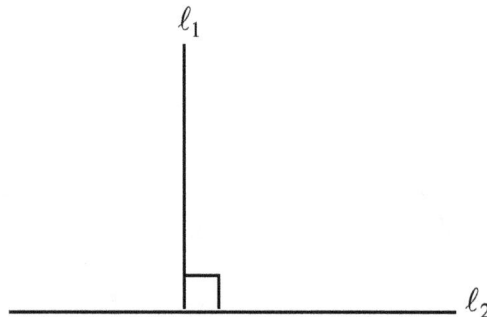

Figure 18

Example 2   In the figure below, find x

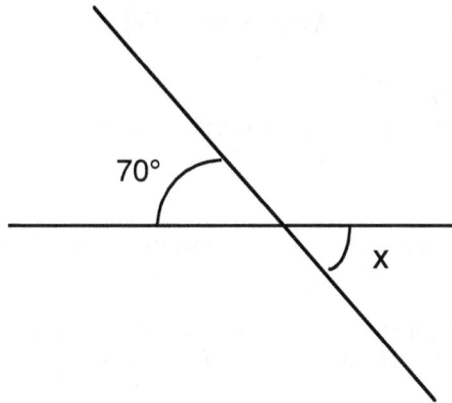

Solution: $x = 70^\circ$ **Reason:** Vertically opposite angles are congruent,
and therefore have equal measures

Lesson 49 Exercises

1. In figure (a) below , find x **2.** In figure (b) below , find x

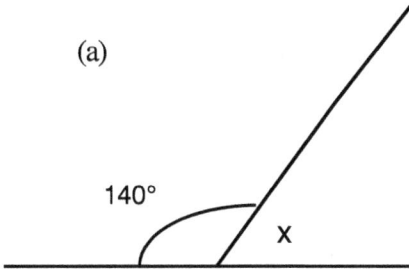

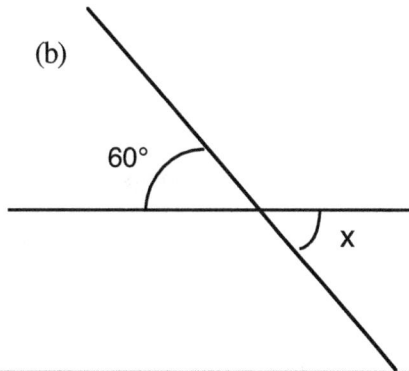

Answers: (a) $x = 40^\circ$; (b) $x = 60^\circ$

Lesson 50

Statements in Geometry, Axioms, Parallel Lines and Transversals; Corresponding Angles, Triangles, Angles of a Triangle

Axioms and postulates (assumptions)

Axioms and postulates are statements which we accept as true, without any proof, in order to deduce other statements.

Axioms are applicable to mathematics in general.
 Example: The statement "If equals are added to equals, the sums are equal " is applicable to both algebra and geometry.

Postulates (in mathematics) are applicable to specific branches of mathematics.
Example: The statement "One and only one straight line can be drawn between any two points" is applicable only in geometry.

Theorems : A theorem is a statement that is proved by deduction.

Some Geometric Postulates

In geometric proofs and in geometric calculations, the following are some of the things that we can do if the need arises.
 (Try to avoid applying more than one postulate in one step, except when one of the postulates is the extension of a straight line; i.e., avoid applying two or more postulates in one step simultaneously, except when a line extension is one of the postulates.)

1. A geometric figure may be moved without changing its size or shape. (motion postulate)

2. Between any two points, one and only one straight line can be drawn.
      (Two points determine a straight line)

3. A straight line segment may be extended indefinitely in both directions.

4. Two straight lines may intersect in one and only one point.

5. The shortest distance between two points is the straight line segment connecting these points.

6. A straight line segment has one and only one midpoint.

7. An angle has one and only one bisector. (We may bisect an angle if the need arises)

8. At a given point on a line, one and only one perpendicular can be drawn to the line.

9. From a given point not on a given line, one and only one perpendicular can be drawn to the line.

10. One and only one circle can be drawn with a given point as center and a given line segment as a radius.

11. Through a given point not on a line, one and only one parallel line can be drawn parallel to the given line.

Transversal

If two lines are cut by a third line, the third line is called a transversal. Angles are formed at where the transversal intersects these lines. In Figure 19, angles 1 and 5 are called corresponding angles; angles 4 and 8 are corresponding angles; angles 2 and 6 are corresponding angles; angles 1, 2, 7, and 8 are called exterior angles; angles 3, 4, 5, and 6 are called interior angles.

Corresponding Angles (F-angles)

In Figure 19, below:
1. $\angle 1$ and $\angle 5$ are corresponding angles.

    (these angles are **above** lines ℓ_1 and ℓ_2 **and to (or on) the right** of the transversal).
2. $\angle 4$ and $\angle 8$ are corresponding angles.

    (these angles are **below** lines ℓ_1 and ℓ_2 **and to (or on) the right** of the transversal).
3. $\angle 2$ and $\angle 6$ are corresponding angles.

    (these angles are **above** ℓ_1 and ℓ_2 **and to (or on) the left** of the transversal).
4. $\angle 3$ and $\angle 7$ are corresponding angles.

    (These angles are **below** lines ℓ_1 and ℓ_2 **and to the left** of the transversal).

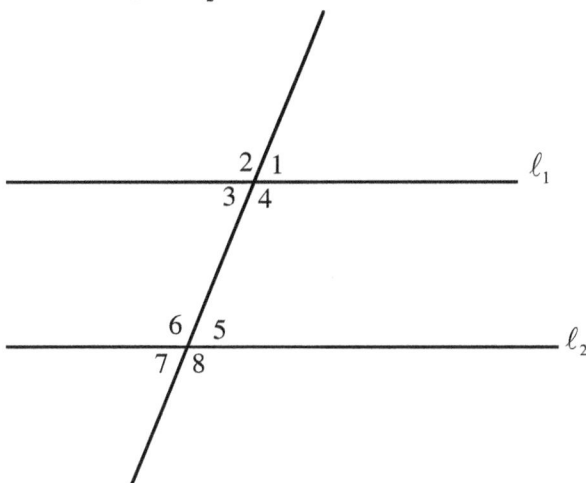

Figure 19

Even though the words " above", "below", "to (on) the right", "to (on) the left" were used to identify corresponding angles, if a given figure does not have the orientation of Figure 19, we can rotate the page to obtain the above orientation and then apply the above descriptions to identify the corresponding angles accordingly. Note that correspondence, as described above, requires that for any two angles to correspond, each angle must satisfy the same two conditions simultaneously; for example, the simultaneous conditions for $\angle 1$ and $\angle 5$ are "above the two lines" and to the right (or on the right) of the transversal."

Alternate interior angles (Z- or N-angles)

$\angle 4$ and $\angle 6$ are alternate interior angles (the lines ℓ_1 and ℓ_2 together with the transversal which form the angles are in the shape of the letter Z or N). $\angle 3$ and $\angle 5$ are alternate interior angles.

Note above that, even though there are corresponding angles and alternate interior angles, we **cannot** say that any of these angles are congruent, since we do not know if the two lines are parallel. In the next section ,we will be able say which angles are congruent, since the two lines are parallel.

Parallel Lines: Two lines l_1 and l_2 in the same plane are parallel if they never meet no matter how far they may be extended. If line l_1 is parallel to l_2, we write $l_1 \; || \; l_2$ (see Fig. 20)

Parallel lines, the transversal, and the pairs of angles formed

In Figure 20, below, line l_1 is parallel to line l_2 .

Theorem: If two parallel lines are cut by a transversal , then:

    1. the corresponding angles (F-angles) are congruent (see Figure 20).

        $\angle 1 \cong \angle 5$
        $\angle 4 \cong \angle 8$
        $\angle 2 \cong \angle 6$
        $\angle 3 \cong \angle 7$

    2. the alternate interior angles (Z- or N-angles) are congruent (see Figure 20).

        $\angle 4 \cong \angle 6$
        $\angle 3 \cong \angle 5$

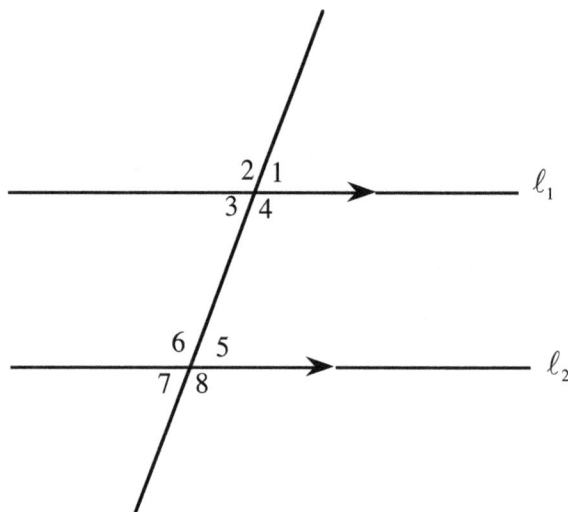

Figure 20

Note also that the vertical angles are congruent (whether or not there are parallel lines).

Theorem: If two lines cut by a transversal have congruent alternate interior angles, then the lines are parallel. (We can use this theorem to prove that two lines are parallel).

Theorem: Any pair of interior angles on the same side of the transversal, as above, are supplementary.

    $m \angle 4 + m\angle 5 = 180°$ (Figure 20)
    $m\angle 3 + m \angle 6 = 180°$

Theorem: The exterior angles on the same side of the transversal, as above, are supplementary.

    $m\angle 1 + m \angle 8 = 180°$ (Figure 20)
    $m \angle 2 + m \angle 7 = 180°$

Example 3 In the figure below, find x given that line ℓ_1 is parallel to line ℓ_2

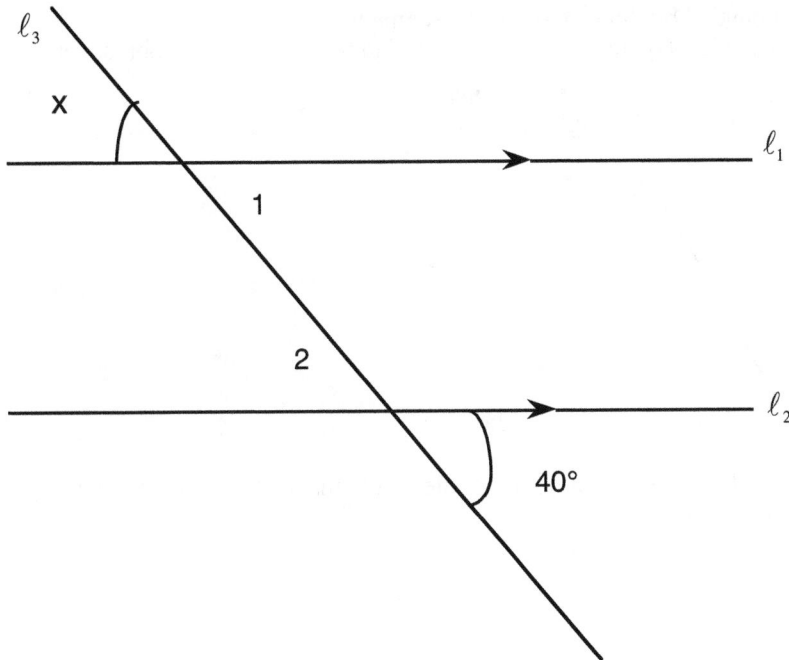

Solution:

$m \angle 2 = 40°$ (Vertical angles are congruent)

$m \angle 2 = x$ (corresponding angles are congruent;

parallel lines ℓ_1, ℓ_2, cut by transversal ℓ_3)

$x = 40°$.

Triangles

A triangle is a closed figure bounded by three straight line segments.
Triangles may be classified by angles. By this classification, a triangle may be acute, obtuse, or right.

Acute triangle: A triangle in which all the angles are acute.

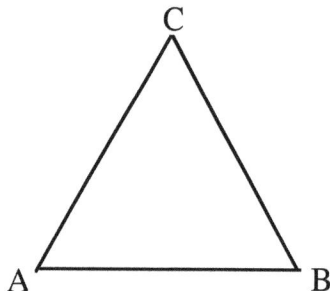

Obtuse triangle: A triangle in which there is an obtuse angle. **Note** that a triangle cannot have more than one obtuse angle.

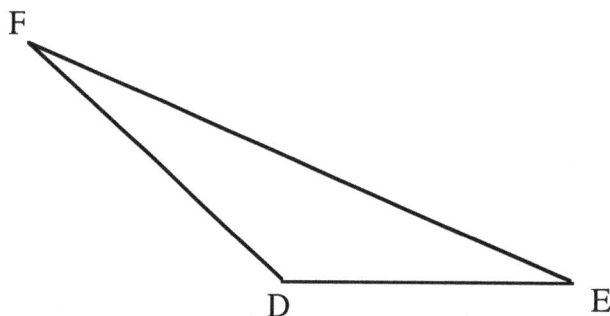

Right triangle (or right-angled triangle): A triangle which has one right angle (90°). **Note** that a triangle cannot have more than one right angle.

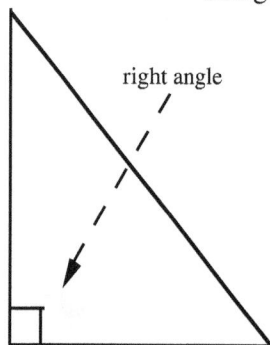

right angle

We can also classify triangles by sides. By this classification, a triangle may be equilateral, isosceles, or scalene.

Equilateral triangle: A triangle in which all three sides are congruent.
(The lengths of all three sides are equal)

All the interior angles of an equilateral triangle are congruent.
(Each angle has a measure of 60°).

Isosceles triangle: A triangle in which two sides are congruent. In an isosceles triangle, the angles opposite to the congruent sides are congruent.

Useful relationships concerning the relative measures of the sides and angles of a triangle

1. In any triangle, the **sum** of the **lengths** of any **two sides** is larger than the length of the **third** side.

2. In any triangle, the **longest side** is opposite to the **largest angle**, and the **shortest side** is opposite to the **smallest angle**.

3. In an **acute triangle**, the square of the length of **longest side is smaller than** the sum of the squares of the lengths of the other two sides.

4. In an **obtuse triangle**, the **square** of the length of the **longest side is larger than** the sum of the **squares** of the lengths of the other two sides.

5. In a **right triangle**, the **square** of the length of the **longest side** (the hypotenuse) **is equal** to the sum of the **squares** of the lengths of the other two sides. (**The Pythagorean theorem**)

Theorem: The sum of the measures of the interior angles of a triangle is 180°.
In the figure below, m ∠ A + m ∠ B + m ∠C = 180°

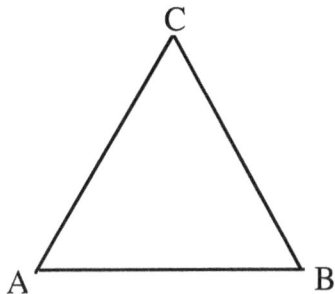

Theorem: If one side of a triangle is extended, the measure of the exterior angle so formed equals the sum of the measures of its remote interior angles.
In the figure below, m ∠ A + m ∠ C = m ∠ CBE
(∠ A and ∠ C are the remote interior angles. ∠ CBE is the exterior angle)

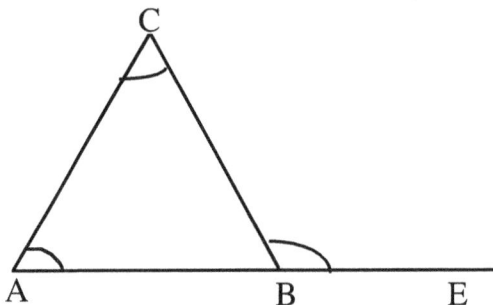

Example 1   In the figure below, find x.

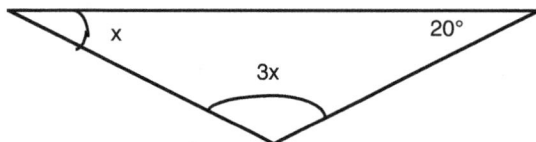

Solution: $x + 3x + 20° = 180°$ (The sum of the measures of the angles of a
                    $4x + 20° = 180°$ triangle is 180°)
                        $4x = 160°$
                         $x = 40°$

Example 2   In the right triangle shown below , find x

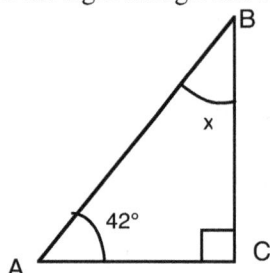

Solution: $m \angle C = 90°$ (A right angle has a measure of 90°)

        $m \angle A + m \angle C + m \angle B = 180°$ (The sum of the measures of the angles of a triangle is 180°)
                    $42°+ 90° + x = 180°$
                        $132° + x = 180°$
                            $x = 180° - 132°$
                            $x = 48°$

Example 3   Find x and y in the figure below.

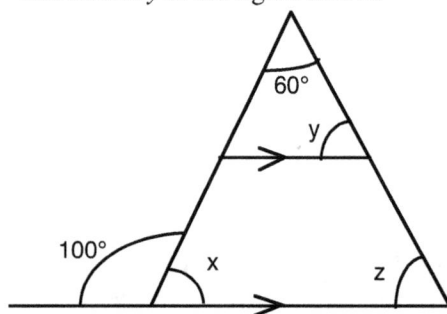

(a) $x + 100° = 180°$ (Supplementary angles)
            $\underline{- 100°   - 100°}$
                $x = 80°$

 (b) Step 1: $x + 60° + z = 180°$
                $80° + 60° + z = 180°$ (The sum of the measures of the angles of a triangle is 180°)
                    $140° + z = 180°$
                $\underline{- 140°   - 140°}$
                        $z = 40°$
Step 2: $y = z = 40°$ (Corresponding angles are congruent.)
            $y = 40$

Lesson 50 Exercises

1. Find x;   **2.** Find m $\angle$ A

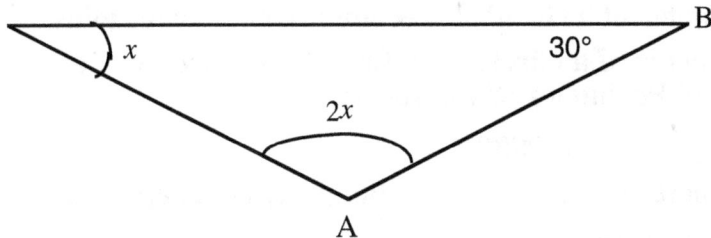

Answers: **1.** $x = 50$;   **2.** m $\angle$ A $= 100$

CHAPTER 19
Areas and Perimeters

Lesson 51: **Area and Perimeter of Triangles, Rectangles, and Trapezoids**

Lesson 52: **Area and Perimeter of a Circle; Arc Length and Sector of a Circle; Area and Perimeter of Composite Figures**

Lesson 51

Area and Perimeter of Triangles, Rectangles, and Trapezoids

Example 1 Find the area of figure (triangle) below.

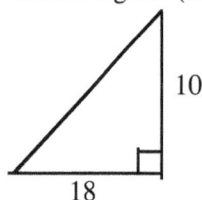

Solution : The figure is a right triangle with base 18 units and altitude 10 units.

$$\text{The area of a triangle} = \frac{1}{2} \text{ base} \times \text{altitude}$$

$$= \frac{1}{2}(18)(10)$$

$$= 90 \text{ square units}$$

Note: The altitude and the base always meet at right angles. In the above problem, it does not matter which of 18 or 10 we call the the base. We could have taken the base to be 10 and then, the altitude would have been 18. The altitude of the triangle is also known as the height of the triangle.

Example 2 In the figure below:
(a) Find the area of the figure.
(b) Find the perimeter of figure.

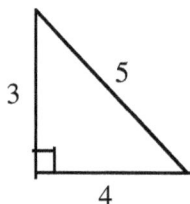

(a) $b = 4, h = 3$

$$\text{Area of the triangle} = \frac{1}{2} bh$$

$$= \frac{1}{2}(4)(3)$$

$$= 6 \text{ sq.units}$$

(b) Perimeter of the triangle $= 3 + 4 + 5$
$$= 12 \text{ units}$$

Example 3   In the figure below:
(a) Find the area of the figure.
(b) Find the perimeter of figure.

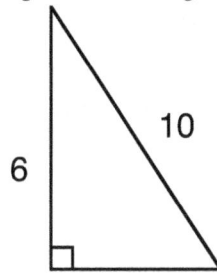

Solution

(a)

Let the base be b, and the altitude h.

Area of the triangle $= \frac{1}{2} bh$.

$h = 6$, but we are **not** given b; however

Step 1: We can find b by using the Pythagorean theorem since the triangle is a right triangle.

$$10^2 = 6^2 + b^2$$
$$100 = 36 + b^2$$
$$64 = b^2$$
$$8 = b$$
$$b = 8$$

Step 2: Now, $b = 8, h = 6$

Area of the triangle $= \frac{1}{2} bh$

$$= \frac{1}{2}(8)(6)$$
$$= 24 \text{ sq. units}$$

The area of the triangle is 24 sq. units

(b)   From figure below,
Perimeter $= 6 + 8 + 10$
$= 24$ units

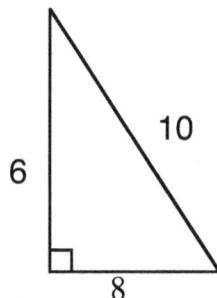

Note: Obtaining the same numerical answers in **(a)** and **(b)** is a coincidence; and therefore, the perimeter is **not** always numerically equal to the area.

Area and Perimeter of a Trapezoid

Example 4   Find the area of the figure shown below.

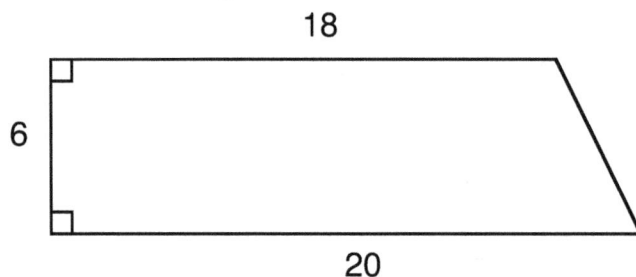

Method 1: The figure is a trapezoid.
The formula for the area, A, of a trapezoid is given
by $A = \frac{1}{2}(b_1 + b_2)h$, where b_1 and b_2 are the bases
(the bases are the parallel sides) and h is the altitude.
$b_1 = 20, b_2 = 18, h = 6$

$$A = \frac{1}{2}(20 + 18)(6)$$
$$= \frac{1}{2}(38)(6)$$
$$= 19(6)$$
$$= 114 \text{ square units}$$

The area of the figure is 114 square units.

Method 2: We break up the figure into a rectangle and a triangle.

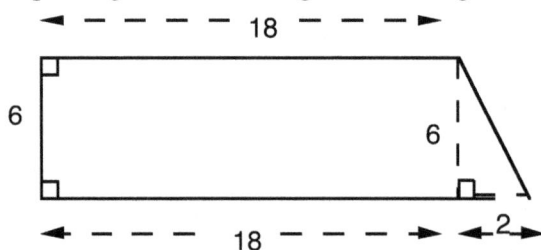

The area of the rectangle $= 6 \times 18$
$= 108$ sq. units

The area of the triangle $= \frac{1}{2}(2)(6)$
$= 6$ sq. units

Area of the figure = Area of the rectangle + Area of the triangle
$= (108 + 6)$ sq. units
$= 114$ sq. units

Again, we obtain the same answer as in Method 1.

Example 5 Find the area of the figure below.

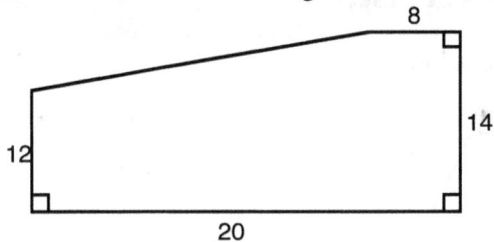

Solution   We break up the figure into a rectangle and a trapezoid.

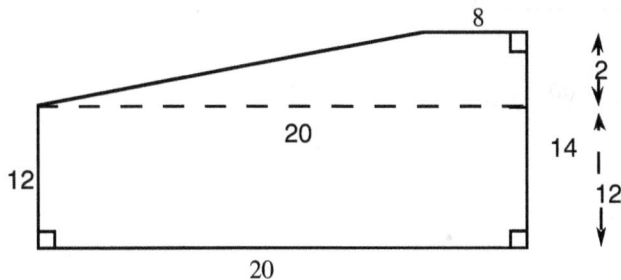

The area of the rectangle = 12 (20)

$$= 240 \text{ sq. units}$$

The area of the trapezoid $= \dfrac{1}{2}(b_1 + b_2)h$

$$= \dfrac{1}{2}(20 + 8)(2) \qquad\qquad b_1 = 20 \ , b_2 = 8, \ h = 2$$

$$= \dfrac{1}{2}(28)(2)$$

$$= 28 \text{ sq. units}$$

Total area = area of the rectangle + area of the trapezoid

$$= (240 + 28) \text{ sq. units}$$

$$= 268 \text{ sq. units}$$

Note in the above problem that the original figure given is **not** a trapezoid,
since it has five sides. By definition, a trapezoid is a quadrilateral (a four-sided figure) which has exactly a
pair of parallel sides.

Note also that in the above problems, given a complex figure for which we do not know any formulas, we
can break up the figure into simpler figures (such as a rectangle, a triangle, or a trapezoid), whose formulas we
know, and then apply the appropriate formulas accordingly to find the areas, which we can then add together
. Sometimes, we can extend the sides of the figures to form "known" figures and then, in this case, subtract
the extensions of the areas.

Lesson 51 Exercises

A Find (a) the area and (b) the perimeter of the figure below.

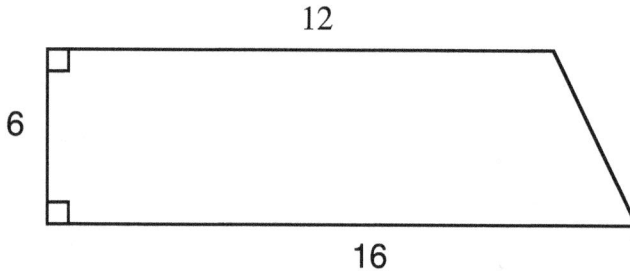

Answers: (a) 84 sq. units; (b) $(34 + 2\sqrt{13})$ units

B

1. Find (a) Area; (b) Perimeter of right triangle ABC.
2. Find (a) Area; (b) Perimeter of right triangle *DEF*.

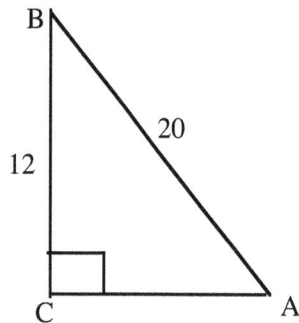

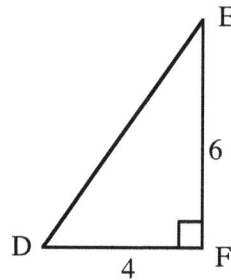

Answers: Triangle ABC:(a). 96 sq. units; (b). 48 units;   Triangle *DEF*: (a) 12 sq. units; (b) $10 + 2\sqrt{13}$

Lesson 52

Area and Perimeter of a Circle; Arc Length and Sector of a Circle; Area and Perimeter of Composite Figures

Area and Perimeter of a Circle

Area of a circle

The area, A , of a circle of radius , r, is given by $A = \pi r^2$.

π is approximately equal to 3.1416; in calculations, we will leave our answers in terms of π, unless instructed otherwise.

Note also that the radius of a circle $= \frac{1}{2}$ of the diameter of the circle.

Area of a semicircle (half-circle) $= \frac{1}{2}$ of the area of the given circle.

Area of a quarter-circle $= \frac{1}{4}$ of the area of the given circle.

Perimeter (or Circumference) of a Circle

The perimeter of a circle is the distance around the circle (i.e., the length of the circle).
The perimeter, P, of a circle of radius, r, is given by $P = 2\pi r$.

The perimeter of a semicircle $= \frac{1}{2}$ of the perimeter of the given circle. However, if there is a diameter

joining the end points of the curved part, then add the diameter. Therefore, the shape of the figure determines whether or not to add the diameter. The circle formula is for the curved part only.

The perimeter of a quarter circle is $\frac{1}{4}$ of the perimeter of the given circle plus $2r$ if two radii

connect the end points of the curved part.

Example Find the area and the perimeter of a circle of radius 6 units.

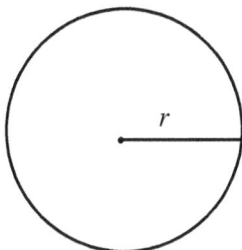

Finding the area

$r = 6$, Area of a circle $= \pi r^2$.
Area of the given circle $= \pi(6)^2 = \pi(36) = 36\pi$ sq. units

Finding the perimeter

$r = 6$ units
Perimeter of a circle $= 2\pi r$
Perimeter of the given circle $= 2\pi(6) = 12\pi$ units

Note: In the above problem, the area of a **semicircle** of radius, $r = 6$ is $\frac{1}{2}(36\pi) = 18\pi$ sq. units;

and the perimeter of the semicircle (curved part only) of radius, $r = 6$, is $\frac{1}{2}(12\pi) = 6\pi$ units.

If the diameter connects the end points of the curved part, then add the diameter, $2(6)$ to obtain $(6\pi + 12)$ units.

Arc Length and Sector of a Circle

Arc length: Arc length is the length of a part of a circle between any two points on the circle. In a circle of radius r, the arc length s cut off by a central angle θ (θ in degrees) is given by

$s = \dfrac{\theta}{360}$ of the circumference of the circle.

$$\boxed{s = \dfrac{\theta}{360}(2\pi r) \text{ or } \dfrac{\theta \pi r}{180}}.$$ (From the proportion, s is to θ as $2\pi r$ is to 360)

However, if the central angle is in radians, replace 360 by 2π and simplify to obtain

$$(360° = 2\pi \text{ radians, or } 180° = \pi \text{ radians})$$

$$\boxed{s = r\theta}$$

Sector of a circle: A sector of a circle is the part of the interior of the circle bounded by two radii and the intercepted arc.

In a circle of radius r, the area A of a sector with central angle θ (θ in degrees) is given by

$A = \dfrac{\theta}{360}$ of the area of the circle.

$$\boxed{A = \dfrac{\theta}{360}(\pi r^2)}$$ (From the proportion, A is to θ as πr^2 is to 360)

However, if θ is in radians, replace 360 by 2π and simplify to obtain

$$\boxed{A = \tfrac{1}{2}r^2\theta}$$

Example In the circle shown below, if the diameter is 6 and the central angle is 30°,
(a) Find the circumference of the circle.
(b) Find the arc length of CB
(c) Find the area of sector COB.

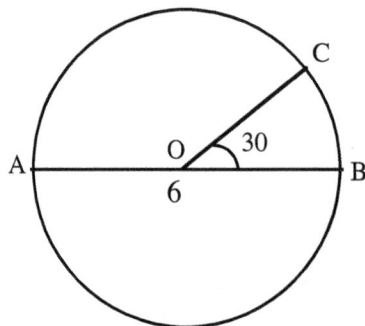

Figure 1

Solution

(a) Circumference of a circle of radius r is given by $2\pi r$.

Radius of a circle $= \dfrac{1}{2}$ of the diameter of the circle.

Therefore, $r = \dfrac{1}{2}(6) = 3$

Circumference of the given circle is $2\pi r. = 2\pi(3) = 6\pi$ units.

(b) arc length, s, of CB $= \dfrac{\theta}{360}$ of the circumference of the circle

$$s = \dfrac{30}{360} \cdot 2\pi(3) = \dfrac{\pi}{2} \text{ units}$$

OR if θ is in radians, $s = r\theta = 3 \cdot \dfrac{\pi}{6}$ 　　　　　　(**Note**: $\theta = 30° = \dfrac{\pi}{6}$; $r = 3$.)

$$s = \dfrac{\pi}{2} \text{ units.}$$

(c) Area, A , of the shaded sector $= \dfrac{30°}{360°}$ of the area of the circle.

$$A = \dfrac{30°}{360°} (\pi r^2)$$

$$= \dfrac{30°}{360°} (\pi \cdot 3^2)$$

$$= \dfrac{30}{360} (9\pi)$$

$$= \dfrac{3\pi}{4} \text{ sq. units}$$

OR if θ is in radians, $A = \dfrac{1}{2} r^2 \theta$ 　　　　　(**Note**: $\theta = 30° = \dfrac{\pi}{6}$; $r = 3$.)

$$= \dfrac{1}{2} \cdot 3^2 \cdot \dfrac{\pi}{6}$$

$$= \dfrac{1}{2} \cdot \dfrac{9}{1} \cdot \dfrac{\pi}{6}$$

$$= \dfrac{3\pi}{4} \text{ sq. units.}$$

Areas and Perimeters of Composite Figures

Example 1   In the figure below:
                    (a) Find the area of the figure.
                    (b) Find the perimeter of figure.

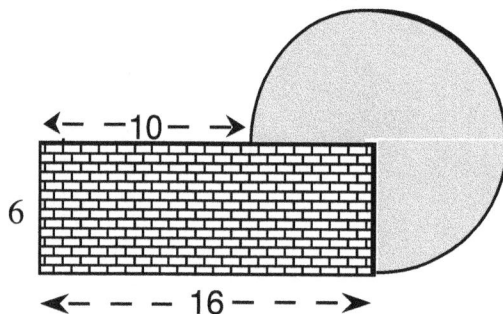

Solution

(a) Finding the area

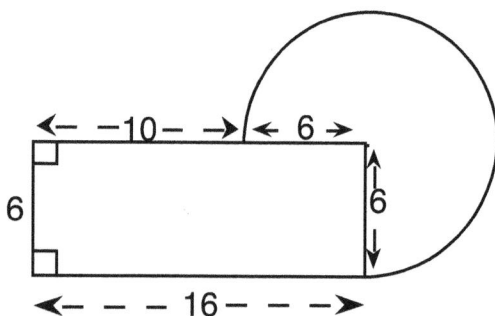

Radius, r, of the three-quarter circle $= 6$

Area of the three-quarter circle $= \dfrac{3}{4}\pi r^2$ ($\dfrac{3}{4}$ of the area of a circle of radius 6 units)

$$= \dfrac{3}{4}\pi(6)^2$$

$$= \dfrac{3}{\cancel{4}}\pi\cancel{(36)}^{9}$$

$$= 27\pi \text{ sq. units}$$

Area of the 16 by 6 rectangle $= 16 \times 6$ sq. units
                             $= 96$ sq. units
Area of the **whole figure** = area of rectangle + area of three-quarter circle
                             $= (96+ 27\pi)$ sq. units

(b) **Finding the perimeter**

Perimeter of the three-quarter circle $= \dfrac{3}{4} 2\pi r$

$$= \dfrac{3}{4}(2\pi)(6)$$

$$= 9\pi \text{ units}$$

Perimeter of the whole figure $= 6 + 16 + 9\pi + 10$

$$= (32 + 9\pi) \text{ units}$$

Begin here and go around counterclockwise

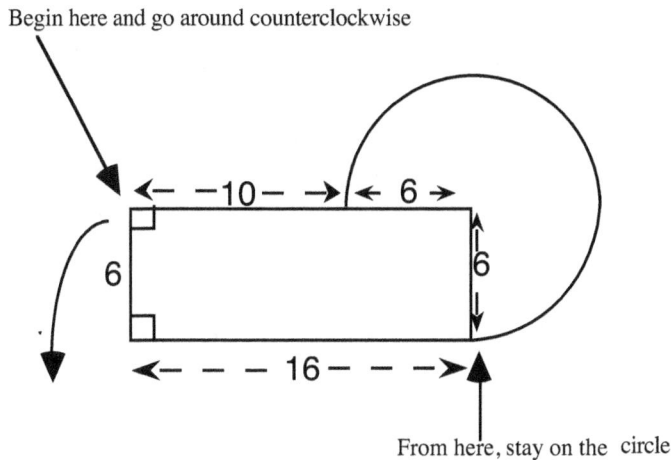

From here, stay on the circle

Note that in finding the perimeter we did **not** add the parts (6 and 6) of the rectangle which are also radii of the three-quarter circle, because the perimeter is the distance around the figure.

Lesson 52 Exercises

A

In the circle below, if the diameter of the circle is 18 units and the central angle has a measure 60°,
(a) Find the circumference of the circle.
(b) Find the area of sector COB.
(c) Find the arc length of CB.

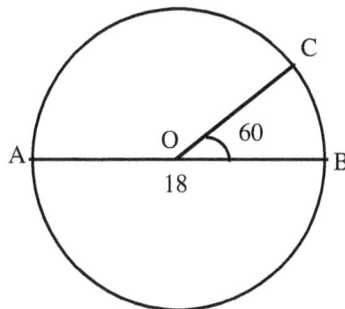

Answers (a) 18π units; (b) $\dfrac{27\pi}{2}$ sq. units; (c) 3π units

B

(a) Find the area of the figure.
(b) Find the perimeter of figure.

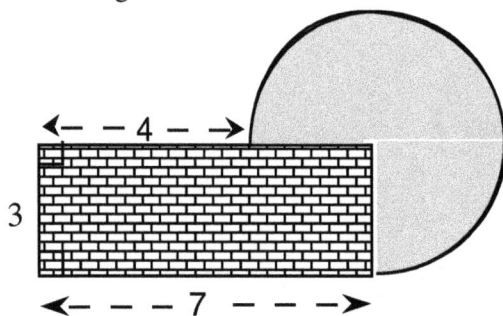

Answers: **(a)** $(21 + \frac{27\pi}{4})$ sq. units; **(b)**. $(14 + \frac{9\pi}{2})$ units

Test # 12 -Student's Self-Test (Always, Test yourself before you are tested)

Attempt all questions on clean sheets of paper, **Do not write in the book** Show all necessary work.

---

1. Simplify:

(a) $\sqrt{12}$; (b) $\sqrt{48}$; (c) $\sqrt{75}$.

Simplify:

2. (a) $\sqrt{162}$;   (b) $\sqrt{128}$;

(c) $\sqrt{96}$;   (d) $\sqrt{\dfrac{16}{81}}$.

---

3. Simplify:

(a) $\sqrt{36x^8}$;   (b) $\sqrt{4x^6y^4}$; (c) $\sqrt{8x^7y^2}$

4. Simplify the following:

(a) $\sqrt{8x^5y^7}$;   (b) $\sqrt{18x^{10}y}$;   (c) $\sqrt{32x^6y^4z^5}$

---

5. Add or subtract:

(a) $5\sqrt{3} + 3\sqrt{3}$

(b) $2\sqrt{6} + \sqrt{18} + \sqrt{24} - 2\sqrt{72}$

(c) $\sqrt{12} + \sqrt{48} + \sqrt{75}$.

6. Add or subtract:

(a) $4\sqrt{50} - \sqrt{24} + 5\sqrt{2}$

(b) $3x\sqrt{25y} - 2x\sqrt{16y}$

---

7. Multiply and simplify:

(a) $4\sqrt{5}$ and $2\sqrt{6}$;   (b) $2\sqrt{5}$ and $3\sqrt{30}$;

(c) $3\sqrt{2}$ and $6\sqrt{2}$

8. Multiply and simplify:

(a) $5\sqrt{7}$ and $6\sqrt{7}$;   (b) $3x\sqrt{2y}$ and $\sqrt{3y}$.

---

9. Find c

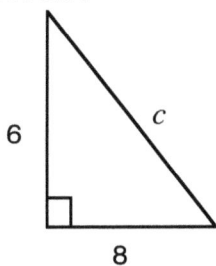

10. Find b

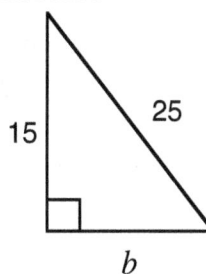

---

11. Fund x:

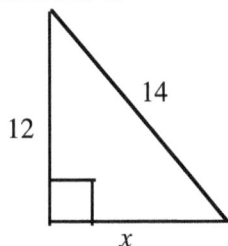

12. Find x:

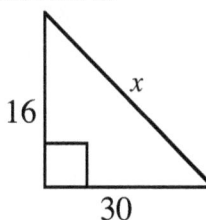

13. Find x:

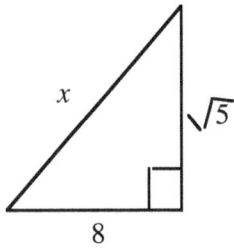

14. Find x:

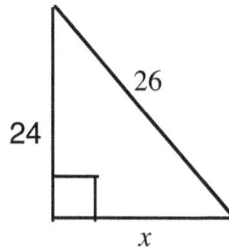

15. Find (a) Area; (b) Perimeter of right triangle ABC

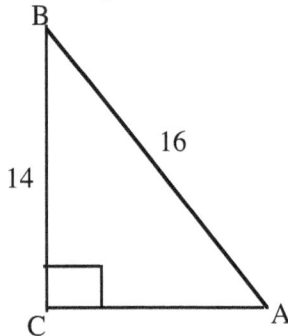

16. Find (a) Area; (b) Perimeter of right triangle DEF.

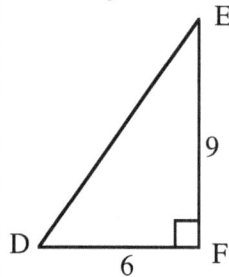

17. Find (a) the area and (b) the perimeter of the figure below.

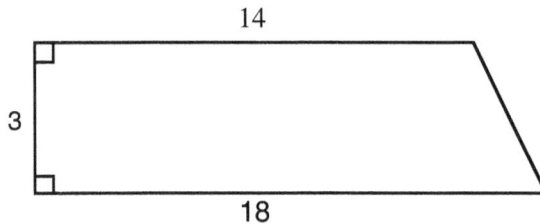

18. a) Find the area of the figure.
(b) Find the perimeter of figure.

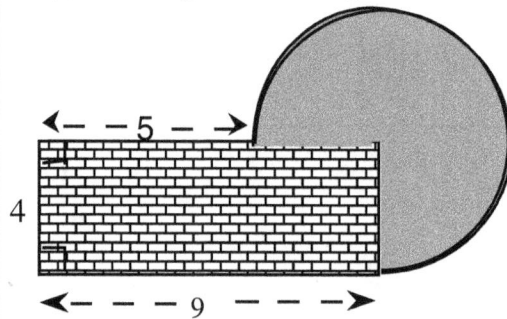

Answers: **1**. (a) $2\sqrt{3}$; (b) $4\sqrt{3}$; (c) $5\sqrt{3}$; **2**. (a) $9\sqrt{2}$; (b) $8\sqrt{2}$; (c) $4\sqrt{6}$; (d) $\frac{4}{9}$;
3. (a) $6x^4$; (b) $2x^3y^2$; **(c)** $2x^3y\sqrt{2x}$; **4**. (a) $2x^2y^3\sqrt{2xy}$; (b) $3x^5\sqrt{2y}$; (c) $4x^3y^2z^2\sqrt{2z}$;
5. (a) $8\sqrt{3}$; (b) $4\sqrt{6}-9\sqrt{2}$; (c) $11\sqrt{3}$; **6**. (a) $25\sqrt{2}-2\sqrt{6}$; (b) $7x\sqrt{y}$;
7. (a) $8\sqrt{30}$; (b) $30\sqrt{6}$; (c) 36; **8**. (a) 210 ; (b) $3xy\sqrt{6}$; **9**. $c=10$; **10** $b=20$;
11. $x=2\sqrt{13}$; **12**. $x=34$; **13**. $\sqrt{69}$; **14**. $x=10$; **15**. (a) $14\sqrt{15}$;
(b) $(30+2\sqrt{15})$ units; **16**. (a) 27 sq. units; (b) $(15+3\sqrt{13})$ units;
17. (a) 48 sq. units; (b) 40 units ; **18**. (a) $(36+12\pi)$ sq. units; (b) $(18+6\pi)$ units.

CHAPTER 20
CONGRUENT TRIANGLES

Lesson 53: **Properties of Congruent Triangles**; **Theorems and Proofs**

Lesson 54: **Applications of Congruency Theorems; Isosceles Triangles**

Before proceeding, review page 216- page 231.

Lesson 53

Properties of Congruent Triangles; Theorems and Proofs

Congruent Figures: Congruent figures have the same shape and the same size.
Congruent figures can be made to coincide.
The symbol for "congruent" or "congruent to" is " $\cong$ ".

Congruent Triangles: Congruent triangles have the same shape and the same size.

Identifying corresponding sides and angles of two congruent triangles

1. Corresponding sides are opposite congruent angles.

How to identify corresponding sides in two congruent triangles say Δ #1 and Δ #2:

Step 1: Pick a side in Δ #1, note the angle this side is opposite to.

Step 2: Go to Δ #2 and locate the angle which is congruent to the angle noted in Δ #1. The side opposite to this located angle in Δ #2 is the corresponding side. Note of course that if the sides in Δ #1 and Δ #2 are given or known to have the same length, then these sides correspond and they are also congruent.

2. Corresponding angles are opposite congruent sides.

How to identify corresponding angles in two congruent triangles say Δ #1 and Δ #2:

Step 1: Pick an angle in Δ #1, note the side this angle is opposite to.

Step 2: Go to Δ #2 and locate the side which is congruent to the side noted in Δ #1. The angle opposite to this located side in Δ #2 is the corresponding angle. Note of course that if the angles in Δ 's #1 and #2 are given or known have the same measure, then these angles correspond and they are also congruent.

Properties of congruent triangles

1. The corresponding angles are congruent (or the measures of corresponding angles are equal).

2. The corresponding sides are congruent (or the lengths of corresponding sides are equal). Also, the ratio of each pair of corresponding sides equals 1.

Conditions or Criteria for Congruency

The conditions or criteria for proving that two triangles are congruent are embodied in the theorems that follow.

Before proceeding to the theorems, review the axioms on pages 176 and 456.

Ordered naming of triangles in stating the congruency of triangles and identification of corresponding parts from the congruency statement

It is good communication to follow a certain order in stating that two triangles are congruent. The logical order is to match the letters of the vertices in stating that two triangles are congruent.

Example Stating that $\triangle ABC \cong \triangle DEF$ (read " triangle ABC is congruent to triangle DEF) implies that the vertex A (of $\triangle ABC$) corresponds to the vertex D (of $\triangle DEF$); the vertex B (of $\triangle ABC$) corresponds to the vertex E (of $\triangle DEF$); and the vertex C (of $\triangle ABC$) corresponds to the vertex F (of $\triangle DEF$).

By matching the vertices, the sides $\overline{AB}$ and $\overline{DE}$ naturally correspond, $\overline{BC}$ and $\overline{EF}$ correspond; and $\overline{AC}$ and $\overline{DF}$ correspond.

$$\triangle ABC \cong \triangle DEF$$

Figure 1

Sometimes, some authors may not follow the above-mentioned order strictly and merely state that some triangles are congruent, and in which case, additional information must be provided either in the form of further statements or in an accompanying diagram to indicate the corresponding parts. **Note** for example that the triangle in Figure 2, below, may be named as either $\triangle ABC$, $\triangle BCA$, $\triangle CAB$, $\triangle ACB$, $\triangle CBA$ or $\triangle BAC$.

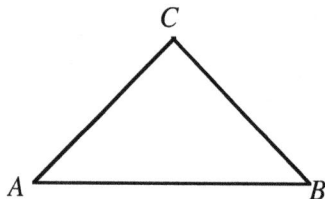

Figure 2

Theorem 1:   If two sides and the included angle of one triangle are congruent to the
corresponding two sides and the included angle of another triangle,
the two triangles are congruent. Abbreviated **SAS** (Side-Angle-Side).

In the following example, we will apply this theorem to prove that two triangles given are congruent.

Example 1   In Figure 1 below, prove that $\triangle ABC \cong \triangle DEF$

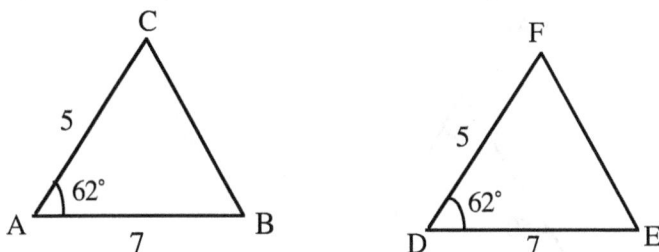

Figure 1

Given: **1.** In $\triangle$ ABC , AB = 7, AC $\cong$ 5 , m$\angle$ A $\cong$ 62°.
   2. In $\triangle$ DEF, DE =7, DF = 5, m$\angle$ D = 62°.

To prove: That $\triangle$ ABC $\cong$ $\triangle$ DEF

Proof: Statements   Reasons
   1. AB = 7   **1.** Given
   2. DE= 7   **2.** Given
   3. AB=DE   **3.** If two quantities are equal to the same quantity, they are equal to each other

   4. $\overline{AB} \cong \overline{DE}$ **4.** If two line segments have equal length, they are congruent .
   5. AC = 5   **5** Given
   6. DF= 5   **6.** Given
   7. AC = DF   **7.** If two quantities are equal to the same quantity, they are equal to each other.

   8. $\overline{AC} \cong \overline{DF}$ **8.** If two line segments have equal length, they are congruent.
   9.   m$\angle$ A = 62°   **9.** Given
   10. m$\angle$ D = 62°   **10.** Given
   11. m$\angle$ A = m $\angle$ D   **11.**   If two quantities are equal to the same quantity, they are equal to each other.
   12. $\angle$ A $\cong$ $\angle$ D   **13.** If two angles have the same measure, they are congruent.
   13. $\angle$ A and $\angle$ D are included angles.
   14. $\triangle$ **ABC** $\cong$ $\triangle$ **DEF**   **14.** SAS
                    *Q.E.D

Some of the steps in the above proof could be skipped or combined with other steps. The thoroughness of the
proof was to show the student how not to take too much for granted and also to show the need to give reasons
for statements made in a geometric proof.

   * **Q.E.D** (or QED) is a Latin abbreviation for **quod erat demonstrandum,** which literally
means "which had to be shown or proved". Written at the end of a proof, it signifies the **end
of the proof**. Some authors signify the end of a proof by writing "and the proof is complete",
"The proof is complete", or "Therefore, the proof is complete".

*Theorem 2: If two angles and the included side of one triangle are congruent to the corresponding two angles and the included side of another triangle, then the two triangles are congruent. Abbreviated **ASA** (Angle-Side-Angle).

Another form of Theorem 2 : If a side and both angles adjacent to that side of one triangle are congruent to the corresponding side and the angles adjacent to that side of another triangle, then the two triangles are congruent.

In the following example, we will apply this theorem to prove that the given triangles are congruent.

Example 2   In Figure 2 below, prove that $\triangle$ ABC $\cong$ $\triangle$ DEF

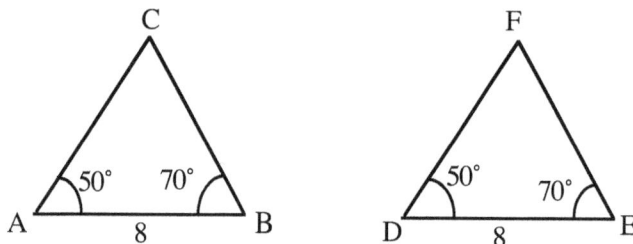

Figure 2

Given: 1. In $\triangle$ ABC, AB = 8, m$\angle$ A = 50°, m$\angle$ B = 70°.
   2. In $\triangle$ DEF , DE = 8, m $\angle$ D = 50°, m$\angle$ E = 70°.

To prove: That $\triangle$ ABC $\cong$ DEF

Proof:

Statements	Reasons
1. AB = 8	1. Given
2. DE = 8	2. Given
3. AB = DE	3. If two quantities are equal to the same quantity, they are equal to each other
4. $\overline{AB}$ $\cong$ $\overline{DE}$	4. If two line segments have equal length, they are congruent
5. $\overline{AB}$ and $\overline{DE}$ are included sides	
6. m$\angle$ A = 50°	6. Given
7. m$\angle$ D = 50°	7. Given
8. m$\angle$ A = m$\angle$ D	8. If two quantities are equal to the same quantity, they are equal to each other.
9. $\angle$ A $\cong$ $\angle$ D	9. If two angles have the same measure, they are congruent.
10. m$\angle$ B = 70°	10. Given
11. m$\angle$ E = 70°	11. Given
12. m$\angle$ B = m$\angle$ E	12. If two quantities are equal to the same quantity, they are equal to each other.
13. $\angle$ B $\cong$ $\angle$ E	13. If two angles have the same measure, they are congruent.
14. $\triangle$ ABC $\cong$ $\triangle$ DEF	14. ASA

Q.E.D.

*Note that another theorem abbreviated **AAS** or **SAA** is sometimes used (as a shortcut in proving two triangles congruent), since if two pairs of corresponding angles are congruent, the third pair of angles are also congruent, recalling that the sum of the measures of the angles of a triangle is 180°.
 In fact, we can easily prove **AAS or SAA** using **ASA.**

Theorem 3: If the three sides of one triangle are congruent to the corresponding three sides of another triangle, then the two triangles are congruent. Abbreviated **SSS** (Side-Side-Side).

Example 3   In Figure 3 below, prove that △ ABC ≅ △ DEF

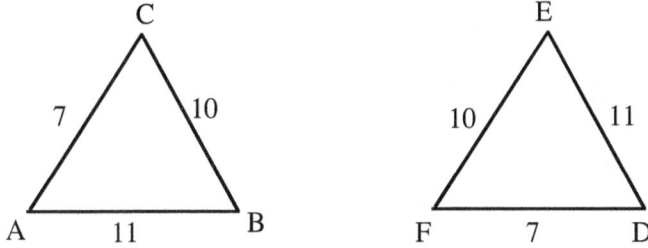

Figure 3

Given: **1.** In △ ABC, AB = 11, BC = 10, CA = 7.
      2. In △ DEF, DE = 11, EF = 10, FD = 7.

To prove: That △ ABC ≅ △ DEF

Proof: Statements   Reasons
      1. AB = 11   **1.** Given
      2. DE = 11   **2.** Given
      3. AB = DE   **3.** If two quantities are equal to the same quantity, they are equal to each other.

      4. $\overline{AB}$ ≅ $\overline{DE}$ **4.** If two line segments have equal length, they are congruent .
      5. BC = 10   **5.** Given
      6. EF = 10   **6.** Given
      7. BC = EF   **7.** If two quantities are equal to the same quantity, they are equal to each other.

      8. $\overline{BC}$ ≅ $\overline{EF}$ **8.** If two line segments have equal length, they are congruent.
      9. CA = 7   **9** Given
      10. FD = 7   **10.** Given
      11. CA = FD   **11.** If two quantities are equal to the same quantity, they are equal to each other.

      12. $\overline{CA}$ ≅ $\overline{FD}$ **12.** If two line segments have equal length, they are congruent.
      13. △ ABC ≅ △ DEF   **13.** SSS
                      Q.E.D.

Lesson 53 Exercises

Determine which of the following triangles are congruent and prove why they are congruent.

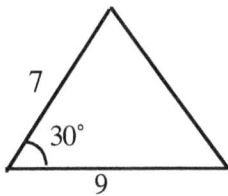

Fig. 1

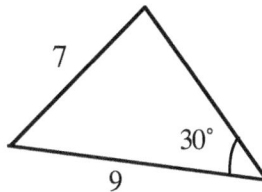

Fig. 2

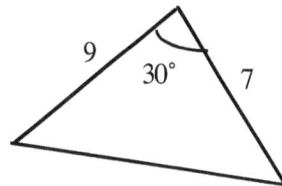

Fig.3

Answer: Fig1 and Fig 3 are congruent.

Lesson 54

More Applications of Congruency Theorems; Isosceles Triangles; Mid-Point Theorem

In proving that two triangles are congruent, it will be sufficient for us to show that any of the congruency theorems is true.

Example 1 In the figure below: (a) Prove that $\triangle BAC \cong \triangle DEC$
(b) Find x and y

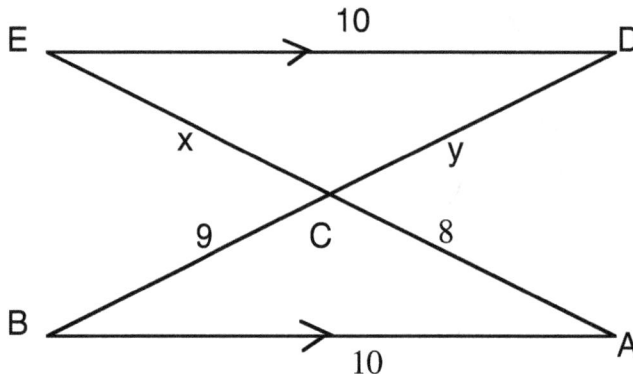

Solution (a): Proving that $\triangle BAC \cong \triangle DEC$

Given: In $\triangle$'s BAC and DEC , ED = 10, BA = 10; $\overline{ED}$ is parallel to $\overline{BA}$.
To prove: That $\triangle BAC \cong \triangle DEC$

Proof:

	Statements		Reasons
1.	ED = 10	**1.**	Given
2.	BA = 10	**2.**	Given
3.	ED = BA	**3.**	If two quantities are equal to the same quantity, they are equal to each other.
4.	$\overline{ED} \cong \overline{BA}$	**4.**	If two line segments have equal length, they are congruent.
5.	$\angle E \cong \angle A$	**5.**	Alt. int. angles, parallel lines $\overline{ED}$, $\overline{BA}$ cut by transversal $\overline{EA}$
6.	$\angle D \cong \angle B$	**6.**	Alt. int. angles, parallel lines $\overline{ED}$, $\overline{BA}$ cut by transversal $\overline{BD}$
7.	$\triangle BAC \cong \triangle DEC$	**7.**	ASA
			Q.E.D

Solution (b): Finding x and y:

$\overline{EC} \cong \overline{AC}$ (Corresponding sides of congruent $\triangle$'s are congruent)

$EC = AC = 8$ ($\overline{EC}$ is opposite $\angle D$; $\overline{AC}$ is opposite $\angle B$; and $\angle D \cong \angle B$)
$x = 8$

Similarly, $y = DC = BC = 9$

$\therefore x = 8, y = 9$

Example 2   Given the parallelogram in the figure below, prove that

(a) $\overline{AE} \cong \overline{EC}$;   (b) $\overline{BE} \cong \overline{ED}$

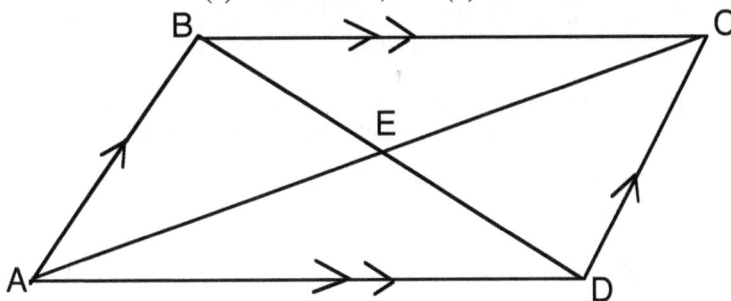

Given: ABCD is a parallelogram.

To prove: (a) That $\overline{AE} \cong \overline{EC}$; (b)   That $\overline{BE} \cong \overline{ED}$

Proof:

Statements	Reasons
1. ABCD is a parallelogram	**1. Given**
2. In $\triangle$'s BCE and DAE,	
$\overline{BC} \cong \overline{DA}$	**2. Opposite** sides of a parallelogram are congruent.
3. $\angle$ BCE $\cong \angle$ DAE	**3. Alternate** interior angles are congruent.
	$\parallel$ lines $\overline{BC}, \overline{AD}$, cut by transversal $\overline{AC}$.
4. $\angle$ EBC $\cong \angle$ EDA	**4.** Alternate interior angles are congruent.
	$\parallel$ lines $\overline{BC}, \overline{AD}$ cut by transversal $\overline{BD}$.
5. $\triangle$ BCE $\cong \triangle$ DAE	**5.** ASA.

(a)   **6.** $\therefore \overline{AE} \cong \overline{EC}$ **6.** Corresponding sides of congruent $\triangle$'s are congruent,

noting that $\overline{AE}$ is the same as $\overline{EA}$

(b)   Similarly, $\overline{BE} \cong \overline{ED}$ Noting that $\overline{ED}$ is the same as $\overline{DE}$
                                                                                          Q.E.D

Alternative proof:

Statements	Reasons
1. ABCD is a parallelogram	**1. Given**
2. In $\triangle$'s BEA and DEC	
$\overline{AB} \cong \overline{CD}$	**2.** Opposite sides of a parallelogram are congruent.
3. $\angle$ ABE $\cong \angle$ CDE	**3.** Alternate interior angles are congruent.
4. $\angle$ EAB $\cong \angle$ ECD	**4.** Alternate interior angles are congruent.
5 $\triangle$ BEA $\cong \triangle$ DEC	**5.** ASA

(a)   **6.** $\therefore \overline{AE} \cong \overline{EC}$ **6.** Corresponding sides of congruent $\triangle$'s are congruent,

noting that $\overline{AE}$ is the same as $\overline{EA}$

(b)   Similarly, $\overline{BE} \cong \overline{ED}$ Noting that $\overline{ED}$ is the same as $\overline{DE}$
                                                                                          Q.E.D

Isosceles Triangles

Definition: An isosceles triangle is a triangle in which two sides are congruent. In an isosceles triangle, the angles opposite the congruent sides are congruent.

Theorem 1: If two sides of a triangle are congruent, the angles opposite to these sides are congruent. (Another form : The base angles of an isosceles triangle are congruent.)

If in $\triangle$ ABC (Figure 1 below), $\overline{AB} \cong \overline{CB}$, then $\angle A \cong \angle C$ (The base is $\overline{AC}$)

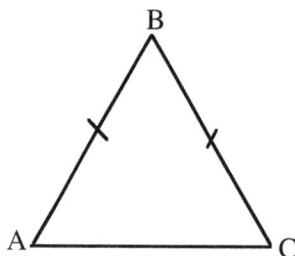

Figure 1

Theorem 2: The bisector of the vertex angle of an isosceles triangle bisects the base.

If in $\triangle$ ABC (Figure 2 below), $\overline{AB} \cong \overline{CB}$, and $\overline{BD}$ bisects $\angle$ B, then $\overline{AD} \cong \overline{CD}$. (The base is $\overline{AC}$)

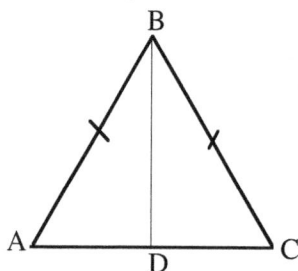

Figure 2

Theorem 3: The bisector of the vertex angle of an isosceles triangle is perpendicular to the base.

If in $\triangle$ ABC (Figure 3 below), $\overline{AB} \cong \overline{CB}$, and $\overline{BD}$ bisects $\angle$ B, then $\overline{BD} \perp \overline{AC}$. (The base is $\overline{AC}$)

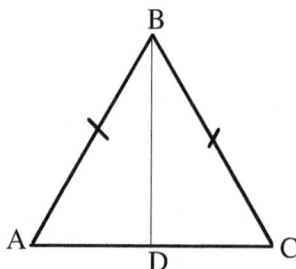

Figure 3

In trigonometry, we will apply a combination of Theorems 2 and 3, above, to find the dimensions of the 30°-60°-90° triangle.(See page 273.)
Note: An equilateral triangle is also isosceles.

Mid-Point Theorem

If a line is drawn from the midpoint of one side of a triangle and the line is parallel to a second side, then the line bisects the third side.

Given: $\triangle ABE$ in which $\overline{BC} \cong \overline{AC}$, $\overline{CD}$ is parallel to $\overline{BE}$.

Required: To prove that $\overline{AD} \cong \overline{ED}$

Construction: Through E, draw $\overline{EF}$ parallel to BA Reason : a line can be drawn parallel to another line

Plan: We first prove that $\overline{AC} \cong \overline{EG}$, followed by proving that $\triangle's\ ACD$ and EGD are congruent

Proof:

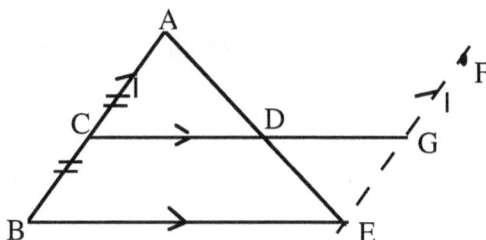

Statements	Reason
1. $\overline{BC}\ \|\|\ \overline{EG}$	1. by construction. $\overline{EF}$ is drawn parallel to BA
2. $\overline{CG}\ \|\|\ \overline{BE}$	2. Given
3 . $\therefore$ quad $BEGC$ is a parallelogram	3. Definition If both pairs of the opposite sides of a quadrilateral are parallel, then the quadrilateral is a parallelogram
4. $\overline{BC} \cong \overline{EG}$	4. Opposite sides of a parallelogram are congruent
5. $\overline{BC} \cong \overline{AC}$	5. Given
6. $\overline{AC} \cong \overline{EG}$	6. Transitive property
7. In $\triangle's\ ACD$ and EGD, $\quad \angle CAD \cong \angle GED$	7. .Alt. int. angles; parallel lines $\overline{BA}$ and $\overline{EF}$ cut by transversal $\overline{AE}$
8. $\angle ACD \cong \angle EGD$	8. Alt. int. angles; parallel lines $\overline{BA}$ and $\overline{EF}$ cut by $\overline{CG}$
9. $\overline{AC} \cong \overline{EG}$	9. Proved in (6) above
10. $\triangle ACD \cong \triangle EGD$	10. ASA (Two angles and the included side)
11. $\overline{AD} \cong \overline{ED}$	11. Sides opposite congruent $\angle's$ are congruent

Q.E.D

Lesson 54 Exercises

In the figure below: (a) Prove that $\triangle\,BAC \cong \triangle\,DEC$; (b) Find x and y

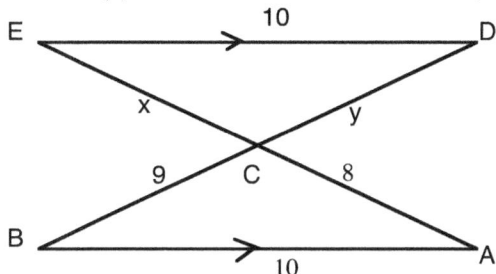

Answers : (b) $x = 8,\ \ y = 9$

CHAPTER 21
Similar Triangles

Lesson 55: **Properties of Similar Triangles; Theorems and Proofs**
Lesson 56: **Comparison of Congruency and Similarity of Triangles and Applications of Similarity Theorems**
Lesson 55

Properties of Similar Triangles; Theorems and Proofs

Preliminaries

Ratio: The ratio of two quantities a and b (written $a{:}b$) is the fraction $\frac{a}{b}$

(i.e., is the first quantity divided by the second quantity). The quantities a and b are called the terms of the ratio.

But the ratio $b{:}a$ is $\frac{b}{a}$. The ratio 3:4 is $\frac{3}{4}$; but the ratio 4:3 is $\frac{4}{3}$.

Note above that order is important.

Proportion : A proportion is a statement that two ratios are equal.

The proportion a is to b as c is to d (symbolically $a{:}b{:}{:}c{:}d$) is the equality

$$\frac{a}{b} = \frac{c}{d} \tag{1}$$

In the above proportion, a and d are called the extremes and b and c are called the means of the proportion.

Proportion Principles

1. We can invert both ratios of the proportion: From (1) above, we would obtain after the inversion:

$$\frac{b}{a} = \frac{d}{c}$$

2. We can interchange the extremes of a proportion: From (1) above we would obtain:

$$\frac{d}{b} = \frac{c}{a}$$

3. We can also interchange the means to obtain: $\frac{a}{c} = \frac{b}{d}$

More (and interesting) proportion principles

4. If $\frac{a}{b} = \frac{c}{d}$, then $\frac{a+b}{b} = \frac{c+d}{d}$;   **5.** If $\frac{a}{b} = \frac{c}{d}$, then $\frac{a-b}{b} = \frac{c-d}{d}$

6. If $\frac{a}{b} = \frac{c}{d} = \frac{e}{f}$, then $\frac{a+c+e}{b+d+f} = \frac{a}{b} = \frac{c}{d} = \frac{e}{f}$

Proportion principles can be used to obtain desired ratios. For example, if we have $\frac{a}{b}$ as the ratio

of one side of a proportion, and we desire the ratio $\frac{b}{a}$, then proportion principle #1 above

(inversion of both sides of a proportion) can be used to obtain this ratio, $\frac{b}{a}$, from $\frac{a}{b}$.

Example: If $\frac{a}{b} = \frac{2}{3}$, find $\frac{b}{a}$. **Solution** By inverting both sides of the equation, we obtain $\frac{b}{a} = \frac{3}{2}$

Identifying corresponding sides and angles of two similar triangles

1. Corresponding sides are opposite congruent angles.
2. Congruent angles are opposite corresponding sides.

To identify corresponding sides in two similar triangles say, Δ #1 and Δ #2:

Case 1: Given or knowing the lengths of two sides and the included angle in each triangle (incl'd angles being congr.).
    The side with the smaller length in Δ #1 corresponds to the side with the smaller length in Δ #2.
    The side with the greater length in Δ #1 corresponds to the side with the greater length in Δ #2.
    The third side (length known or unknown) in Δ #1 corresponds to the third side of Δ #2.
Case 2: Given the lengths of all three sides of each triangle.
    The side with the smallest length in Δ #1 corresponds to the side with smallest length in Δ #2.
    The side with the greatest length Δ #1 corresponds to the side with the greatest length in Δ #2.
    The side with intermediate length in Δ #1 corresponds to the side with intermediate length in Δ #2,
Case 3: Given the measures of two pairs of congruent angles in the two triangles.
    Step 1: Pick a side in Δ #1, and note the angle this side is opposite to.
    Step 2: Go to Δ #2 and locate the angle which is congruent to the angle noted in Δ #1. The side opposite to this
        located angle in Δ #2 is the corresponding side.

To identify corresponding angles in two similar triangles say Δ's #1 and #2:

Given or knowing the corresponding sides
Step 1: Pick an angle in Δ #1, note the side this angle is opposite to.
Step 2: Go to Δ #2 and locate the side which corresponds to the side noted in Δ #1. The angle opposite to this
    located side in Δ #2 is the corresponding angle. Note of course that if the angles in Δ #1 and Δ #2 have
    the ame measure, these angles correspond and they are also congruent.

Note: The above identification guidelines assume that the two triangles are given to be or have been proved to be
    similar. Do not therefore apply them until you know that the two triangles are similar. However, a modified
    form of this guide can be helpful in picking corresponding sides, even if the triangles are not similar.
    See also Examples 1 & 2, pages 258 & 260 respectively: the "Plan "parts.

Example Identify the corresponding lengths and corresponding angles in the Δ's given below

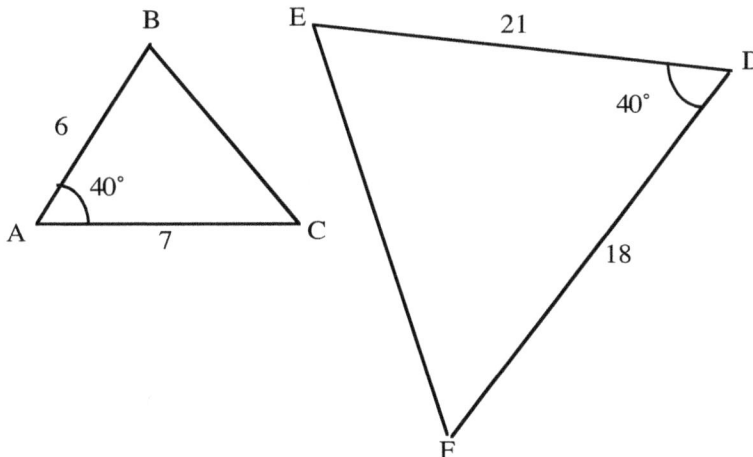

Solution: Smaller length of one triangle corresponds to the smaller length of the other triangle;
the larger length of one triangle corresponds to the larger length of the other triangle. Therefore
6 corresponds to 18; 7 corresponds to 21. ∠ D corresponds to ∠ A; ∠ C to ∠ E; and ∠ B to ∠ F.

Definition: Two triangles are similar ($\sim$) if and only if the corresponding angles are congruent and the ratio of corresponding sides is the same for each pair of corresponding sides.

Corresponding angles being congruent implies that corresponding angles have equal measures. The statement that "the ratio of corresponding sides is the same for each pair of corresponding sides" implies that the corresponding sides are in proportion.

Properties of Similar Triangles

1. Each pair of corresponding angles are congruent (i.e. have equal measure).
2. Corresponding sides are in proportion (that is, the ratio of the lengths of any pair of corresponding sides is equal to the ratio of the lengths of any other pair of corresponding sides).

Ordered naming of triangles in stating the similarity of triangles and identification of corresponding parts from the similarity statement

As it is in the case of the congruency of triangles, it is also good communication to follow a certain order in stating that two triangles are similar. The logical order is to match the letters of the vertices in stating that two triangles are similar.

Example   Stating that $\triangle ABC \sim \triangle DEF$ (read "triangle ABC is similar to triangle DEF"), implies that the vertex A (of $\triangle ABC$) corresponds to the vertex D (of $\triangle DEF$); the vertex B (of $\triangle ABC$) corresponds to the vertex E (of $\triangle DEF$); and the vertex C (of $\triangle ABC$) corresponds to the vertex F (of $\triangle DEF$).

By matching the vertices, the sides $\overline{AB}$ and $\overline{DE}$ naturally correspond, $\overline{BC}$ and $\overline{EF}$ correspond; and $\overline{AC}$ and $\overline{DF}$ correspond.

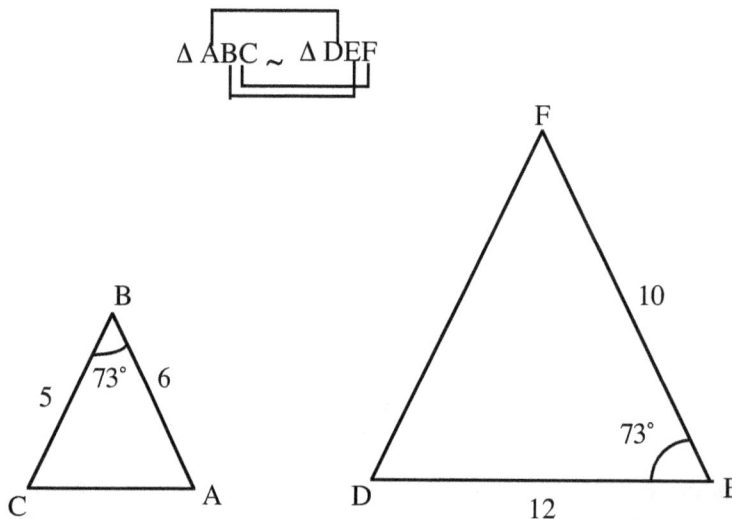

Theorems and Proofs

Conditions or criteria for proving similarity of triangles

* The conditions or criteria for proving that two triangles are similar are embodied in th following theorems:

Theorem 1: If two pairs of corresponding sides of two triangles are in proportion **and** the included angles (the angles between the sides involved) are congruent, then the two triangles are similar. Abbreviated **SAS.**

In the figure below, Δ ABC ~ Δ DEF (read "triangle ABC is similar to triangle DEF"), since the conditions in Theorem 1 have been satisfied. In the next example we will prove the similarity of these two triangles.

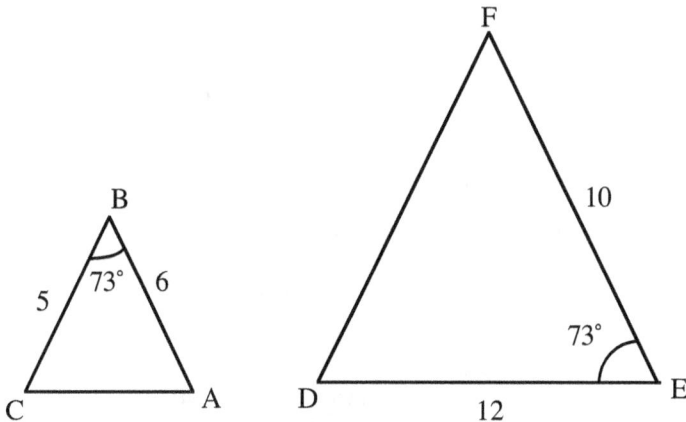

* In the Appendix, all the theorems on similar triangles covered have been stated in terms of the lengths of the sides, and the measures of the angles. These forms of the theorems may be useful when dealing with the lengths of the sides of the triangles.

Example 1 In the triangles shown (**Figure 1**), prove that $\triangle ABC \sim \triangle DEF$

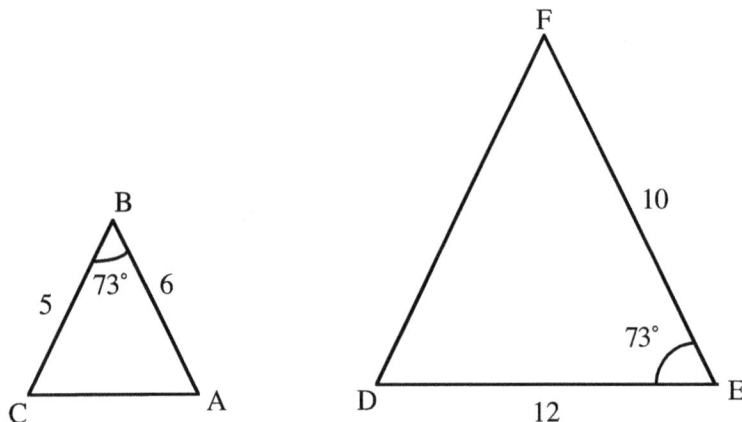

Figure 1

Given: 1. In $\triangle ABC$, $m\angle B = 73°$, $AB = 6$, $BC = 5$.
 2. In $\triangle DEF$, $m\angle E = 73°$, $DE = 12$, $EF = 10$.

To prove: That $\triangle ABC \sim \triangle DEF$

Plan: Arrange the dimensions of each triangle in increasing order (or decreasing order) ,i.e. smaller length first, etc.
 For $\triangle ABC$, we have 5,6; for $\triangle DEF$, we have 10 ,12.
 Divide the smaller length of $\triangle ABC$ by the smaller length of $\triangle DEF$, and note the line segments
 involved in the ratio. Next, divide the larger length of $\triangle ABC$ by the larger length of $\triangle DEF$ and again note
 the sides involved.. If we obtain the same ratio (same quotient), the sides of the triangles are in
 proportion, otherwise, they are not in proportion. We will also check to see if the given angles have
 the same measure (or are congruent) **and** are also included (i.e., the angles are also between the sides
 involved in the ratios).

Proof:

Statements	Reasons
1. The ratio, $5{:}10 = \dfrac{5}{10} = \dfrac{1}{2} = \dfrac{BC}{EF}$	1. The ratio of two quantities is the first quantity divided by the second quantity
2. The ratio, $6{:}12 = \dfrac{6}{12} = \dfrac{1}{2} = \dfrac{AB}{DE}$	2. The ratio of two quantities is the first quantity divided by the second quantity.
3. $\dfrac{AB}{DE} = \dfrac{BC}{EF}$	3. If quantities are equal to the same quantity, they are equal to each other.
4. AB, DE, BC and EF are in proportion.	4. A proportion is the equality of two ratios.
5. $m\angle B = 73°$	5. Given
6. $m\angle E = 73°$	6. Given
7. $m\angle B = m\angle E$	7. If quantities are equal to the same quantity, they are equal to each other.
8. $\angle B \cong \angle E$	8. If two angles have the same measure, they are congruent.
9. $\angle B$ and $\angle E$ are included angles.	
10. $\therefore$ $\triangle ABC \sim \triangle DEF$	10. SAS

Q.E.D.

How to find an unknown side (Knowing that the triangles involved are similar)

Pick the unknown side from one triangle and then pick the corresponding side in the other triangle and form a ratio. Go back to the first triangle and pick a known side whose corresponding side in the second triangle is known, and form another ratio. Equate the two ratios and solve for the unknown.

You may also pick corresponding sides and set up a proportion as "a is to b as c is to d" and translate to $\frac{a}{b}=\frac{c}{d}$ (where $a, b, c,$ and d are the lengths of the corresponding sides).

Example 1b If the two triangles below are similar, find x.

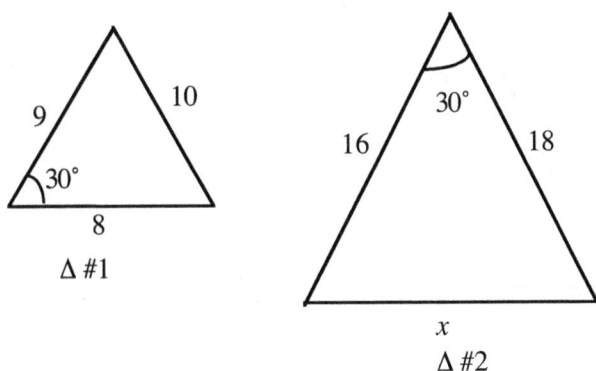

Δ #1

Δ #2

Solution
Step 1: Pick the x in say, Δ #2; note that x is opposite the angle whose measure is 30°

Step 2 : Go to the other triangle say, Δ #1; pick the side opposite the angle whose measure is 30°. The length of this corresponding side is 10.

Step 3: Form a ratio using these corresponding sides x and 10 to obtain $\frac{x}{10}$.

Step 4: Go back to Δ #2, pick a known side whose corresponding side in Δ #1 is also known. In this problem, we can pick either the "18" or the "16". We pick 16; and pick its corresponding side in Δ #1 which is 8.

Step 5: Form another ratio using 16 and 8 to obtain $\frac{16}{8}$.

Step 6 : Equate the ratios from Step 3 and Step 5, and solve for x.
$$\frac{x}{10}=\frac{16}{8}$$
$$8x = 10(16)$$
$$x = 20$$

Theorem 2: If all three pairs of corresponding sides of two triangles are in proportion (i.e., the ratio of corresponding sides is the same for each pair of corresponding sides) , then the two triangles are similar. Abbreviated **SSS** (Side-Side-Side: but note that implied are the **ratios of sides,** and **not** equality of sides)

In **Figure 2** below, Δ ABC ~ Δ FDE (read "triangle ABC is similar to triangle FDE", since the conditions in Theorem 2 have been satisfied. We will prove this similarity in the following example.

Example 2 In Figure 2 below, prove that Δ ABC ~ Δ FDE

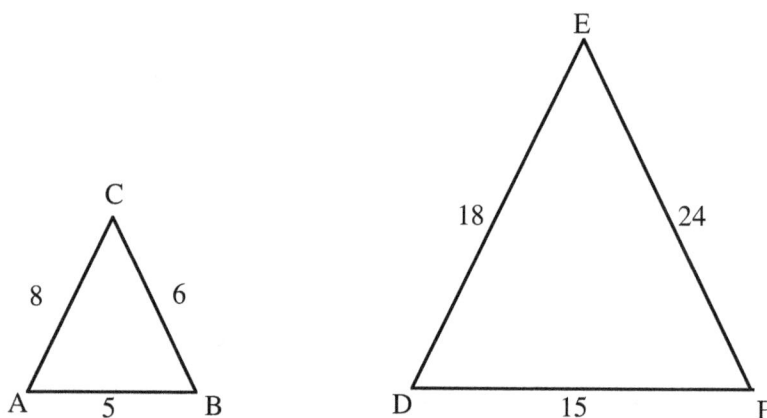

Figure 2

Given: **1**. In Δ ABC, AB = 5, BC = 6, CA = 8.
            2. In Δ FDE, DF = 15, FE = 24, ED = 18.

To prove: That Δ ABC ~ Δ FDE

Plan: Arrange the dimensions of each triangle in increasing order (or decreasing order) ,i.e. smallest length first, etc. For Δ ABC, we have 5, 6, 8; for Δ FDE, we have 15 ,18, 24.
Divide the smallest length of Δ ABC by the smallest length of Δ FDE, and note the line segments involved in the ratio. Next, divide the larger length of Δ ABC by the larger length of Δ FDE and note the sides; and similarly divide the largest length of Δ ABC by the largest length of Δ FDE, and again note the sides.. If we obtain the same ratio (same quotient) for all three cases, the three pairs of sides of the triangles are in proportion, and the triangles are similar; otherwise, the sides are not in proportion, and the triangles are not similar.

Proof: Statements   Reasons

1. The ratio, $5{:}15 = \dfrac{5}{15} = \dfrac{1}{3} = \dfrac{AB}{DF}$ **1**.The ratio of two quantities is the first quantity divided by the second quantity.

2. The ratio, $6{:}18 = \dfrac{6}{18} = \dfrac{1}{3} = \dfrac{BC}{DE}$ **2**.The ratio of two quantities is the first quantity divided by the second quantity.

3. The ratio, $8{:}24 = \dfrac{8}{24} = \dfrac{1}{3} = \dfrac{AC}{FE}$ **3**.The ratio of two quantities is the first quantity divided by the second quantity.

4. $\dfrac{AB}{DF} = \dfrac{BC}{DE} = \dfrac{AC}{FE}$ **4**. If quantities are equal to the same quantity, they are equal to each other.

                Also all the three pairs of corresponding sides are in proportion.

5. Δ ABC~ Δ FDE   **5**. SSS
                Q.E.D

Theorem 3: If two angles of one triangle are congruent to two corresponding angles of another triangle, then the two triangles are similar. Abbreviated **AA** (Angle-Angle).

Note that if two pairs of angles are congruent, the third pair are congruent, and therefore, AA implies AAA.

In **Figure 3** below, Δ ABC ~ Δ DEF (read "triangle ABC is similar to triangle DEF"), since the conditions in Theorem 3 have been satisfied. In the following example, will prove that the two triangles given are similar .

Example 3 In Figure 3 below, prove that Δ ABC ~ Δ DEF.

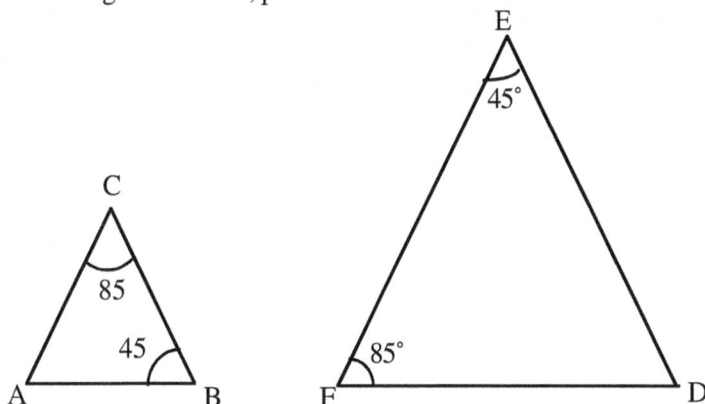

Figure 3

Given: **1.** In Δ ABC, m∠ B = 45°, m∠ C = 85°
2. In Δ DEF, m∠ E = 45° , m∠ F = 85°

To prove: That Δ ABC ~ Δ DEF

Proof: Statements   Reasons

1. m∠ B = 45°   1. Given
2. m∠ E = 45°,   2. Given
3. m∠ B = m∠ E   **3.** If quantities are equal to the same quantity, they are equal to each other.
4. ∠ B ≅ ∠ E   **4.** If two angles have the same measure, they are congruent.
5. m∠ C = 85°   5. Given
6. m∠ F = 85°   6. Given
7. m∠ C = m∠ F   **7.** If quantities are equal to the same quantity, they are equal to each other.
8. ∠ C ≅ ∠ F   **8.** If two angles have the same measure, they are congruent.
9. Δ ABC ~ Δ DEF   **9.** AA

Q.E.D

Lesson 55 Exercises

A   Determine which of the triangles in Fig.1, Fig.2 and Fig 3 are similar; and state the reason.

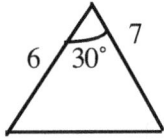   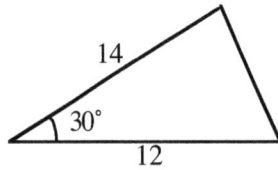   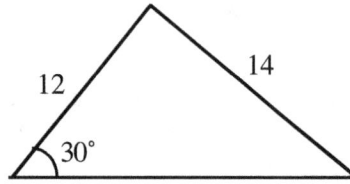

Fig.1   Fig. 2   Fig. 3

Answers: Fig 1 and Fig.2 by SAS

B In the figures below, (a) Show that the Δ's are similar.
(b) Find x.

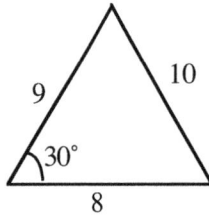   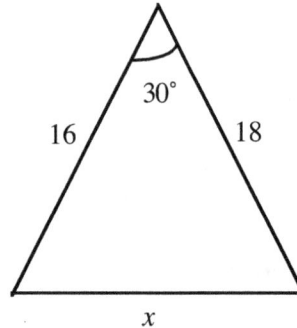

Solution: (a) See lesson and imitate; (b) 20.

C Determine if the two triangles in each problem with the give dimensions are similar:

1. Triangle #1 of sides 12, 16, 24 ; Triangle #2 of sides 8, 6, 12

2. Triangle #1 of sides 4, 6, 8; Triangle #2 of sides 8, 18, 12.

Answers: **1.** Similar; **2.** Not similar

D
Determine if the two triangles in each of the following problems are similar according to the data given.

1. Δ #1 of sides 3 and 4; included angle 75°; Δ #2 of sides 8 and 6, included angle 25°.

2. Δ #1 of sides 3, and 4; included angle 65°; Δ #2 of sides 8, and 6, included angle 65°

Answers: **1.** Not similar; **2.** Similar

Lesson 56

Comparison of Congruency and Similarity of Triangles; Applications of Similarity Theorems

Comparison of congruency and similarity of triangles

1. Corresponding angles: For **both** congruency and similarity, the **corresponding angles** are **congruent.**

2. Corresponding sides: For congruency we have **equality of sides** (equality of lengths)
   For similarity, we have **equality of ratios of sides** (equality of ratio of lengths)

3. For congruency, each ratio of corresponding sides = 1. (Ratio of similitude = 1.)
   For similarity, each ratio of corresponding sides could be 1 or another number.
   (Ratio of similitude = 1, or another number)

Note 1: If two triangles are congruent, they are also similar; but if two triangles are similar, they are not necessarily congruent.

Note 2: There is no need for **ASA** for similar triangles (the "**S**" in **ASA** is redundant), since **ASA** implies **AA** which is sufficient for two triangles to be similar.

Example 4   In Figure 4 below, $\overline{DE}$ is parallel to $\overline{BC}$.
             (a) Prove that $\triangle$ ADE $\sim$ $\triangle$ ABC. (b) If BC = 12, DE = 4, AD = 6, Find AB.

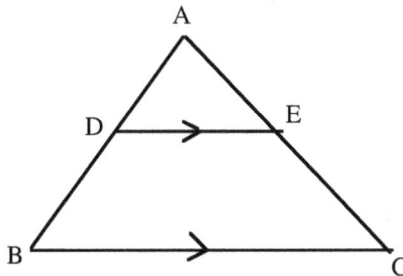

Figure 4

(a)   Given : $\triangle$ ABC in which DE || BC.
      To prove: That $\triangle$ ADE $\sim\triangle$ ABC.

Proof: Statements   Reason

   1.$\angle$ ADE $\cong$ $\angle$ ABC   **1.** Corresponding angles, || lines $\overline{DE}$, $\overline{BC}$ cut by transversal $\overline{AB}$.

   2.$\angle$ AED $\cong$ $\angle$ ACB   **2.** Corresponding angles, || lines $\overline{DE}$, $\overline{BC}$ cut by transversal $\overline{AC}$.

   3. $\triangle$ ADE $\sim\triangle$ ABC   **3. AA**

(b)
 BC = 12 ,DE = 4, AD = 6.
      Let AB = x
Since $\triangle$ ADE $\sim\triangle$ ABC, the corresponding sides are in proportion:

Therefore, $\dfrac{AD}{x} = \dfrac{DE}{BC}$ (AD corresponds to AB. They are opposite congruent angles)

Substituting: AD = 6, DE = 4, BC =12.

$$\frac{6}{x} = \frac{4}{12}$$
$$4x = 72$$
$$x = 18$$
$$\therefore AB = 18$$

Example 5 A tree cast a 18 ft. shadow at noon when a nearby upright pole of length 7 ft. cast a shadow of length 3 ft. Find the height of the tree assuming that both the tree and pole meet the ground at right angles.

Solution

We will assume that the measure of the angle between the sun's ray and the ground is the same for both the tree and the nearby pole.

Step 1: The tree, its shadow and the sun's ray form a right triangle. Similarly, the pole, its shadow together with the sun's ray form another right triangle. The two triangles formed are similar by AA. See Figure 1 below.

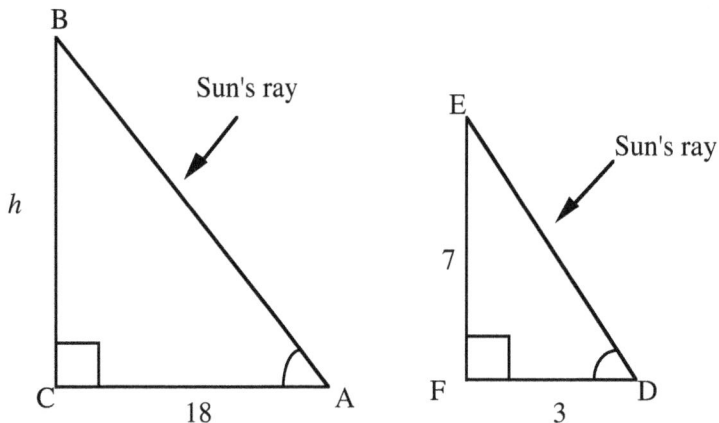

Figure 1

Step 2:

Since $\triangle ABC \sim \triangle DEF$,

$$\frac{h}{7} = \frac{18}{3}$$ ($\overline{BC}$ and $\overline{EF}$ are corresponding sides; $\overline{CA}$ and $\overline{FD}$ are corresponding sides)

and $h = \frac{18(7)}{3}$

$h = 42$

The height of the tree is 42 ft.

Extra If you study calculus in the future, you will apply similarity of triangles when covering problems on related rates and when finding volumes of solids of known cross-sections.

Lesson 56 Exercises

A tree cast a 18 ft. shadow at noon when a nearby upright pole of length 7 ft. cast a shadow of length 3 ft. Find the height of the tree assuming that both the tree and pole meet the ground at right angles.

Answer: 42 ft.

CHAPTER 22

Lesson 57: **Radian-Degree Conversions**
Lesson 58: **Right Triangle Trigonometry and Applications**
Lesson 59: **Special Δ 's: The 30°-60°-90° and 45°-45°-90° triangles**

Angle: In **geometry**, we define an angle as being formed by two rays (or line segments) meeting at a common point called the vertex of the angle. The main unit for measuring angles in geometry is the **degree**.

In **trigonometry**, we define an **angle** as being formed by rotating a line segment (or ray) about an end point, from an initial position to a final or terminal position. We call the initial position , the initial side and the terminal position the terminal side. The angle formed is measured by the amount of rotation. If the rotation is in a counterclockwise direction, then the measure of the angle formed is positive. If the rotation is in a clockwise direction, the measure of the angle formed is negative. In trigonometry, we measure angles in degrees as well as in **radians**.

Lesson 57
Radian-Degree Conversions

Converting from Degrees to Radians
Converting from one unit to another unit is a direct proportion problem, and as such any of the methods of solving direct proportion problems are applicable.

Example 1   Convert 240° to radians.

Solution $360° = 2\pi$ radians, or $\boxed{180° = \pi \text{ radians}}$

Method 1   Let x radians $= 240°$............(1)
π radians $= 180°$ ....(2)

Divide equation (1) by equation (2)

Then $\dfrac{x \text{ radians}}{\pi \text{ radians}} = \dfrac{240°}{180°}$, from which

$$x = \frac{\pi(240)}{180}$$

$$x = \frac{\pi(\cancel{240°})^4}{\cancel{3}_{180°}}$$

$$x = \frac{4\pi}{3}; \qquad \therefore 240° = \frac{4\pi}{3} \text{ radians.}$$

Method 2   Using the units label method

Step 1: $\dfrac{240 \text{ degrees}}{1} \times \dfrac{?}{?}$

Step 1: $\dfrac{240 \text{ degrees}}{1} \times \dfrac{\pi \text{ radians}}{180 \text{ degree}}$

$= \dfrac{240}{1} \times \dfrac{\pi \text{ radians}}{180}$

$= \dfrac{4\pi}{3} \text{ radians}$

Method 3 We will use simple proportion to solve the problem. Let x radians correspond to 240°.
Then x radians is to 240° as π radians is to 180°.

Step 1: " x radians is to 240° as π radians is to 180°" translates to:

$$\frac{x \text{ radians}}{240°} = \frac{\pi \text{ radians}}{180°}$$

Step 2: Solve for x:

$$x = \frac{\pi(240)}{180}$$

$$x = \frac{\pi(\cancel{240°})^4}{\cancel{180°}_3} = \frac{4\pi}{3} \; ; \quad \text{Therefore} \quad 240° = \frac{4\pi}{3} \text{ radians.}$$

Converting from Radians to Degrees

Example 2 Convert $\frac{7\pi}{5}$ radians to degrees.

Method 1 Let $x° = \frac{7\pi}{5}$ radians .... (1)

$180° = \pi$ radians .... (2)

Divide equation (1) by equation (2): $\dfrac{x°}{180°} = \dfrac{\frac{7\pi}{5} \text{ radians}}{\pi \text{ radians}}$

$$\frac{x°}{180°} = \frac{7\pi}{5} \frac{1}{\pi}$$

$$x = \frac{7\pi}{5} \frac{1(180)}{\pi}$$

Solving for x,

$$x = \frac{7\cancel{\pi}}{\frac{5}{1}} \frac{1\cancel{(180)}^{36}}{\cancel{\pi}}$$

$$x = 252$$

$$\therefore \frac{7\pi}{5} \text{ radians} = 252°.$$

Method 2

Step 1: $\dfrac{\frac{7\pi}{5} \text{ radians}}{1} \times \dfrac{?}{?}$

Step 1: $\dfrac{\frac{7\pi}{5} \text{ radians}}{1} \times \dfrac{180 \text{ degrees}}{\pi \text{ radians}}$

$$= \frac{7}{5} \times \frac{180 \text{ degrees}}{1}$$

$$= 252 \text{ degrees}$$

$\frac{7\pi}{5}$ radians $= 252°$

Method 3

Let $x°$ correspond to $\frac{7\pi}{5}$ radians.

Then, "$x°$ is to $\frac{7\pi}{5}$ radians as 180° is to π radians." translates to:

$$\frac{x°}{\frac{7\pi}{5}\text{ radians}} = \frac{180°}{\pi \text{ radians}}$$

Solving for x,

$$x = \frac{180}{\pi}\frac{7\pi}{5}$$
$$x = 252°$$
$$\therefore \frac{7\pi}{5}\text{ radians} = 252°.$$

We could have solved Examples 1 and 2 using proportion, knowing that π radians = 180°.

In Example 1: If 180° = π

Then, $240° = \frac{\pi \text{ radians}}{1} \cdot \frac{\overset{4}{\cancel{240}}}{\underset{3}{\cancel{180}}}$

$\therefore 240° = \frac{4\pi}{3}$ radians.

Lesson 57 Exercises

1. Convert 270° to radians
2. Convert $\frac{6\pi}{5}$ radians to degrees

Answers: **1**. $\frac{3\pi}{2}$ radians ; **2**. 216°

Lesson 58
Right Triangle Trigonometry and Applications

Trigonometric Ratios (Trigonometric Functions)
sine (**sin**); cosine (**cos**); tangent (**tan**); cosecant (**csc**); secant (**sec**); and cotangent (**cot**).

Definitions: The following definitions are applicable to **right** triangles only.

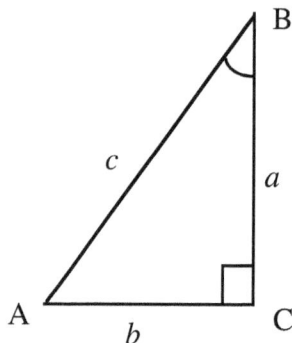

1. $\sin A = \dfrac{\text{opposite side}}{\text{hypotenuse}} = \dfrac{a}{c}$ (sin A is abbreviation for the sine of A)

2. $\cos A = \dfrac{\text{adjacent side}}{\text{hypotenuse}} = \dfrac{b}{c}$ (cos A is abbreviation for the cosine of A)

3. $\tan A = \dfrac{\text{opposite side}}{\text{adjacent side}} = \dfrac{a}{b}$ (tan A is abbreviation for the tangent of A)

4. $\csc A = \dfrac{1}{\sin A} = \dfrac{\text{hypotenuse}}{\text{opposite side}} = \dfrac{c}{a}$ (csc A is abbreviation for the cosecant of A)

5. $\sec A = \dfrac{1}{\cos A} = \dfrac{\text{hypotenuse}}{\text{adjacent side}} = \dfrac{c}{b}$ (sec A is abbreviation for the secant of A)

6. $\cot A = \dfrac{1}{\tan A} = \dfrac{\text{adjacent side}}{\text{opposite side}} = \dfrac{b}{a}$ (cot A is abbreviation for the cotangent of A)

Note above that $\sin B = \dfrac{b}{c}$; $\cos B = \dfrac{a}{c}$; $\tan B = \dfrac{b}{a}$

Note also that for ∠ B, the opposite side is $\overline{AC}$, and the adjacent side is $\overline{BC}$.

For ∠ A, the opposite side is $\overline{BC}$ and the adjacent side is $\overline{AC}$
Some mnemonic devices will be discussed after Example 1

Example 1   Solve the right triangle ABC for the unknown parts, given
that $b = 4$, m $\angle$ B = 30°

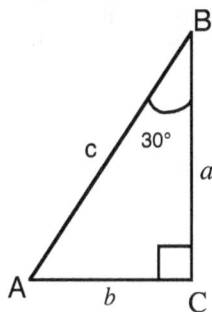

Solution

(a) m $\angle$ A $= 180° - (30° + 90°)$ (m$\angle$ C = 90°; sum of the measures of the $\angle$'s of a Δ is 180)

$\qquad\qquad = 180° - 120°$

$\quad$ m $\angle$ A $= 60°$

(b) To find c:

$\qquad c$ is the hypotenuse, $b = 4$ is opposite $\angle$ B.

$\qquad \sin 30 = \dfrac{4}{c}$ $\qquad\qquad\qquad$ (m$\angle$ B = 30°,b = 4)

$\qquad c \sin 30° = 4$

$\qquad c = \dfrac{4}{\sin 30°} = \dfrac{4}{\frac{1}{2}} = 8$ $\qquad$ $(\sin 30° = \frac{1}{2})$

$\qquad$ now $c = 8$, $b = 4$

Apply the Pythagorean theorem (We may apply this theorem since Δ ABC is a right triangle)

$\qquad\qquad 8^2 = 4^2 + a^2$

$\qquad\qquad 64 = 16 + a^2$

$\qquad\qquad 48 = a^2$

$\qquad\qquad a = \sqrt{48} = \sqrt{16}\sqrt{3} = 4\sqrt{3}$

Note: In (b), we could have found a first using the tangent relationship

$\tan A = \dfrac{a}{4}$ $\qquad \left(\dfrac{\mathbf{O}\text{pposite}}{\mathbf{A}\text{djacent side}} \text{ "TOA" } \longleftarrow \text{-------- mnemonic device}\right)$

$a = 4 \tan A$ $\qquad\qquad\qquad\qquad$ **Also**, $\cos 30° = \dfrac{a}{c}$

$a = 4 \tan 60°$ $\qquad\qquad\qquad\qquad\qquad$ $\dfrac{\sqrt{3}}{2} = \dfrac{a}{8}$ and $a = \dfrac{8(\sqrt{3})}{2} = 4\sqrt{3}$

$a = 4\sqrt{3}$

Note: The determination of which trigonometric relationship to use depends on what we are given and what we want to find. A good practice is to scribble down the following three relationships:

$\sin \theta = \dfrac{\text{opposite side}}{\text{hypotenuse}}$ $\quad \left(\dfrac{\mathbf{O}\text{pposite}}{\mathbf{H}\text{ypotenuse}} : \text{"SOH" } \longleftarrow \text{-------- mnemonic device}\right)$

$\cos \theta = \dfrac{\text{adjacent. side}}{\text{hypotenuse.}}$ $\quad \left(\dfrac{\mathbf{A}\text{djacent}}{\mathbf{H}\text{ypotenuse}} : \text{"CAH" } \longleftarrow \text{-------- mnemonic device}\right)$

$\tan \theta = \dfrac{\text{opposite side}}{\text{adjacent side}}$ $\quad \left(\dfrac{\mathbf{O}\text{pposite}}{\mathbf{A}\text{djacent}} : \text{"TOA" } \longleftarrow \text{-------- mnemonic device}\right)$

and use the relationship in which you know all but one quantity. Sometimes, you may use two relationships simultaneously to find two unknowns.

Applications of Right Angle Trigonometry

Definition: For an observer sighting (looking) at something above the observer, **the angle of elevation** is the angle between the horizontal line (x-axis) from the observer's eye and the line of sight to the object, the angle being measured from the horizontal to the line of sight.

Example Find the height reached by a kite, if the length of the kite is 400 ft, and the angle of elevation of the kite is 60°.

Step 1: Sketch a diagram for the system under consideration. Assume that the hand holding on to the kite and the eye coincide at one point.

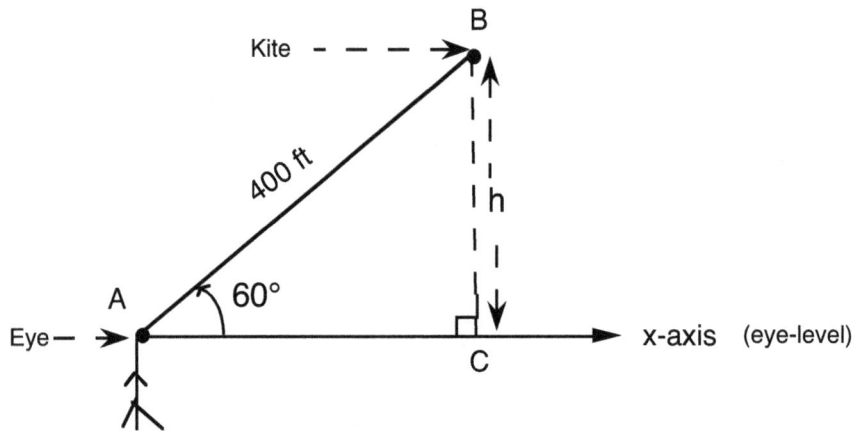

AB = Length of kite = 400 ft.
BC = h = height of kite (vertical distance from eye-level to the kite)

Step 2: $\triangle$ ABC is a right triangle, and therefore we may apply the trigonometric relationships.

$$\sin 60° = \frac{h}{400} \qquad \left(\text{SOH} : \frac{\text{Opposite}}{\text{Hypotenuse}} \right)$$

$$h = 400 \sin 60°$$

$$= 400 \left(\frac{\sqrt{3}}{2} \right) \text{ ft.}$$

$$= 200 \sqrt{3} \text{ ft.} \qquad\qquad (\sqrt{3} \approx 1.73)$$

$$\approx 346.41 \text{ ft.}$$

The height reached is $200 \sqrt{3}$ ft. ($\approx$346.41 ft.)

Angle of depression

Definition:

For an observer sighting (looking) at something below the observer, the angle of depression is the angle between the horizontal line (x-axis) from the observer's eye and the line of sight to the object, the angle being measured from the horizontal to the line of sight.

In the figure below: $\angle \alpha$ is the angle of elevation of B from A.

$\quad\quad\quad\quad\quad\quad$ $\angle \beta$ is the angle of depression of A from B.

The angle of elevation is congruent to the angle of depression.

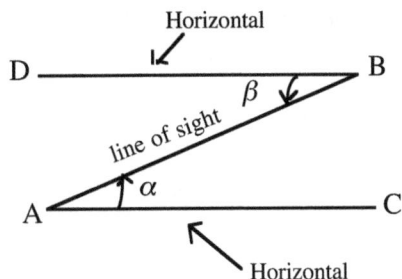

Lesson 58 Exercises

A Solve the right triangle ABC for the unknown parts, given
that $b = 6$, and $m \angle A = 60°$

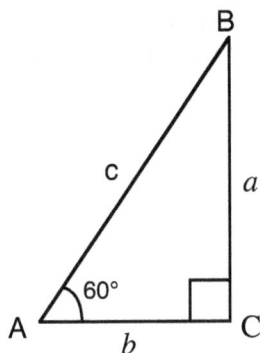

$\quad\quad\quad\quad\quad$ Answers: $m \angle B = 30°$; $a = 6\sqrt{3}$; $c = 12$

B Find the height reached by a kite, if the length of the kite is 600 ft, and the angle of elevation of the kite is 30°.

$\quad\quad\quad\quad\quad\quad$ Answer: 300ft.

Lesson 59

Special triangles: **The 30º-60º-90º triangle and the 45º-45º-90º triangle**

The trigonometric functions for 30º, 45º, 60º, and 90º occur very frequently in mathematics that students are usually required to memorize them.

The 45º-45º-90º triangle

By sketching an isosceles right triangle with measures of 45º, 45º, and 90º (Figure 1) we will be able, using the basic definitions, write down the trigonometric functional value for 45º.

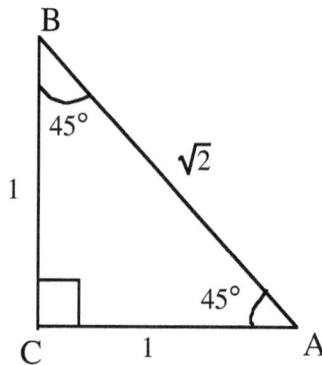

Figure 1

In an isosceles triangle, the sides opposite the congruent angles have the same length. If we let the length of the sides opposite the congruent angles be 1 unit each, we can calculate the length of the hypotenuse using the Pythagorean theorem.

$$AB^2 = AC^2 + BC^2$$
$$AB^2 = 1^2 + 1^2$$
$$AB^2 = 2$$
$$AB = \sqrt{2}.$$

Example $\sin 45° = \dfrac{\text{opposite side}}{\text{hypotenuse}} = \dfrac{1}{\sqrt{2}} = \dfrac{1}{\sqrt{2}} \cdot \dfrac{\sqrt{2}}{\sqrt{2}} = \dfrac{\sqrt{2}}{2}$ (Rationalizing the denominator)

$\cos 45° = \dfrac{\text{adjacent side}}{\text{hypotenuse}} = \dfrac{1}{\sqrt{2}} = \dfrac{1}{\sqrt{2}} \cdot \dfrac{\sqrt{2}}{\sqrt{2}} = \dfrac{\sqrt{2}}{2}$

$\tan 45° = \dfrac{\text{opposite side}}{\text{adjacent side}} \dfrac{1}{1} = 1.$

The 30⁰-60⁰-90⁰ triangle

For the 30⁰-60⁰-90⁰ triangle (Figure 2), the dimensions are obtained by considering an equilateral triangle whose sides are each 2 units long (see page 252, Theorems 2 and 3). From geometry, the length of the hypotenuse is twice the length of the shortest side. The third side is calculated by the Pythagorean theorem.

Thus if $AB = 2$, then $AC = 1$ and $2^2 = 1^2 + BC^2$ and from which $BC = \sqrt{3}$. Usually, after having gone through the derivation, using geometric considerations and the Pythagorean theorem, students become familiar with the dimensions $1, 2, \sqrt{3}$. The problem of recall that remains then is how to place these dimensions on the 30⁰-60⁰-90⁰ triangle, if you do not want to draw the equilateral triangle, but draw only $\triangle ABC$. The following mnemonic device will be helpful:

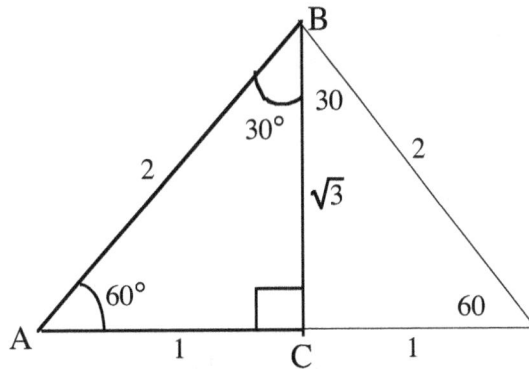

Figure 2

Mnemonic device: If you can remember the numbers $1, 2, \sqrt{3}$ as being the lengths of the sides of this triangle, and also remember that in any triangle, the longest side is opposite the largest angle, and that the shortest side is opposite the smallest angle, you should be able to place these dimensions accordingly on the triangle. Thus, 1 is opposite the 30⁰ angle, 2 is opposite the 90⁰ angle (the right angle) , and $\sqrt{3}$ is opposite the 60⁰ angle. ($\sqrt{3} \approx 1.73$)

By applying the mnemonic devices, SOH , CAH, TOA (see p.269) we can write down some functional values.

Examples

$\sin 30^\circ = \dfrac{\text{opposite side}}{\text{hypotenuse}} = \dfrac{1}{2}$

$\cos 30^\circ = \dfrac{\text{adjacent side}}{\text{hypotenuse}} = \dfrac{\sqrt{3}}{2}$

$\sin 60^\circ = \dfrac{\text{opposite side}}{\text{hypotenuse}} = \dfrac{\sqrt{3}}{2}$.

Lesson 59 Exercises

In Figure 2, above, find the following:

1. $\tan 30^\circ$;   **2.** $\tan 60^\circ$;   **3.** $\csc 30^\circ$; **4.** $\csc 60^\circ$;   **5.** $\sec 30^\circ$;   **6.** $\sec 60^\circ$.

Answers: 1. $\frac{\sqrt{3}}{3}$; 2. $\sqrt{3}$; 3. 2; 4. $\frac{2\sqrt{3}}{3}$; 5. $\frac{2\sqrt{3}}{3}$; 6. 2

CHAPTER 23

Trigonometric Functional Value of any Angle

Lesson 60

Introduction: Reference Angle

Angle in standard position

Definition: An angle is in standard position in an *x-y* rectangular coordinate system if it's vertex is at the origin and its initial side is along positive *x* -axis.

Reference Angle

Definition: If an angle, θ, is standard position, then the reference angle is the positive acute angle between the terminal side of θ and the *x*- axis (the horizontal axis).

Example 1   Find reference angle for 230°.

Solution

Step 1:   Draw (or sketch) θ = 230° in standard position.
To draw the arc, begin from the positive *x*-axis and draw the arc counterclockwise.

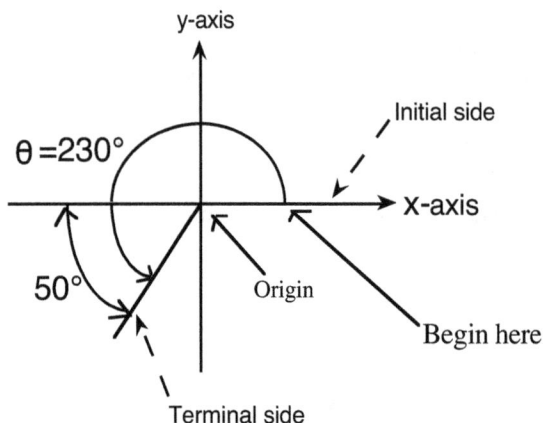

Step 2:   Determine the reference angle by difference. The reference angle for 230° is 230 - 180
i.e. 50°
The reference angle for 230° is 50° .

Note: The reference angle is always between the terminal side and the *x*-axis.

Example 2 Find the reference angle for 460º

Solution

Step 1: Draw (sketch) θ = 460º in standard position.

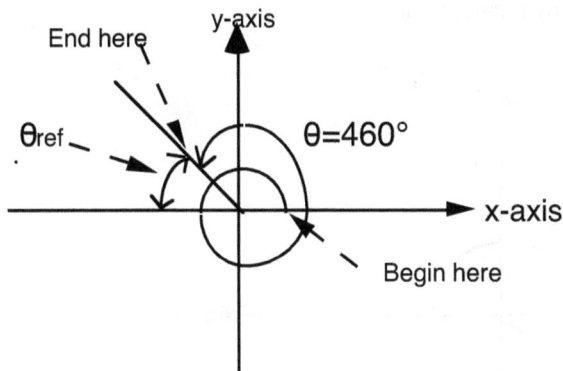

Step 2: The reference angle is the difference between 180º and 100º.

i.e., 180º - 100º = 80º

Note that 360º will bring you back to the initial side, leaving 460 - 360 = 100 to consider,

Lesson 60 Exercises

Find the reference angles for the following: **1.** 250˚ ; **2.** 295˚; **3.** 560˚; **4.** 75˚; **5.** 360˚

Answers. **1.** 70˚ ; **2.** 65˚ ; **3.** 20˚; **4.** 75˚ ; **5.** 0˚

Lesson 61

Trigonometric Functional Value (Given the measure of the angle)

Example   (a) Find the value of sin 210° without using a calculator: use trigonometric tables.
(b) Verify your answer using a calculator.

Solution (a)

Step 1:   Draw (or sketch) the given angle in standard position.

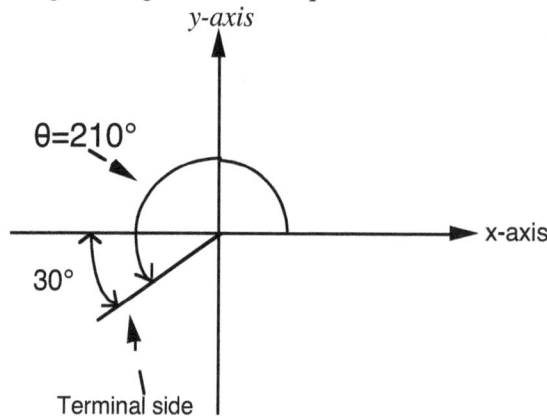

Figure 1

Step 2:   Find the reference angle.

The reference angle is 210° - 180° = 30°

Step 3:   From tables or from memory, find sin 30°. This value will be positive, and we must then determine the sign of sin 210° according to the quadrant in which the terminal side of this angle lies. (See Figure 2 below) The terminal side of 210° is in the 3rd quadrant where the sine function is negative. $\therefore \ \sin 210 = -\sin 30 = -\frac{1}{2}$

Signs of Trigonometric Functions According to Quadrants

Mnemonic device " **CAST**" for remembering in which quadrant a trigonometric function is positive or negative

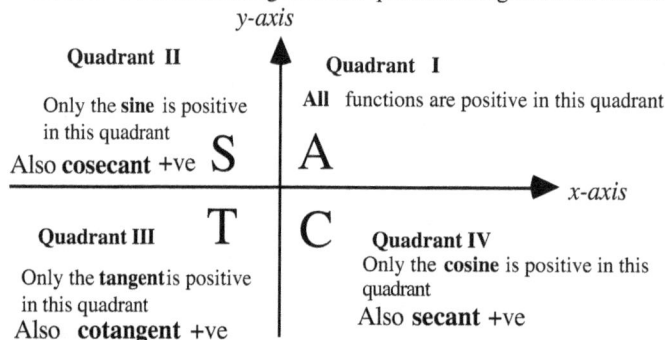

Figure 2

In the above device: The **A** in "CAST" means **All** functions are positive in this quadrant; **S** for only sine is positive in this quadrant; and similarly **T** for tangent; and **C** for cosine.

Note: The order of the above"CAST" is counterclockwise. There are other mnemonic devices such as "**All Students Take Calculus**".

(b) With the calculator in degree mode, press 210; then press "the sin" key and read -0.5

Note above: The sign of a trigonometric function and its reciprocal are the same.

Lesson 61 Exercises

In the following problems, first use tables; and then verify your answers using a calculator.

Find the values of the following: **1**. sin 250° ; **2.** cos 250°; **3.** sin 460°; **4.** sin 75°

Answers: **1.** - 0.94 ; **2.** -0.34; **3.** 0.98; **4.** 0.96

Lesson 62
Finding Trigonometric Functional Value
(Given the coordinates of a Point on the Terminal Side of Angle)

Example Find the six trigonometric values of θ , (if θ is an angle in standard position) given that the terminal side of θ passes through the point (-3, $2\sqrt{3}$).

Step 1:   Draw (or sketch) θ with terminal side passing through the point (-3, $2\sqrt{3}$)
          (x-coordinate = -3; y-coordinate = $2\sqrt{3}$)

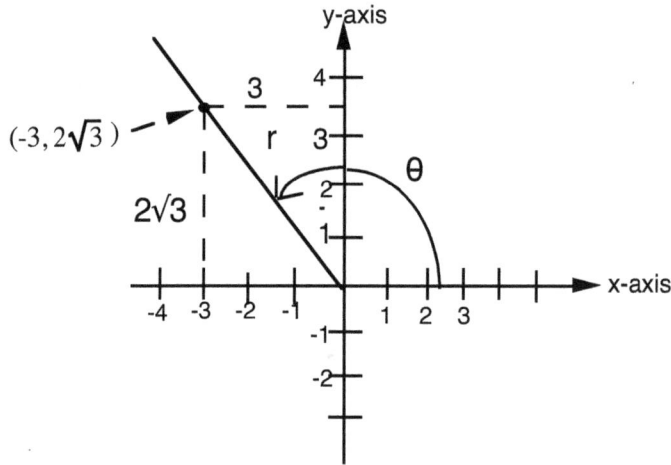

Step 2: Find r using the Pythagorean Theorem.
$$r^2 = (-3)^2 + (2\sqrt{3})^2 \quad (r^2 = x^2 + y^2)$$
$$r^2 = 9 + 4\sqrt{9}$$
$$r^2 = 9 + 12 \qquad \text{Note: } 4\sqrt{9} = 4(3) = 12)$$
$$r^2 = 21$$
$$r = \sqrt{21}$$

Step 3:   Now, $r = \sqrt{21}$, $x = -3$, $y = 2\sqrt{3}$

(a) $\cos \theta = \dfrac{x}{r}$

$$= \dfrac{-3}{\sqrt{21}}$$

$$= -\dfrac{3}{\sqrt{21}} \cdot \dfrac{\sqrt{21}}{\sqrt{21}} \quad \text{(Rationalizing the denominator)}$$

$\therefore \cos \theta = -\dfrac{3\sqrt{21}}{21}$

(b) $\sin \theta = \dfrac{y}{r}$ $\qquad\qquad (y = 2\sqrt{3}, \ r = \sqrt{21})$

$$= \dfrac{2\sqrt{3}}{\sqrt{21}}$$

$$= \frac{2\sqrt{3}}{\sqrt{21}} \cdot \frac{\sqrt{21}}{\sqrt{21}} \qquad \text{(Rationalizing the denominator)}$$

$$= \frac{2\sqrt{63}}{21}$$

$$= \frac{2\sqrt{9}\sqrt{7}}{21}$$

$$= \frac{2\cancel{(3)}\sqrt{7}}{\cancel{21}\,7} \qquad\qquad (\sqrt{9} = 3)$$

$$\sin \theta = \frac{2\sqrt{7}}{7}$$

(c) $\quad \tan \theta = \dfrac{y}{x} \qquad\qquad (y = 2\sqrt{3}, x = -3)$

$$= \frac{2\sqrt{3}}{-3}$$

$$\therefore \tan \theta = -\frac{2\sqrt{3}}{3}$$

(d) $\quad \sec \theta = \dfrac{1}{\cos \theta}$

$$= \frac{1}{\dfrac{x}{r}} = \frac{r}{x} = \frac{\sqrt{21}}{-3}$$

$$\therefore \sec \theta = -\frac{\sqrt{21}}{3}$$

(e) $\quad \csc \theta = \dfrac{1}{\sin \theta}$

$$= \frac{r}{y}$$

$$= \frac{\sqrt{21}}{2\sqrt{3}}$$

$$= \frac{\sqrt{21}}{2\sqrt{3}} \cdot \frac{\sqrt{3}}{\sqrt{3}} \qquad \text{(Rationalizing the denominator)} \qquad = \frac{\sqrt{63}}{2\sqrt{9}}$$

$$= \frac{\sqrt{63}}{2(3)} \qquad\qquad (\sqrt{9} = 3)$$

$$= \frac{\sqrt{9}\sqrt{7}}{6}$$

$$= \frac{3\sqrt{7}}{6}$$

$$\therefore \csc \theta = \frac{\sqrt{7}}{2}$$

(f) $\quad \cot \theta = \dfrac{1}{\tan \theta}$

$$= \frac{1}{\dfrac{y}{x}}$$

$$= \frac{x}{y}$$

$$= \frac{-3}{2\sqrt{3}} \qquad (x = -3, y = 2\sqrt{3})$$

$$= -\frac{3}{2\sqrt{3}} \cdot \frac{\sqrt{3}}{\sqrt{3}} \qquad \text{(Rationalizing the denominator).}$$

$$= -\frac{3\sqrt{3}}{2\sqrt{9}}$$

$$= -\frac{3\sqrt{3}}{2(3)} \qquad (\sqrt{9} = 3)$$

$$\therefore \cot \theta = -\frac{\sqrt{3}}{2}$$

Note in the last two examples that:
1. In Lesson 61,, we had to determine the sign of the function according to the quadrant (in which the terminal side of the angle lies) if we obtain the values of the acute angles from tables or from memory.

2. In Lesson 62, we did not have to determine the sign of the function. The signs were obtained solely from the signs of the x- and y-coordinates; but note also that r is always positive.

Lesson 62 Exercises

Find the six trigonometric values of θ , if θ is an angle in standard position, given that the terminal side of θ passes through the point (5,2).

Answers: $\sin \theta = \frac{2\sqrt{29}}{29}$; $\cos \theta = \frac{5\sqrt{29}}{29}$; $\tan \theta = \frac{2}{5}$; $\csc \theta = \frac{\sqrt{29}}{2}$; $\sec \theta = \frac{\sqrt{29}}{5}$; $\cot \theta = \frac{5}{2}$

Lesson 63

Finding Other trigonometric functional values
(Given one functional value and the quadrant of the angle)

Example   If $\sin \theta = -\dfrac{4}{5}$ and the terminal side of θ is in the fourth quadrant (θ being in standard position)

(a) Find $\cos \theta$, (b) $\tan \theta$, (c) $\csc \theta$

We will determine r, x and y and apply the definitions of the trigonometric functions

Step 1: Since we know that the terminal side is in the 4th quadrant, we will draw θ accordingly.

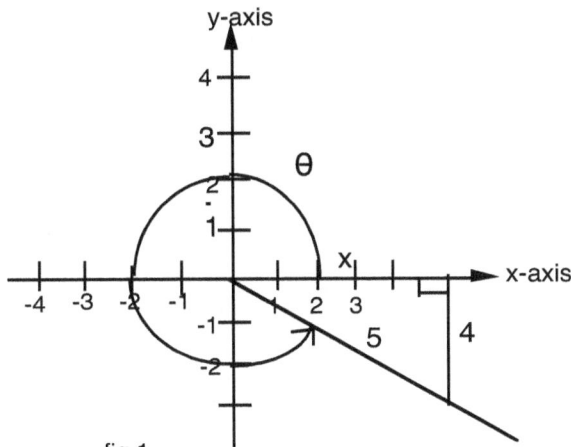

fig.1

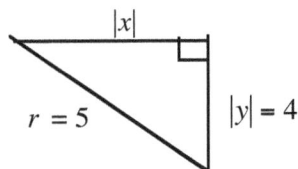

fig.2

Step 2: In this quadrant , the y-coordinate is negative and the x-coordinate is positive

$$\sin \theta = \frac{y}{r} = -\frac{4}{5}$$

$y = -4$ since r is always positive (We give the minus sign to the "4") .
$r = 5$

Step 3:   To find $x,$ we use the Pythagorean theorem.

$$5^2 = (-4)^2 + x^2$$
$$25 = 16 + x^2$$
$$9 = x^2$$
$$3 = x$$
$$\therefore \quad x = 3, \ y = -4, \ r = 5$$

Step 4:

(a) $\cos \theta = \dfrac{x}{r}$

   $\cos \theta = \dfrac{3}{5}$ $\qquad\qquad$ $(x = 3, r = 5)$

(b) $\tan \theta = \dfrac{y}{x}$

   $= \dfrac{-4}{3}$ $\qquad\qquad$ $(y = -4, x = 3)$

   $= -\dfrac{4}{3}$

(c) $\csc \theta = \dfrac{1}{\sin \theta}$

   $= \dfrac{r}{y}$

   $= \dfrac{5}{-4}$

   $= -\dfrac{5}{4}$

Note: We could have obtained $\csc \theta$ from $\sin \theta = -\dfrac{4}{5}$ by inversion.

Lesson 63 Exercises

If $\sin \theta = -\dfrac{3}{7}$ and the terminal side of θ is in the third quadrant (θ being in standard position)
(a) Find $\cos \theta$, (b) $\tan \theta$, (c) $\csc \theta$

Solutions: $\cos \theta = \dfrac{-2\sqrt{10}}{7}$; (b) $\tan \theta = \dfrac{3\sqrt{10}}{20}$; (c) $\csc \theta = -\dfrac{7}{3}$

CHAPTER 24

Trigonometry of Oblique Triangles

Lesson 64: **The Laws of Cosines and Sines**

Lesson 65: **Applications of the Laws of Cosines and Sines**

An **oblique triangle** is a triangle in which no angle is a right angle. We may consider the right triangle as a special case of the oblique triangle and that the laws which are applicable to oblique triangles are also applicable to the right triangle.

In solving oblique triangles, we are usually given three parts of the oblique triangle and we are required to find the remaining three parts. We will usually seek a unique solution (that is only one solution). There may be a unique solution, two distinct solutions or no solutions at all depending on the relative lengths of the sides and the measures of the angles

Lesson 64

The Laws of Cosines and Sines

Use of the Law of Cosines.

Direct use of the Law of Cosines applies to the following cases:

(1) Given two sides and an included angle (abbreviated SAS) we are therefore to find the remaining side and two angles.

(2) Given all three sides (abbreviated SSS) we are therefore to find the remaining three angles. The above cases yield unique solutions. A mnemonic device for the applicability of this law of cosines is **Co-SAS -SSS** - pronounced Kosaszz.

Statement of the Law of Cosines

Using standard notation for corresponding angles and sides, the law of cosines states that :

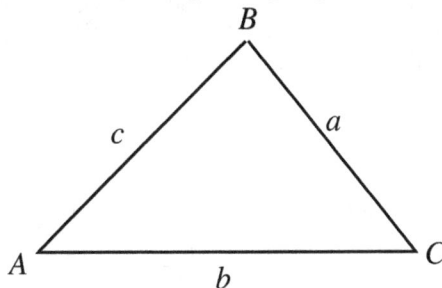

$$a^2 = b^2 + c^2 - 2bc \cos A$$

or

$$b^2 = a^2 + c^2 - 2ac \cos B$$

or

$$c^2 = a^2 + b^2 - 2ab \cos C$$

In words, the square of the length of one side is equal to the sum of the squares of the other two sides minus twice the product (of the lengths) of the other two sides multiplied by the cosine of the angle between these two sides.

Law of Cosine for the Right Triangle (a special case)

For a right triangle, let $m\angle A = 90°$. Then the law of cosines, $a^2 = b^2 + c^2 - 2bc \cos A$ becomes

$$a^2 = b^2 + c^2 - 2bc \cos 90°$$

$$a^2 = b^2 + c^2 - 2bc\,(0) \qquad (\cos 90° = 0)$$

$$a^2 = b^2 + c^2 \leftarrow --- \text{The Pythagorean theorem}$$

We may conclude that the law of cosines is a generalization of the Pythagorean theorem.

The Law of Sines

The **direct use** of the Law of Sines applies in the following cases:

(1) Given **two angles** and an **included side** (ASA) and therefore, we are to find the remaining sides and angle. This case yields a unique solution.

(2) Given **two sides** and an **angle** (ASS or SSA) there may be only one solution (one triangle), two solutions (two triangles) or there may be no solution at all (no triangle) depending upon the relative lengths of the sides and the measures of the angles. This case is commonly known as the ambiguous case of the law of sines.

A mnemonic device for its application is:
(1) sin ASA - pronounced "sinasa" **unique case**
(2) ASS - t**he ambiguous case**

Statement of the Law of Sines

The law of sines states that in any triangle, the sides are proportional to the sines of the angles opposite (facing) these sides. That is, the ratio of the length of a side to the sine of the angle opposite this side is equal to the ratio of the length of any other side to the sine of the angle opposite that side.

Using standard notation for angles and sides,

$$a \text{ is to } \mathbf{sin\ A} \text{ as } b \text{ is to } \mathbf{sin\ B} \text{ as } c \text{ is to } \mathbf{sin\ C}$$

Symbolically, $\dfrac{a}{\sin A} = \dfrac{b}{\sin B} = \dfrac{c}{\sin C}$ (equivalent to $\dfrac{\sin A}{a} = \dfrac{\sin B}{b} = \dfrac{\sin C}{c}$)

Lesson 64 Exercises

1. State the Law of Sines from memory.

2. State the Law of Cosines from memory

Lesson 65

Applications of the Law of sines, Law of cosines

Example 1: Solve $\triangle ABC$ given that, $m\angle C = 150°$, $b = 3$, $c = 8$

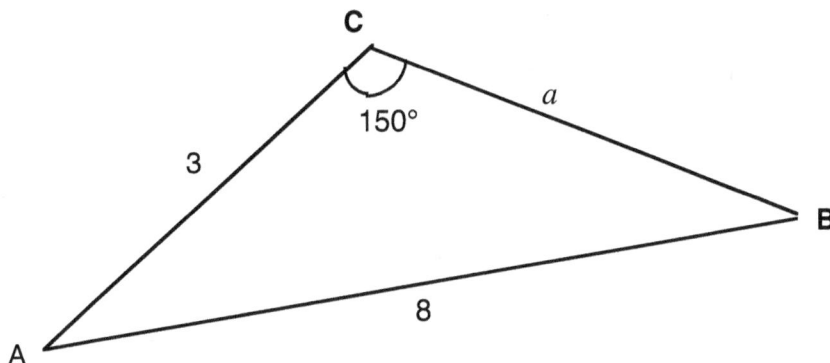

To **solve the triangle**, we will find the remaining three parts, namely the length a, $m\angle A$, and $m\angle B$.

Solution: We **cannot** apply the Pythagorean theorem, since $\triangle ABC$ is **not** a right triangle.

We will try the law of sines, or the law of cosines.

We apply the law of sines: $\dfrac{a}{\sin A} = \dfrac{b}{\sin B} = \dfrac{c}{\sin C}$

Step 1:   We apply $\dfrac{c}{\sin C} = \dfrac{b}{\sin B}$

$$\dfrac{8}{\sin 150°} = \dfrac{3}{\sin B}$$

$$\dfrac{8}{\sin 30°} = \dfrac{3}{\sin B} \qquad\qquad (\sin 150° = \sin 30° = \tfrac{1}{2})$$

$$\dfrac{8}{\frac{1}{2}} = \dfrac{3}{\sin B} \qquad (b = 3; c = 8; \sin 150° = \sin 30° = \tfrac{1}{2})$$

$$8 \sin B = \dfrac{3}{2} \qquad\qquad \text{(by cross-multiplication or otherwise)}$$

$$\sin B = \dfrac{3}{2} \times \dfrac{1}{8} = \dfrac{3}{16}$$
$$= .1875$$

Step 2:   Find the angle whose sine is .1875 (From tables or calculator)
        then, $m\angle B = 10.81°$
    $m\angle A = 180° - (150° + 10.8°) = 19.19°$

Finding a

Method 1 Given: $c = 8$, m $\angle$ C$= 150°$, $b = 3$

Using the law of cosines:

$$c^2 = a^2 + b^2 - 2ab \cos C \quad (\angle \text{ C is an included angle})$$

$$8^2 = a^2 + (3)^2 - 2a(3) \cos 150°$$

$$64 = a^2 + 9 - 2a(3)\left(-\frac{\sqrt{3}}{2}\right) \qquad (\cos 150° = -\cos 30° = -\frac{\sqrt{3}}{2})$$

$$64 = a^2 + 9 + 2a(3)\frac{\sqrt{3}}{2}$$

$$64 = a^2 + 9 + 3\sqrt{3}\, a$$

$$0 = a^2 + 3\sqrt{3}\, a - 55 \;\text{<-------------quadratic equation}$$

Solve by the quadratic formula.
$$a^2 + 3\sqrt{3}a - 55 = 0$$

$$a = \frac{-3\sqrt{3} \pm \sqrt{(3\sqrt{3})^2 - 4(1)(-55)}}{2} \qquad (a = 1, b = 3\sqrt{3}, c = -55)$$

$$= \frac{-3\sqrt{3} \pm \sqrt{9(3) + 220}}{2}$$

$$= \frac{-3\sqrt{3} \pm \sqrt{247}}{2}$$

$$= \frac{-3\sqrt{3} \pm 15.71}{2} = \frac{-3\sqrt{3} + 15.71}{2} \;\text{or}\; \frac{-3\sqrt{3} - 15.71}{2}$$
$$a = 5.26 \text{ or } -10.45$$

we reject the negative root since the length of a side of a triangle is always positive.
$\therefore a = 5.26$.

Finding a

Method 2   We could also have found a, using the law of sines after having found.

m ∠ A = 19.19°

$$\frac{a}{\sin A} = \frac{c}{\sin C}$$

$$\frac{a}{\sin 19.19°} = \frac{8}{\sin 150°}$$ (sin 19.19 = .3287)

(sin 150° =sin 30° = .5)

$$a = \frac{8 \sin 19.19}{\sin 150}$$

$$= \frac{8(.32870)}{.5}$$

$$a = 5.26$$

Again, we obtain the same value for a. Method 2 is less involved than Method 1. However, the disadvantage of Method 2, in this particular problem, is that we used a value, m ∠ A = 19.19°, which we calculated; and if there were error in m ∠ A, there would be error in the value of a. The advantage of Method 1 is that we used only given values and no calculated values ($c = 8$, m ∠ C = 150, $b = 3$); but then, we had to solve a quadratic equation.

Example 2   Solve Δ ABC given that   m ∠ C = 130° , $a = 6$, $b = 9$.

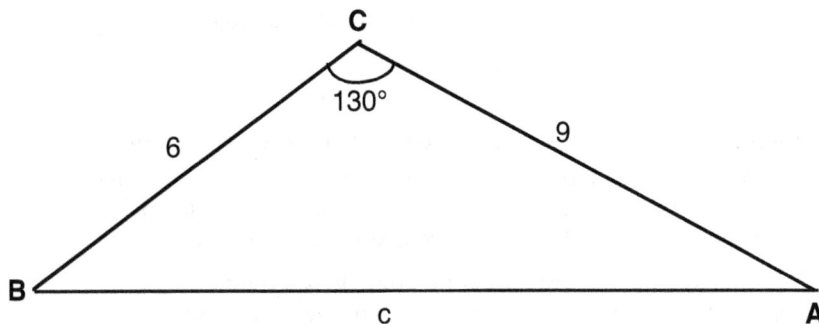

We will find m ∠ B, M ∠A and length c.

Finding c

$$c^2 = a^2 + b^2 - 2ab \cos C$$ <-------Law of cosines

$$c^2 = 6^2 + 9^2 - 2\,(6)(9) \cos 130°$$

$$= 36 + 81 - 108(-.642)$$ (cos 130° = - .642)

$$= 36 + 81 + 69.42$$

$$c^2 = 186.42$$

$$c = \sqrt{186.42}$$

$$c = 13.65$$

(Note above that if we had applied the law of sines to find c, we would have had one equation with two unknowns; and we would not be able to obtain a numerical value for c without an additional equation with the same unknowns.)

Finding $M \angle A$:

$$\frac{c}{\sin 130°} = \frac{a}{\sin A} \qquad \text{(law of sines).}$$

$$\frac{13.65}{\sin 130} = \frac{6}{\sin A} \quad (c = 13,65, a = 6)$$

We solve for sine A.

$$\sin A = \frac{6 \sin 130}{13.65}$$

$$= \frac{6(.766)}{13.65}$$

$$\sin A = .3367$$

Find inverse sine of .3367   (from tables or a calculator)
(That is, find the angle whose sine is .3367)
$$m \angle A = 19.68$$

$$m \angle B = 180 - (130 + 19.68)$$
$$m \angle B = 30.32$$

(We could also use the law of sines to find $m \angle B$, and then check that the sum of the measures of the angles equals 180°)

$$\therefore \quad c = 13.65 \; ; \; m \angle A = 19.68° \; ; \; m \angle B = 30.32°$$

When do we apply the Pythagorean theorem, the law of cosines, or the law of sines?

We have learned how to solve triangles using the Pythagorean theorem, the law of sines and the law of cosines. Given a triangle to solve, how do we know which theorem or law to apply? Which theorem or law to apply depends on what we are given and what we want to find.

1. We may apply the Pythagorean theorem only if the triangle is a right triangle. However, if by construction, we are able to obtain right triangles, then we can apply the theorem to any such triangles obtained.

2. For the law of cosines:
   (a) Given **two sides and an included angle** (SAS), use of the law of cosines yields a unique solution (i.e. one triangle). In this case, we will find the remaining side and the measures of two angles.

   (b) Given all **three sides** (SSS), the application of the law of cosines yields a unique solution. In this case, we will find the measures of all the angles.

3. For the law of sines:
   (a) Given **two angles** and **any side** (ASA, AAS, SAA), the use of the law of sines will yield a unique solution (one triangle).

   Note that given any two angles, we can easily find the third angle, since the sum of the measures of the angles of a triangle equals 180°.

(b) **Ambiguous Case:** Given **two sides** and **an angle opposite one of the sides** (ASS or SSA), there may be only one solution (one triangle), two solutions (two Δ's), or there may be no solution at all (no triangle) depending on relative lengths of the sides and the measures of the angles,

Note also that if the measure of any angle given is 180º or more there is no solution, since the sum of the measures of the three angles of a triangle is 180º.

Also, if the sum of the measures of any two angles given is 180° or more there is no solution.

Note also that if the triangle is a **right triangle**, we can consider applying any of the six trigonometric functions (ratios): sine, cosine, tangent, cosecant, secant, and cotangent.

Question: How do we remember the above cases?
Answer: By practice, repetition and finding mnemonic devices.

Lesson 65 Exercises

A Solve the triangle:

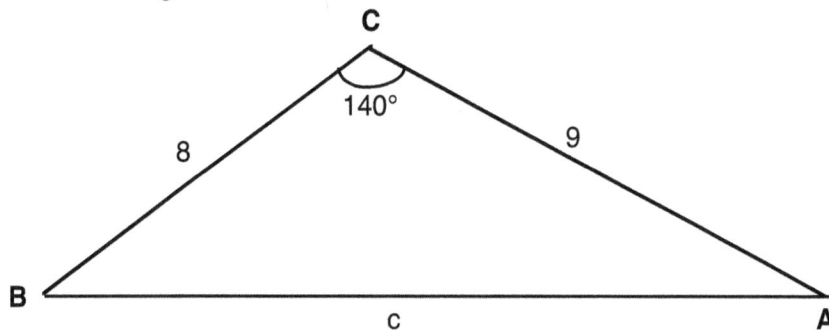

Solutions: c = 15.99;   m ∠ B = 21.2° ; m ∠ A = 18.8°
            ≈ 16;   ≈ 21°;   ≈ 19°

B Solve the following triangles:

1.

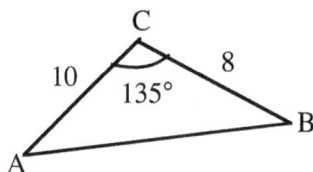

2.

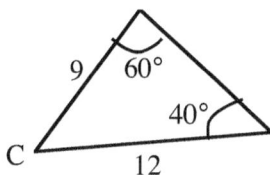

Solve the following triangles using the given dimensions:
3. a = 15;   m∠A = 50°;   m∠B = 110°;   **4.** b = 8 ;   m∠B = 75°;   m∠C = 55°

Answers: 1. c = 17, m∠A = 20°;   m∠B = 25°; **2.** c = 13.7, m∠C = 80°
      3. m∠C = 20°; b = 18;   c = 6.7;   **4.** m∠A = 50°;   a = 6.3;   c = 6.8

C Determine the measure of the largest angle to the nearest degree.

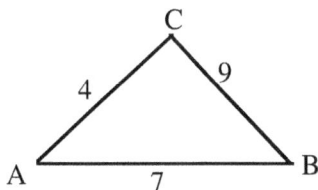

Answer: m∠A = 107°   (Hint: Largest angle is opposite longest side)

CHAPTER 25

Lesson 66: **Trigonometric Identities**
Lesson 67: **Proving Trigonometric Identities**
Lesson 66
Trigonometric Identities

An identity is an equation that is true for all permissible replacements of the independent variable.

Trigonometric identities are based on eight fundamental identities which are grouped as reciprocal identities, ratio identities and the Pythagorean identities.

Reciprocal identities

1. $\sec\theta = \dfrac{1}{\cos\theta}$ $(\cos\theta \neq 0)$

2. $\csc\theta = \dfrac{1}{\sin\theta}$ $(\sin\theta \neq 0)$

3. $\cot\theta = \dfrac{1}{\tan\theta}$ $(\tan\theta \neq 0)$

Ratio identities

1. $\tan\theta = \dfrac{\sin\theta}{\cos\theta}$ $(\cos\theta \neq 0)$

2. $\cot\theta = \dfrac{\cos\theta}{\sin\theta}$ $(\sin\theta \neq 0)$

Pythagorean identities

1. $\sin^2\theta + \cos^2\theta = 1$
2. $1 + \tan^2\theta = \sec^2\theta$
3. $1 + \cot^2\theta = \csc^2\theta$

For the Pythagorean identities, identities **2** and **3** can be derived from identity **1** by dividing identity **1** by $\cos^2\theta$ and by $\sin^2\theta$ respectively. Keep this in mind when trying to memorize these identities.

Addition formulas

1. $\cos(A - B) = \cos A \cos B + \sin A \sin B$

2. $\cos(A + B) = \cos A \cos B - \sin A \sin B$

3. $\sin(A + B) = \sin A \cos B + \cos A \sin B$

4. $\sin(A - B) = \sin A \cos B - \cos A \sin B$

5. $\tan(A - B) = \dfrac{\tan A - \tan B}{1 + \tan A \tan B}$

6. $\tan(A + B) = \dfrac{\tan A + \tan B}{1 - \tan A \tan B}$

Double angle identities

1. $\sin 2\theta = 2\sin\theta\cos\theta;$ **3.** $\cos 2\theta = 2\cos^2\theta - 1$

2. $\cos 2\theta = \cos^2\theta - \sin^2\theta;$ **4.** $\cos 2\theta = 1 - 2\sin^2\theta;$ **5.** $\tan 2\theta = \dfrac{2\tan\theta}{1 - \tan^2\theta}$

Lesson 66 Exercises

From memory, write down the reciprocal identities, ratio identities, and the Pythagorean identities

Lesson 67
Proving Trigonometric Identities

In proving trigonometric identities, the following guidelines will be helpful:

1. Express all functions in terms of sines and cosines and simplify.
   (However, sometimes, this may not be necessary.)

2. Work **separately** on the left-hand-side and **separately** on the right-hand side.

   Do **not** work on the equation as you would operate on both sides of a conditional equation
   This is not an "if-then statement" proof.

3. Apply the appropriate trigonometric identities or formulas (from above). For example, if there are double angles, use the double angle identities to change to single angles.
   (However, sometimes, this may not be necessary.)

4 . If there are indicated operations such as addition or subtraction of fractions, combine these fractions and simplify. (You may have to go back and review how to add rational fractions)
   You may sometimes do the opposite of **4** as in **5** below.

5. If the numerator of a fraction consists of two or more terms and the denominator consists of only one term (especially, one function only), divide every term in the numerator by the denominator.

6. If there is a binomial in either the numerator or the denominator, multiply the binomial by its conjugate

7. Only practice will make the above guidelines meaningful. Do not be afraid to try and err.
   You will learn from your mistakes. After some mastery, you may even begin to enjoy proving trigonometric identities.

Example 1 Prove the following identity: $\dfrac{\sin x}{\tan x} + \dfrac{\cos x}{\cot x} = \cos x + \sin x$.

(**Note** that the right-hand side (RHS) is in terms sines and cosines. Therefore, we do not have to do any work on the RHS. We therefore work on the LHS and express the LHS in terms of sines and cosines and simplify.)

$$\frac{\sin x}{\tan x} + \frac{\cos x}{\cot x} \stackrel{?}{=} \cos x + \sin x \qquad \text{(We will remove the question mark" ? " when we have}$$

exactly the same expression on both sides of the equation)

$$\frac{\sin x}{\dfrac{\sin x}{\cos x}} + \frac{\cos x}{\dfrac{\cos x}{\sin x}} \stackrel{?}{=} \cos x + \sin x \qquad \left(\tan x = \frac{\sin x}{\cos x} \; ; \cot x = \frac{\cos x}{\sin x}\right)$$

$$\frac{\sin x}{1} \cdot \frac{\cos x}{\sin x} + \frac{\cos x}{1} \cdot \frac{\sin x}{\cos x} \stackrel{?}{=} \cos x + \sin x \qquad \text{(Inverting the divisor and multiplying)}$$

$$\frac{\sin x}{1} \cdot \frac{\cos x}{\sin x} + \frac{\cos x}{1} \cdot \frac{\sin x}{\cos x} \stackrel{?}{=} \cos x + \sin x$$

$$\cos x + \sin x = \cos x + \sin x.$$

QED

Example 2   Prove that $\dfrac{\sin 2x}{\sin x} = \sec x + \dfrac{\cos 2x}{\cos x}$

(In this problem, we will work on both sides.)

$$\frac{\sin 2x}{\sin x} \overset{?}{=} \sec x + \frac{\cos 2x}{\cos x}$$

$$\frac{2\sin x \cos x}{\sin x} \overset{?}{=} \frac{1}{\cos x} + \frac{\cos^2\theta - \sin^2\theta}{\cos x} \qquad (\sin 2x = 2\sin x \cos x \text{ for the LHS})$$

$$\frac{2\cancel{\sin x}\cos x}{\cancel{\sin x}} \overset{?}{=} \frac{1}{\cos x} + \frac{\cos^2 x - \sin^2 x}{\cos x} \qquad (\cos 2\theta = \cos^2\theta - \sin^2\theta \text{ for the RHS})$$

$$2\cos x \overset{?}{=} \frac{1 + \cos^2 x - \sin^2 x}{\cos x} \qquad (\text{Adding the fractions})$$

$$2\cos x \overset{?}{=} \frac{1 - \sin^2 x + \cos^2 x}{\cos x} \qquad (\text{Rewriting})$$

$$2\cos x \overset{?}{=} \frac{\cos^2 x + \cos^2 x}{\cos x} \qquad (1 - \sin^2\theta = \cos^2\theta)$$

$$2\cos x \overset{?}{=} \frac{2\cos^2 x}{\cos x}$$

$$2\cos x \overset{?}{=} \frac{2\cos x \cancel{\cos x}}{\cancel{\cos x}}$$

$$2\cos x = 2\cos x \qquad (\text{Note the LHS} = \text{the RHS})$$

Lesson 67 Exercises

Prove the following identities:

1. $\cot x + \tan x = \sec x \csc x$

2. $\tan x + \csc x = \dfrac{\sin x + \cot x}{\cos x}$

3. $\dfrac{\cos 2x}{\cos x} + \sec x = \dfrac{\sin 2x}{\sin x}$;

4. $\dfrac{\cos x}{\sec x} - \dfrac{\sin x}{\cot x} = \dfrac{\cos x \cot x - \tan x}{\csc x}$

CHAPTER 26

Lesson 68
Solutions of Trigonometric Equations

A trigonometric equation is an equation containing a trigonometric function of an unknown angle. A trigonometric equation is a conditional equation. By conditional we mean the equation is true for some but not all permissible replacements of the variable. We can solve a trigonometric equation either graphically or algebraically. We will cover only algebraic methods although on occasion we may complement the algebraic solutions by graphical methods. In solving trigonometric equations, we will make use of trigonometric identities as well as inverse trigonometric relations.

We will cover two types of trigonometric equations, namely,

1. The trigonometric equation which is linear in one trigonometric function of one variable.

2. The trigonometric equation which is quadratic in one variable of the function.

We will restrict our solutions for an unknown angle x such that $0 \le x \le 360°$ $(0 \le x \le 2\pi)$.

General Procedure for solving trigonometric equations:

Step 1: Collect all terms on one side of the equation, preferably on the left-hand side of the equation.

Step 2: Solve the equation for the trig function as exemplified below:

Examples in which the functions have been solved for: **1**. $\sin x = .5$

           2. $\tan x = 1$

Examples in which the functions have **not** been solved for: **1**. $4 \sin^2 x = 1$

           2. $\cos^2 x = .5$

In trying to solve for the functions we may have to apply trigonometric identities if the equation contains multiple angles and / or more than one type of a trig function. If there are multiple angles such as double angles, we may apply the double angle identities or we may substitute a new variable for the multiple angle, solve for the function in terms of the new variable, and reconvert to the original variable after having found the measure of the angle.

Step 3: Solve for the angle: Use inverse trigonometric functions (from tables, memory or calculator) to find the measures of the angles involved.

Step 4: Check the solutions if the equation was squared or raised to a power, since we might have introduced extraneous solutions.

Example 1   Solve for x: $\sin x = \frac{1}{2}$ or .5 $0 \le x < 360°$

Solution

Step 1: Find the angle whose sine is $\frac{1}{2}$ (from tables or calculator). The reference angle is $30°$

Step 2: Determine the quadrants in which the sine is positive.

The sine is positive in quadrants I and II.   ($\frac{1}{2}$ is positive)

Step 3 : Specify the measures of the angles (in standard position).

      $x = 30°$ or $150°$ (The angles are measured from the positive x-axis counterclockwise)

The solution set is $\{30°, 150°\}$

Example 2 Solve for x: $2\sin x - 1 = 0$ $0 \le x < 360°$

Step 1: Solve for $\sin x$

$$2\sin x = 1$$
$$\sin x = \frac{1}{2} \quad ....(1)$$

Step 2: Equation (1) of Step 1 is the same as Example 1 above. We therefore repeat the solution:
The solution set is {30°, 150°}

Example 3   Solve for x: $\cos 2x = \frac{1}{2}$.

Method 1

Step 1: Let $2x = \theta$

Then, $\cos 2x = \frac{1}{2}$ becomes

$$\cos \theta = \frac{1}{2}$$

Step 2: Determine the reference angle for θ.

The reference angle is 60°   (From tables, memory or calculator)

Step 3: Determine the quadrant in which $\cos \theta$ is positive.

The cosine is positive in quadrant I and IV.   ($\frac{1}{2}$ is positive)

Step 4: Specify the measures of the angles   (in standard position).

$\theta = 60°, \theta = 300°$

Step 5:   The general solutions are $\theta = 60° + 360n$ or $\theta = 300° + 360n$ ($n = 0, 1$ for this problem)

Step 6: Change back to x. That is, replace θ by $2x$

$$2x = 60° + n \cdot 360° \quad \text{or} \quad 2x = 300° + n \cdot 360°$$

$$x = 30° + n \cdot 180° \quad \text{or} \quad x = 150° + n \cdot 180° \qquad \text{(Dividing by 2 and solving for } x\text{)}$$

For $n = 0, x = \mathbf{30°}$ or $x = \mathbf{150°}$ (When $n = 0, n \cdot 180 = 0$)

For $n = 1, x = 30° + 180°$ or $x = 150° + 180°$

$\qquad\qquad x = \mathbf{210°}$ or $x = \mathbf{330°}$

The solution set is {30°, 150° , 210°, 330°}

Note that in Step 6, we considered two values of n ($n = 0, 1$) because of the "2" in $\cos 2x$.
If we were given $\cos 3x$, we would have considered three values of n ($n = 0, 1, 2$) because
of the "3" in $\cos 3x$.

Method 2: Using the double angle identity, $\cos 2x = 2\cos^2 x - 1$

The given equation $\cos 2x = \frac{1}{2}$ becomes

$$2\cos^2 x - 1 = \frac{1}{2}$$
$$2\cos^2 x = \frac{1}{2} + 1$$
$$2\cos^2 x = \frac{3}{2}$$
$$\cos^2 x = \frac{3}{4}$$
$$\cos x = \pm \sqrt{\frac{3}{4}}$$
$$\cos x = \pm \frac{\sqrt{3}}{2} \quad \text{(That is, } \cos x = +\frac{\sqrt{3}}{2} \quad \text{or } \cos x = -\frac{\sqrt{3}}{2}\text{)}$$

For $\cos x = +\dfrac{\sqrt{3}}{2}$, the reference angle is 30°.

$\cos x$ is positive in quadrants I and IV.

Therefore, $x =$ **30° or 330°**   (angles in standard position).

Similarly, for $\cos x = -\dfrac{\sqrt{3}}{2}$, the reference angle is 30°.

$\cos x$ is negative in quadrants II and III.

Therefore, $x =$ **150° or 210°**   (angles in standard position).

The solution set is {30°, 150°, 210°, 330°}.

Again, we obtain the same solution set as by Method 1. You decide which method you prefer.

Example 4 Solve for x: $2 \cos^2 x + 3 \sin x = 3 \quad 0 \le x < 360°$

Solution $2 \cos^2 x + 3 \sin x - 3 = 0$

Step 1: $2(1 - \sin^2 x) + 3 \sin x - 3 = 0$ ($\cos^2 x = 1 - \sin^2 x$)

$\qquad\qquad 2 - 2 \sin^2 x + 3 \sin x - 3 = 0$

$\qquad\qquad\quad -2 \sin^2 x + 3 \sin x - 1 = 0$

$\qquad\qquad\quad\ 2 \sin^2 x - 3 \sin x + 1 = 0$ (1)

*Step 2: Let $\sin x = t$, Then equation (1) becomes

$\qquad 2t^2 - 3t + 1 = 0$ <--------quadratic equation.

We solve by factoring since the factors are easily recognizable.

$\qquad\quad (2t - 1)(t - 1) = 0$

$\qquad\quad 2t - 1 = 0$ or $t - 1 = 0$

$\qquad\quad t = \dfrac{1}{2}$ or $t = 1$

Now, replace t by $\sin x$.

Then $\sin x = \dfrac{1}{2}$ or $\sin x = 1$

Step 3:   Solve for the angles:

$\qquad$ When $\sin x = \dfrac{1}{2}$,the reference angle is 30°,

$\qquad$ and $x =$ **30°** or **150°**

$\qquad$ When $\sin x = 1, x =$ **90°**

The solution set is {30°, 90°, 150°}.

* Note that in Step 2, we could avoid substitution by doing the following: $(2 \sin x - 1)(\sin x - 1) = 0$.
Then either $2 \sin x - 1 = 0$ or $\sin x - 1 = 0$.

Solving, $\sin x = \dfrac{1}{2}$ or $\sin x = 1$

Note:

To check the solutions to the above equations, substitute in the original equations

Lesson 68 Exercises

Solve for x: 1. $4\cos^2 x - 8\cos x + 3 = 0$;   **2.** $2\cos^2 x - \cos x = 1$

Answers: **1** {60°, 300°}   **2**. {0°, 120°, 240°}

CHAPTER 27

Lesson 69: **Distance Formula** (Review)
Lesson 70: **Equation and Graph of a Circle**

Lesson 69
Distance Formula (Review)

The distance, d, between the points $P_1(x_1, y_1)$ and $P_2(x_2, y_2)$ on a line in a plane (Fig.1) is given by

$$d = \sqrt{(x_2 - x_1)^2 + (y_2 - y_1)^2}$$ (By applying the Pythagorean theorem to the right triangle, and solving for d)

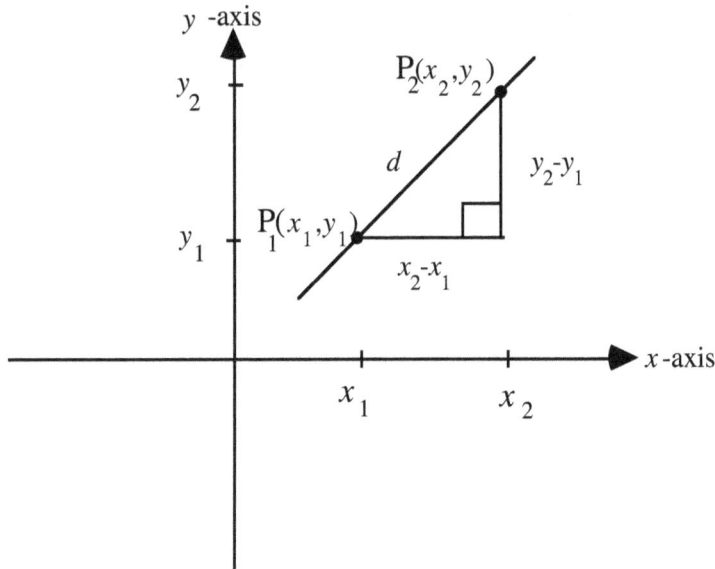

Figure 1

Example   Find the distance between the points $(-2, 3)$ and $(4, -5)$.

Solution: Apply the distance formula:

$$d = \sqrt{(x_2 - x_1)^2 + (y_2 - y_1)^2} \qquad (1)$$

(where d is the distance between the points $P_1(x_1, y_1)$ and $P_2(x_2, y_2)$).

Substituting $x_1 = -2, y_1 = 3, x_2 = 4, y_2 = -5$ in equation (1) above,

$$d = \sqrt{(4 - (-2))^2 + (-5 - 3)^2}$$
$$d = \sqrt{(4 + 2)^2 + (-8)^2}$$
$$= \sqrt{(6)^2 + (-8)^2}$$
$$= \sqrt{36 + 64}$$
$$= \sqrt{100}$$
$$= 10$$

$\therefore$ the distance between the given points is 10 units.

Lesson 69 Exercises

1. Find the distance between the points (3,4) and (5, -1) .

2. Find the distance between the points (-4, 2) and (-6, -3)

Answers: **1.** $\sqrt{29}$; **2.** $\sqrt{29}$

Lesson 70

Equation and Graph of a Circle

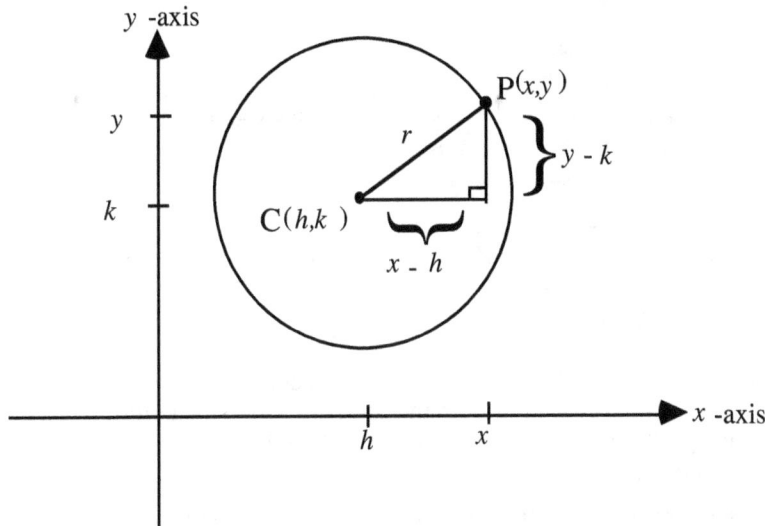

Figure 2

We will derive an equation of a circle with its center at the point $C(h, k)$ (Figure 2, above) and having radius, r, by applying the Pythagorean theorem.

Let the distance between the center $C(h, k)$ of the circle and a point $P(x, y)$ on the circle be r. Applying the Pythagorean theorem (since we have a right triangle),

$$r^2 = (x - h)^2 + (y - k)^2 \text{ or rewriting,}$$
$$(x - h)^2 + (y - k)^2 = r^2 \text{ <--------- center-radius form of the equation of a circle}$$

This equation could also be obtained from the distance formula as follows:
Replace d by r; x_2 by x; x_1 by h; y_2 by y , and y_1 by k in

$$d = \sqrt{(x_2 - x_1)^2 + (y_2 - y_1)^2} \quad \text{<--------- distance formula.}$$

Then we obtain $r = \sqrt{(x - h)^2 + (y - k)^2}$

Squaring both sides of the equation, $r^2 = (x - h)^2 + (y - k)^2$
Rewriting, $(x - h)^2 + (y - k)^2 = r^2$ <--------- center-radius form

Example 1   (a) Write an equation for the circle with radius 5 and center at (2, -3).
        (b) Sketch the graph of the circle.
    The center-radius form of the equation of a circle is given by

      $(x - h)^2 + (y - k) = r^2$, where r is the radius of the circle, h = the x- coordinate of the center, and k = the y-coordinate of the center of the circle.

Step 1: Substituting $r = 5$, $h = 2, k = -3$ in
        $(x - h)^2 + (y - k)^2 = r^2$, we obtain
        $(x - 2)^2 + (y - (-3))^2 = 5^2$
        $(x - 2)^2 + (y + 3)^2 = 5^2$ <-------center-radius form.
   or $(x - 2)^2 + (y + 3)^2 = 25$.

(b) **Graph**   Step 1: Plot the coordinates (2,-3) of the center of the circle.

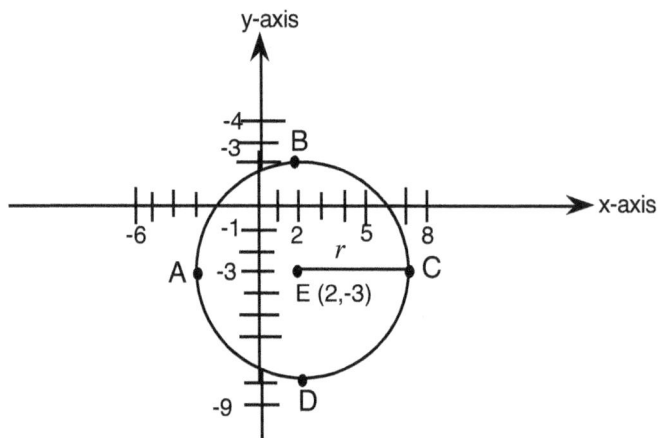

Step 2: Since the radius is 5, count 5 units each , horizontally, from the center (2,-3) to the points A, and C; and also count 5 units each, vertically, from the center (2, -3) to the points B and D. Connect these four points by a smooth "circular " curve to obtain the circle.

Note: Use the same scale for the x- and y-axes.

Note also that for a circle, the distance from the center of the circle to any point on the circumference is the same as the distance from the center to any other point on the circumference. Keep this in mind when drawing a circle.

Example 2   Find an equation of the circle with center at the point (4, -2) and passing through the point (-3, 6).

If we know the radius, r, and the coordinates h, k of the center of a circle we can immediately write down an equation for the circle by applying

$$(x - h)^2 + (y - k)^2 = r^2. \qquad (1)$$

We are given $h = 4, k = -2$, but we do not know r yet; however, we can find it.

Finding r:

Since the point (-3, 6) is given to be on the circle, we may substitute this ordered pair in equation (1).

Step 1: Substitute $h = 4, k = -2, x = -3, y = 6$

in equation (1) and solve for r.

$$(-3 - 4)^2 + (6 - (-2))^2 = r^2$$
$$(-7)^2 + (6 + 2)^2 = r^2$$
$$(-7)^2 + 8^2 = r^2$$
$$49 + 64 = r^2$$
$$113 = r^2,$$
$$\sqrt{113} = r$$
$$\text{and } r = \sqrt{113}$$

Step 2: Substitute $h = 4, k = -2, r = \sqrt{113}$ (or $r^2 = 113$) in equation   (1) above.

∴ an equation for the given circle is $(x - 4)^2 + (y + 2)^2 = 113$ <-------center-radius form.

Finding the center and radius of a circle (Given the center-radius form of the equation of a circle)

Example 1   Find the center and radius of the circle whose equation is given by
$$(x + 4)^2 + (y - 3)^2 = 36$$
Procedure: The center-radius form of the equation of the circle with center at the point (h, k) and with radius r is given by:
$$(x - h)^2 + (y - k)^2 = r^2 \qquad\qquad (1)$$
Compare $(x + 4)^2 + (y - 3)^2 = 36$ with equation (1) above.

Then $(x - h)^2 = (x + 4)^2$; $(y - k)^2 = (y - 3)^2$; and $r^2 = 36$
$$x - h = x + 4 ; \qquad y - k = y - 3 \qquad\qquad r = \sqrt{36} = 6$$
$$-h = 4 ; \qquad\qquad -k = -3$$
$$h = -4 \qquad\qquad k = 3$$
Therefore $h = -4$; $k = 3$; and $r = 6$.

The center of the circle is at (-4,3); and its radius is 6 units.

Shortcut method: Step 1: Set $x + 4 = 0$ and solve to obtain $x = -4$ (Thus $h = -4$
                            Step 2: Set $y - 3 = 0$ and solve to obtain $y = 3$ (Thus $k = 3$
                            The center of the circle is at (-4,3);
With some practice, we shall be able to read, by comparison with equation (1) above, the coordinates of the center of the circle without performing any calculations.

Finding the center and radius of a circle given the general form of the equation of a circle.

Example 2   Find the center and radius of a circle whose equation is given by
$$x^2 + y^2 + 8x - 6y - 11 = 0$$

Procedure: We will group the x-terms, group the y-terms, and complete the square on the x-terms, and on the y-terms. (see page 131 and review how to complete the square.)
Step 1: $x^2 + y^2 + 8x - 6y - 11 = 0$
$$x^2 + 8x + y^2 - 6y = 11 .... (1)$$
Step 2: Complete the squares:
$$* \qquad x^2 + 8x + 16 + y^2 - 6y + 9 = 11 + 16 + 9 ....(2)$$

$$x^2 + 8x + 16 + y^2 - 6y + 9 = 36$$
(For the x-terms: $x^2 + 8x + (4)^2$)
(For the y-terms : $y^2 - 6y+ (-3)^2$)

$(x + 4)^2 + (y - 3)^2 = 36$ <---From this center-radius form, follow the steps exactly as in Example 1 above.

The center of the circle is at (-4, 3); and its radius is 6 units.

* Whenever we add any quantity to the left-hand side of equation (1), we must add the same quantity to the right-hand side of the equation: First, we added 16 (i.e., 4^2) for the x-terms, to both sides of the equation; then we added 9 (i.e.,$(-3)^2$) for the y-terms to both sides of the equation.

Lesson 70 Exercises

A (a) Write an equation for the circle with radius 4 and center at (-2, 3).

(b) Sketch the graph of the circle.

Solution: $(x + 2)^2 + (y - 3)^2 = 16$<------- center-radius form.

B Find an equation of the circle with center at the point $(2, 3)$ and passing through the point $(-4, 1)$.

Answer: $(x - 2)^2 + (y - 3)^2 = 40$

C Find the center and radius of the circle whose equation is given by
$$(x - 5)^2 + (y + 2)^2 = 16$$

Answer: Center at $(5, -2)$; and radius, 4 units.

D Find an equation for each circle with the given properties.

1. Radius 4 units with center at the origin.

2. Radius 3 units and center at $(-3, 2)$.

3. Radius 5 units and center at $(-2, -4)$

Problems 4-6: Find the center and radius of each of the given circle

4. $x^2 - 2x + y^2 - 4y - 11 = 0$;   **5.** $x^2 + y^2 + 4x - 6y = 12$;   **6.** $x^2 + y^2 - x - \dfrac{y}{2} = \dfrac{63}{16}$.

Answers: **1.** $x^2 + y^2 = 16$;   **2.** $(x + 3)^2 + (y - 2)^2 = 9$;   **3.** $(x + 2)^2 + (y + 4)^2 = 25$
4. Center at $(1, 2)$, radius 4 units;   **5.** Center at $(-2, 3)$, radius 5 units;
6. Center at $(\frac{1}{2}, \frac{1}{4})$, radius $\dfrac{\sqrt{17}}{2}$ units.

APPENDIX A

EXTRA: Factoring Perfect Square Trinomials

In factoring, knowing that a given trinomial is a **perfect square** trinomial will speed-up the factoring process. However, note that the methods we have used, so far, in factoring trinomials, can be used to factor perfect square trinomials.

Case 1: The coefficient of the leading term is 1 (assuming the trinomial is properly arranged)

Examples: **1.** $x^2 + 6x + 9 = (x + 3)^2$ same as $(x + 3)(x + 3)$

Observe that (a) the first and third terms are perfect squares, and

(b) $\left(\frac{6}{2}\right)^2 = (3)^2 = 9$ (that is the square of half the coefficient of the x-term is 9)

2. $x^2 - 6x + 9 = (x - 3)^2$ same as $(x - 3)(x - 3)$

Observe that (a) the first and third terms are perfect squares, and

(b)$\left(\frac{-6}{2}\right)^2 = (-3)^2 = 9$ (that is, the square of half the coefficient of the x-term is 9)

3. $x^2 - 4xy + 4y^2 = (x - 2y)^2$ same as $(x - 2y)(x - 2y)$

Observe that (a) the first and third terms are perfect squares, and

(b) $\left(\frac{-4y}{2}\right)^2 = (-2y)^2 = 4y^2$ (that is, the square of half the coefficient of the x-term is $4y^2$)

Case 2: The coefficient of the leading term **is not 1** (assuming the trinomial is properly arranged)

Here, if the first and third terms are perfect squares, the trinomial may or may not be a perfect square trinomial, Obtain the binomial factor using the square roots of the first and third terms and check the factorization by multiplication to see if you obtain the middle term. **Note** that even if the first and third terms are perfect squares, the trinomial may **not** be a perfect square trinomial as in Examples **8** and **9** below.

We can also check for a perfect square trinomial as follows:

Step 1: Find the square root of the first term (the leading term)

Step 2: Divide the middle term by twice the root from step 1:

Step 3: Square the quotient from step 2. If this square equals the third term, then the trinomial is a perfect square trinomial. Factor by taking the square roots of the first and third terms

Simply, if $\left(\dfrac{\text{middle term}}{2(\text{square root of first term})}\right)^2$ = Third term, then the trinomial is a perfect square trinomial

Examples The following are perfect square trinomials

To factor, find the square roots of the first and third terms and use the roots to   form the binomial, with a plus or minus sign between the terms according as the middle term has a plus sign or a minus sign.

1. $4x^2 - 4xy + y^2 = (2x - y)(2x - y) = (2x - y)^2$

2. $4x^2 + 4xy + y^2 = (2x + y)(2x + y) = (2x + y)^2$

3. $25x^2 - 60xy + 36y^2 = (5x - 6y)(5x - 6y) = (5x - 6y)^2$

4. $25x^2 + 60xy + 36y^2 = (5x + 6y)(5x + 6y) = (5x + 6y)^2$

5. $9a^2 - 12ab + 4b^2 = (3a - 2b)^2$

6. $a^2 + 18a + 81 = (a + 9)^2$

7. $4a^2 - 36a + 81 = (2a - 9)^2$

The following are **not** perfect square trinomials, even though the first and third terms are perfect squares. Here, we use the usual methods for factoring.

8. $9x^2 + 30x + 16$ whose factorization is $(3x + 2)(3x + 8)$; but $9x^2 + 24x + 16$ is perfect square. Why?

9. $4x^2 + 15xy + 9y^2$ whose factorization is $(x + 3y)(4x + 3y)$; but $4x^2 + 12xy + 9y^2$ is perfect square. Why?

APPENDIX B

The following theorems were proposed by the author in the Summer of 1993, and may be helpful when dealing with problems in which the lengths of the sides and the measures of the angles of the triangles are given.

Similar Triangles

Theorems on similar triangles in terms of the **lengths** of the sides and the **measures** of the **angles**

Theorem 1 (Another form of the **SSS** theorem for similar triangles)

If the lengths of the sides one triangle are x_1, x_2, x_3, where $x_1 \le x_2 \le x_3$, and the lengths of the sides of another triangle are y_1, y_2, y_3, where $y_1 \le y_2 \le y_3$, and if

$\dfrac{x_1}{y_1} = \dfrac{x_2}{y_2} = \dfrac{x_3}{y_3}$, then the two triangles are similar.

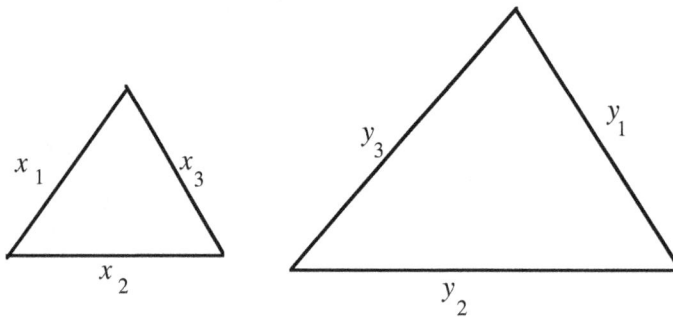

* Note: The "=" in "$\le$" is for the cases of isosceles or equilateral triangles.

Converse

If two triangles are similar and the lengths of the sides of one triangle are x_1, x_2, x_3, where $x_1 \le x_2 \le x_3$, and the lengths of the sides of the other triangle are y_1, y_2, y_3, where $y_1 \le y_2 \le y_3$,

then $\dfrac{x_1}{y_1} = \dfrac{x_2}{y_2} = \dfrac{x_3}{y_3}$.

Note above that x_1 and y_1 are corresponding lengths; x_2 and y_2 are corresponding lengths, and x_3 and y_3 are corresponding lengths. (i.e., the smallest length of Δ #1 corresponds to the smallest length of Δ #2, and the largest length of Δ #1 corresponds to the largest length of Δ #2 , and the remaining lengths correspond.)

* 1. If two triangles are similar and one triangle is isosceles, the other triangle is also isosceles.

* 2. If two triangles are similar and one triangle is equilateral, the other triangle is also equilateral.
* 3. Also, all equilateral triangles are similar.

Application of Theorem 1

In the figures below:
(a) Find the corresponding lengths .
(b) Form ratios using the corresponding lengths.
(c) Are the two triangles similar?

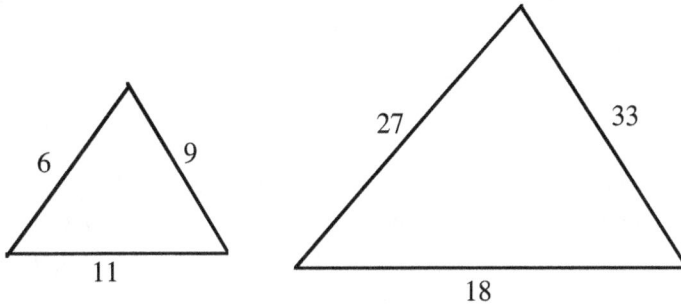

Solutions:
(a) $6 < 9 < 11$ and $18 < 27 < 33$
  6 and 18 correspond; 9 and 27 correspond; and 11 and 33 correspond.

(b) $6{:}18 = \dfrac{6}{18} = \dfrac{1}{3}$; $9{:}27 = \dfrac{9}{27} = \dfrac{1}{3}$; $11{:}33 = \dfrac{11}{33} = \dfrac{1}{3}$.

(c) Since the ratios are the same for each pair of corresponding sides, the corresponding side s are in proportion and the two triangles are similar by SSS.

Theorem 2 (Another form of the **SAS** theorem for similar triangles)

> If the lengths of two sides one triangle are x_1, x_2, where $x_1 \le x_2$, with an included angle whose measure is α, and the lengths of the sides of another triangle are y_1, y_2, where
>
> $y_1 \le y_2$, with an included angle whose measure is β; and if $\dfrac{x_1}{y_1} = \dfrac{x_2}{y_2}$ and $\alpha = \beta$, then the
>
> two triangles are similar.

Theorem 3 (Another form of **AA** theorem for similar triangles)

> If the measures of two angles of a triangle are α_1, α_2, where $\alpha_1 \le \alpha_2$ and the measures of
> two angles of another triangle are β_1, β_2, where $\beta_1 \le \beta_2$; and if $\alpha_1 = \beta_1$ and $\alpha_2 = \beta_2$, then
> the two triangles are similar.

Congruent Triangles

Theorems on congruent triangles in terms of the lengths of the sides and the measures of the angles

Theorem 1 (Another form of **SSS** for congruent triangles)

If the lengths of the sides of one triangle are x_1, x_2, x_3, where $x_1 \le x_2 \le x_3$, and the lengths of the sides of another triangle are y_1, y_2, y_3, where $y_1 \le y_2 \le y_3$; and if $x_1 = y_1, x_2 = y_2, x_3 = y_3$, then the two triangles are congruent.

Theorems 2 (Another form of the **SAS** theorem for congruent triangles)

If the lengths of two sides one triangle are x_1, x_2, where $x_1 \le x_2$, with an included angle whose measure is α, and the lengths of the sides of another triangle are y_1, y_2, where $y_1 \le y_2$, with an included angle whose measure is β, and if $x_1 = y_1, x_2 = y_2$ and $\alpha = \beta$, then the two triangles are congruent.

Theorems 3 (Another form of the **ASA** theorem for congruent triangles)

If the measures of two angles of one triangle are α_1, α_2, where $\alpha_1 \le \alpha_2$, with an included side of length x, and the measures of two angles of another triangle are β_1, β_2, where $\beta_1 \le \beta_2$, with an included side of length y; and if $\alpha_1 = \beta_1, \alpha_2 = \beta_2$ and $x = y$, then the two triangles are congruent.

Note:

1. If two triangles are congruent and one triangle is isosceles, the other triangle is also isosceles.

2. If two triangles are congruent and one triangle is equilateral, the other triangle is also equilateral

Absolute Value Equations

Example 4 (From p.185)   Solve for x: $|x - 2| = |x + 6|$. We consider four cases.

Case 1: Both expressions within the absolute value bars are positive.

Then $x - 2 = x + 6$ (A)

Case 2: Both expressions within the absolute value bars are negative.

Then $-(x - 2) = -(x + 6)$ (B)

Case 3: The expression within the absolute value bars on the left is positive and that on the right is negative.

Then $x - 2 = -(x + 6)$ (C)

Case 4: The expression within the absolute bars on the left is negative and that on the right is positive.

Then $-(x - 2) = x + 6$ (D)

On simplifying, (B) reduces (A). Similarly, (C) and (D) are equivalent, ignoring the domains of the definitions.

We therefore solve only Case 1, $x - 2 = x + 6$

and Case 3, $x - 2 = -(x + 6)$

Solving for Case 1, $x - 2 = x + 6$

$-2 = 6$, which is a contradiction.

There is no solution for Cases 1 and 2.

For Cases 3 and 4, $x - 2 = -(x + 6)$

$x - 2 = -x - 6$, and from which $x = -2$

We check for $x = -2$.

Then $|-2 - 2| \overset{?}{=} |-2 + 6|$; $|-4| \overset{?}{=} |4|$; $4 \overset{?}{=} 4$. Yes.

The solution is -2.

APPENDIX C

Lesson 71: One-to-One Functions, Composite Functions
Lesson 72: Inverse Functions and Inverse Relations

Lesson 71

One-to-One Functions, Composite Functions

One-to-One Functions

(In Lesson 72, we will learn that if a function is one-to-one, then it has an inverse function.)

We consider three main cases according to how the function is specified.

Case 1: **Given a set of ordered pairs** (or a table of x- and y-values)

A one-to-one (1-1) function is a set of ordered pairs in which for any two ordered pairs, the first elements are different from each other and the second elements are also different from each other.

Example The set, $A = \{(3,2),(4,7),(1,5),(2,3)\}$ is a one-to-one function. However.

The set, $B = \{(3,2),(4,7),(1,5),(5,2)\}$ is **not** a one-to-one function because the first and the last ordered pairs have the same second elements, namely, 2.

Case 2: Given the graph of the function

By the so called **horizontal line test**, the graph of a function is that of a one-to-one function if every possible horizontal line drawn to intersect the graph cuts (intersects) the graph only once (at one point only). **Figures** 2 and 3, p. 223, are graphs of one-to-one functions but **Figure 1** is not one-to-one.

Case 3: Given the equation of the function

A function $f(x)$ is one-to-one if whenever $f(x_1) = f(x_2), x_1 = x_2$. If we let $y_1 = f(x_1)$, and $y_2 = f(x_2)$, then $y_1 = y_2$ **must** imply that $x_1 = x_2$, otherwise, $f(x)$ is not one-to-one.

Example 2: Determine if $f(x) = 2x + 3$ is one-to-one.
Solution

Step 1: $f(x_1) = 2x_1 + 3$	**Step 2**: Solve for x_1. (You may also solve for x_2)
$\qquad f(x_2) = 2x_2 + 3$	$\qquad 2x_1 + 3 = 2x_2 + 3$
Equate RHS of $f(x_1)$ to RHS of $f(x_2)$	$\qquad 2x_1 = 2x_2$
(That is, let $f(x_1) = f(x_2)$):	$\qquad x_1 = x_2 \qquad (2)$
$\qquad 2x_1 + 3 = 2x_2 + 3 \qquad (1)$	

Since from above, whenever $f(x_1) = f(x_2)$ (from equation (1)) $x_1 = x_2$ (from equation (2))

$f(x) = 2x + 3$ is one-to-one. (You may check by sketching its graph and using the horizontal line test)

Example 3 Determine if $f(x) = \sqrt{25 - x^2}$ is one-to-one.
Solution

Step 1: $f(x_1) = \sqrt{25 - x_1^2}$

$f(x_2) = \sqrt{25 - x_2^2}$

Equate RHS of $f(x_1)$ to RHS of $f(x_2)$
(That is, let $f(x_1) = f(x_2)$.)

$\sqrt{25 - x_1^2} = \sqrt{25 - x_2^2}$ (1)

Step 2: Solve for x_1.

$$\sqrt{25 - x_1^2} = \sqrt{25 - x_2^2}$$
$$25 - x_1^2 = 25 - x_2^2$$
$$-x_1^2 = -x_2^2$$
$$x_1^2 = x_2^2$$
$$x_1 = \pm\sqrt{x_2^2}$$
$$x_1 = +x_2 \text{ or } -x_2 \quad (2)$$

Since from above, whenever $f(x_1) = f(x_2)$, $x_1 = -x_2$, (That is for the same y-value, x is not
unique: $x_1 = +x_2$, and **also** $x_1 = -x_2$ (from equation (2))

$f(x) = \sqrt{25 - x^2}$ is **not** one-to-one. (You may check by sketching its graph and using the horizontal line test)

Example 4: Determine if $f(x) = |x - 2|$ is one-to-one.
Solution

Step 1: $f(x_1) = |x_1 - 2|$

$f(x_2) = |x_2 - 2|$

Equate RHS's of $f(x_1)$ to $f(x_2)$
(That is, let $f(x_1) = f(x_2)$):
$$|x_1 - 2| = |x_2 - 2| \quad (1)$$

Step 2: Solve for x_1:
If both $x_1 - 2$ and $x_2 - 2$ are positive,
$x_1 - 2 = x_2 - 2$ and from which $x_1 = x_2$
(same result as if both $x_1 - 2$ and $x_2 - 2$ are negative)

However, if $x_1 - 2$ is positive and $x_2 - 2$ is negative,
$$x_1 - 2 = -(x_2 - 2)$$
$$x_1 - 2 = -x_2 + 2$$
$$x_1 = -x_2 + 4. \text{ That is, } x_1 \neq x_2$$

Since from above, whenever $f(x_1) = f(x_2)$, $x_1 \neq x_2$
$f(x) = |x - 2|$ is **not** one-to-one. (You may check by sketching its graph and using the horizontal line test)

Another method for determining if a function is one-to-one

The author proposes the following definition for a one-to-one function.

Definition : A function is one-to-one if its inverse relation is a function.
This definition provides another method for determining if a given function is one-to-one.

Example 5

Determine if $f(x) = 2x + 6$ is one-to-one
Solution
Given: $f(x) = 2x + 6$
Required: To determine if $f(x) = 2x + 6$ is one-to-one.
Plan: If it can be shown that $f(x) = 2x + 6$ has an inverse relation which is function, then
$$f(x) = 2x + 6 \text{ is one-to-one.}$$
Determination:
Step 1: Let $f(x) = y$ to obtain $y = 2x + 6$.
Step 2: Interchange x and y to obtain the inverse relation $x = 2y + 6$
Step 3: Solve for y.
$$2y = x - 6$$
$$y = \tfrac{1}{2}x - 3$$
Since clearly, for a given value of x there is exactly one corresponding value of y.

the inverse relation, $y = \tfrac{1}{2}x - 3$, is a function

and therefore, $f(x) = 2x + 6$ is one-to-one.

(You may also check that $y = \tfrac{1}{2}x - 3$ is a function by sketching its graph and applying the vertical line test)

Example 6

Determine if $f(x) = x^2$ is one-to-one.
Solution
Given: $f(x) = x^2$
Required: To determine if $f(x) = x^2$ is one-to-one.

Plan: If it can be shown that $f(x) = x^2$ has an inverse relation which is function, then
$$f(x) = x^2 \text{ is one-to-one.}$$
Determination
Step 1: Let $f(x) = y$ to obtain $y = x^2$
Step 2: Interchange x and y to obtain the inverse relation $x = y^2$
Step 3: Solve for y.
$$\text{If } y^2 = x$$
$$y = \pm\sqrt{x} \quad (\text{that is, } y = +\sqrt{x} \text{ or } y = -\sqrt{x})$$
Clearly, for a given value of x there are two different corresponding y-values. Therefore, y is

not a function of x, and the inverse relation $y^2 = x$ is **not** a function and therefore , the given function $f(x) = x^2$ is **not** one-to-one.

(You may also check that $y^2 = x$ or $y = \pm\sqrt{x}$ is **not** a function by sketching its graph and applying the vertical line test)

Composite Functions

Some authors refer to composite functions as "product functions". This alternative terminology may be misleading because a reader might be inclined to multiply the given functions. Perhaps, a better alternative terminology for a composite function is " a function within another function".

Use of Composition of Functions:

The principle of composition of functions can be used to determine algebraically if two given functions are inverses of each other.

Definition: If two functions f and g are such that the range of g is in the domain of f, then the composite function of f with g, symbolized $f \circ g$, is specified by $(f \circ g)(x) = f[g(x)]$ which is read "f of g of x". That is, the output of g becomes the input for f.

Similarly, if the range of f is in the domain of g, then the composite function of g with f, symbolized $g \circ f$ is specified by $g \circ f = g[f(x)]$ which is read "g of f of x." That is, the output of f becomes the input for g.

We must **note** that generally, $f \circ g \neq g \circ f$ (i.e., generally, $f \circ g$ is not equal to $g \circ f$)

Example 1

If $f(x) = x^2$, and $g(x) = x + 1$, find (a) $f[g(x)]$; (b) $g[f(x)]$.

Solution

Recall that if, for example,

$$f(x) = x^2$$

then $f(3) = (3)^2$.

Similarly, (a) $f[g(x)] = f[x + 1]$

$$= [x + 1]^2$$

$$f[g(x)] = x^2 + 2x + 1$$

Thus, (in the above problem) wherever there is x in the equation for $f(x) = x^2$, write (substitute) $x + 1$.

(b) $g[f(x)]$

$$= g[x^2]$$

$$= (x^2) + 1 \quad \text{(substitute } x^2 \text{ for } x \text{ in the equation for } g(x) = x + 1\text{)}$$

$$= x^2 + 1$$

Thus, wherever there is x in the equation for $g(x) = x + 1$, we substitute x^2.

Example 2: If $f(x) = 2(x + 10)$, and $g(x) = x - 2$, find (a) $f[g(x)]$; (b) $g[f(x)]$.

Solution

(a) $f[g(x)] = 2[(x - 2) + 10$

$$= 2[x - 2 + 10]$$

$$= 2[x + 8]$$

$$f[g(x)] = 2x + 16$$

(Substituting $x - 2$ in the equation for $f(x)$.

(b) $g[f(x)] = 2(x + 10) - 2$

$$= 2x + 20 - 2$$

$$g[f(x)] = 2x + 18$$

(Substituting $2(x + 10)$ in the equation for $g(x)$.

Domains of Composite Functions   3 1 1

The approach here is similar to that for domains of algebra of functions. However, here, in the first step, we check only for the domain of the "inside" function. In the second step, we check for the domain of the resulting composite function. We add any excluded values to those from Step 1 and if there is an overlap in the domains we use the intersection of the domains from Step 1 and Step 2. Pay attention to radical functions such as $f(x) = \sqrt{x-2}$,.

Given $f(x) = \frac{3}{x}$ and $g(x) = \frac{3x-4}{x+2}$, find the domains of the following:

find $(a)\ f[g(x)];\ (b)\ g[f(x)].$

(a) Step 1: We check the domain of

the "inside function", $g(x) = \frac{3x-4}{x+2}$

Domain of $g : \{x \mid x$ is a real number and $x \neq -2\}.$
Step 2: Form the composite function and
check its domain.

$f[g(x)] = \dfrac{3}{\frac{3x-4}{x+2}} \qquad (f(x) = \frac{3}{x})$

$\qquad = \dfrac{3}{1} \cdot \dfrac{x+2}{3x-4}$

$f[g(x)] \quad = \dfrac{3(x+2)}{3x-4}$

Set $3x - 4 = 0$ to obtain $x = \frac{4}{3}$. Therefore $x \neq \frac{4}{3}$.

Combining the domains from Step 1 and Step 2,

the excluded values are -2, and $\frac{4}{3}$.; and

the domain of $f[g(x)] = \dfrac{3(x+2)}{3x-4}$ is

$\left\{ x \mid x \text{ is a real number and } x \neq -2 \text{ and } x \neq \frac{4}{3} \right\}$

(a) Step 1: We check the domain of

the "inside function", $f(x) = \frac{3}{x}$

Domain of $f : \{x \mid x \neq 0\}.$
Step 2: Form the composite function and
check its domain.

$g[f(x)] = \dfrac{3(\frac{3}{x})-4}{\frac{3}{x}+2} \qquad (g(x) = \frac{3x-4}{x+2})$

$g[f(x)] \quad = \dfrac{-4x+9}{2x+3}$

Set $2x + 3 = 0$ to obtain $x = -\frac{3}{2}$.

Therefore $x \neq -\frac{3}{2}$.

Combining the domains from Step 1 and Step 2,

the excluded values are 0, and $-\frac{3}{2}$.; and

the domain of $g[f(x)] = \dfrac{-4x+9}{2x+3}$ is

$\left\{ x \mid x \text{ is a real number and } x \neq -0 \text{ and } x \neq -\frac{3}{2} \right\}$

Application of Composition of Functions

If two functions f_1 and f_2 are inverses of each other, then, the following two conditions must be satisfied simultaneously. **1.** $f_1[f_2(x)] = x$ and **2.** $f_2[f_1(x)] = x$. For examples, see page 319.

Lesson 71 Exercises

A Show and determine which of the following functions are one-to-one

1. The set $\{(4,5),\ (3,4),\ (1,2)\}$; 2. The set $\{(2,2),\ (4,3),\ (5,7)\}$; 3. $f(x) = 3x - 2$;

4. $f(x) = \sqrt{x+3}$; 5. $f(x) = \sqrt{16-x^2}$; 6. $f(x) = |x|$; 7. $f(x) = |x-2|$; 8. $f(x) = x^3 + 2.$

Answers: **1.** Yes;   **2.** Yes; **3.** Yes; **4. Yes**; **5.** No; **6.** No; **7.** No.; **8.** Yes

B 1. Given that $f(x) = x + 2$, $g(x) = x - 1$,   **2.** Given that $f(x) = 3(x+2)$, $g(x) = \frac{1}{x}$

find $(a)\ f[g(x)];\ (b)\ g[f(x)]$ find $(a)\ f[g(x)];\ (b)\ g[f(x)]$

3. Given that $f(x) = (x+1)^2 + 3$, $g(x) = -x$,

find $(a)\ f[g(x)];\ (b)\ g[f(x)]$

Answers: **1.** (a) $x + 1$; (b) $x + 1$; **2.** (a) $\frac{3}{x} + 6$; (b) $\dfrac{1}{3x+6}$; **3.** (a) $x^2 - 2x + 4$; (b) $-x^2 - 2x - 4$

Lesson 72
Inverse Functions and Inverse Relations

Recalling the definitions of a relation and a function, a relation is a set of ordered pairs and function is a relation in which no two distinct ordered pairs have the same first elements (components).

Inverse Relation

Given a relation which is specified by the set of ordered pairs (x, y), the inverse relation of this given relation is the set of ordered pairs (y, x). This inverse relation is obtained by interchanging the first and second elements of each ordered pair.

Example 1: Find the inverse relation of the function specified by the set A:
$$A = \{(1, 3), (3, 2), (5, 7)\}$$

Solution   The inverse relation is obtained by interchanging the first and second elements of each ordered pair
The inverse relation of A is the set
$$\{(3,1),(2,3),(7,5)\}$$

From the definition of the inverse, we can conclude that when we form the inverse of a relation, the domain and the range of the original relation are interchanged. Thus, the domain of the original relation becomes the range of the inverse, and the range of the original relation becomes the domain of the inverse. The inverse may be a relation or a function. (See page 139)

Inverse function

Let a function be specified by the set of ordered pairs $\{(x, y)\}$. Then the inverse relation of this function is the set of ordered pairs $\{(y, x)\}$. If this inverse relation is also a function, then for the inverse relation ,we symbolize $f^{-1}(x)$, which is read "f inverse of x" and we say that $f(x)$ and $f^{-1}(x)$ are inverse functions of each other.

Example 2   Find the inverse function of the function specified by the set $B_,$:
$$B = \{(a, b), (c, d), (e, f)\}$$

Solution: The inverse function of B is the set, denoted by B^{-1}, is given by
$$B^{-1} = \{(b, a), (d, c), (f, e)\}$$

Sometimes, authors ask the question :" Does this function have an inverse?" Such a question may sometimes be misunderstood, since the inverse is found by interchanging the roles of x and y, which is always possible. What is implied in such a question is whether or not the inverse relation obtained is **also** a function. Perhaps, an unambiguous form of the question should be "Does this function have an **inverse function**?" If the inverse relation of a given function is also a function, then we also say that the given function is invertible

Note that it is possible that a given relation which is not a function may have an inverse relation which is a function.

Let us elucidate how the terms, " relation, function and inverse", are connected by using the terms "inverse relation" and "inverse function" in the following statements:

Every relation has an inverse relation. This inverse relation may or may not be a function.
Every function has an inverse relation. Some functions have inverse relations which are (inverse) functions.
A function is also a relation , but a relation is not necessarily a function.

Determining if a function has an inverse function

Necessary and sufficient condition for a function to have an inverse function: A function has an inverse function if and only if it is a one-to-one function.

We will consider the different forms in which the rules for specifying a function may be given and determine if the function has an inverse function. We will consider the set form, the tabular form, the graphical form, and the equation form.

Case 1: Given the set form of the function

A function has an inverse function if for any two different ordered pairs, the second elements are different. If a function has an inverse function it is said to be invertible.

The necessary and sufficient condition is a consequence of the definitions of a function (page 139) and of an i nverse function. We may note here that the condition for invertibility refers to the differences in the second components while the condition for being a function refers to the differences in the first components.

Example 3 Given the function specified by the set

$A = \{(1, 2),\ (2, 3), (4, 5), (6, 7)\}$, determine if the inverse relation of this set is a function.

Solution Since the second components are all different from one another, the inverse relation is a function.

In fact, the inverse of the set A is given by $A^{-1} = \{(2,1),(3,2),(5,4),(7,6)\}$,
which is clearly a function, since the first components are all different from one another.

An example of a function whose inverse relation is **not** a function is the set $\{(3,4),(5,4),(6,5)\}$, because the first and the second ordered pairs have the same second component , 4.

Case 2: Given the tabular form of the function

If any two y-values are the same, then the inverse relation is not a function (but only a relation) otherwise, it is a function. This case corresponds to the set form of a function.

Example 4   **Table 3** represents a function whose inverse relation is a function, while **Table 4** represents a function whose inverse relation is **not** a function.

Tables 3

x	y
1	2
2	3
3	4
4	5
5	6
6	7

Table 4

x	y
1	2
2	3
3	3
4	5
5	7
6	8

$\}$ same two y-values

Case 3: Given the graphical form of the function

The horizontal line test

If any **horizontal line** drawn to intersect the graph intersects the graph at only one point, then the inverse relation of the function (graph) is a function. However, if a horizontal line meets the graph in more than one point, then the inverse of the graph is not a function but only a relation. See Figs 1, 2 and 3 below.

Example 3 Figure **1** is **not** the graph of a one-to-one function, Figures **2** and **3** are the graphs of one-to-one functions.

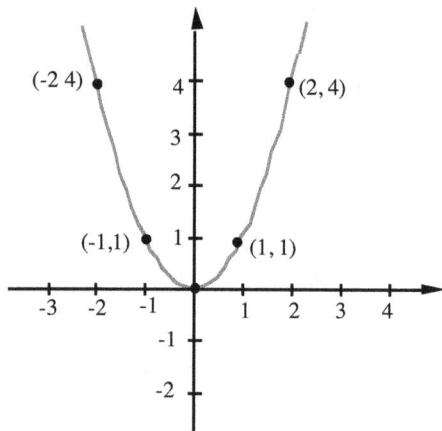

Figure 1: The graph of $y = x^2$,
   The inverse relation of this graph is **not** a function
(This is the graph of a one-to-one function)

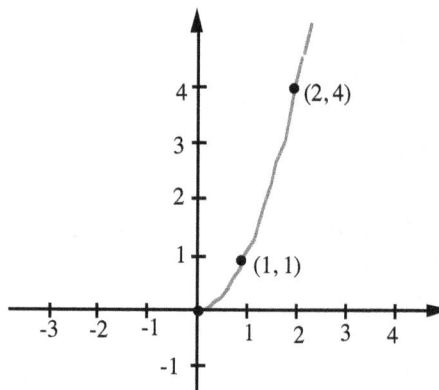

Figure 2: The graph of $y = x^2$, $x \geq 0$.
The inverse relation of this graph is a function

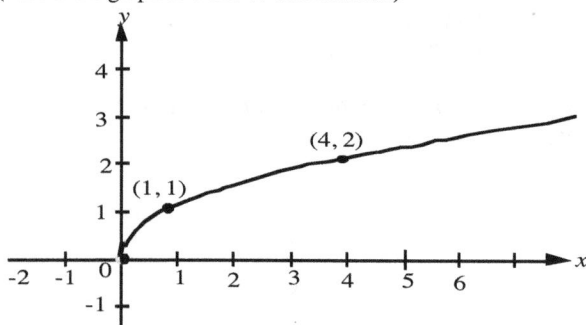

Figure 3: The inverse relation of this graph is a function. (This is the graph of a one-to-one function)

Case 4: Given the equation of the function

By definition, y is a function of x if there is a rule which gives only one corresponding value of y for a given value of x.

Example Determine if the inverse relation of $y = x^2$ is a function.
   Method 1: By interchanging the roles of x and y in the given equation, we obtain the inverse relation

$$x = y^2 \text{ or } y^2 = x. \text{ Solving } y^2 = x \text{ for } y, \text{ we obtain, } y = +\sqrt{x} \text{ or } y = -\sqrt{x}$$

Since for the same x-value (say, 4) we have two different y-values (+2 and -2),. the inverse relation of $y = x^2$ is not a function. However, we could make the inverse relation become a function by restricting the domain of this function (see Figures 1 and 2, above)
For some functions, we will graph the function and use the graphical method.

Method 2: Let $y = f(x)$

Step 1: $f(x_1) = x_1^2$

$\qquad f(x_2) = x_2^2$

$\qquad$ Equate RHS of $f(x_1)$ to RHS of $f(x_2)$ $\qquad$ (That is, let $f(x_1) = f(x_2)$):

$\qquad x_1^2 = x_2^2$ $\qquad$ (1)

Step 2: Solve for x_1. $\qquad\qquad\qquad$ (You may also solve for x_2)

$\qquad x_1^2 = x_2^2$

$\qquad x_1 = \pm\sqrt{x_2^2}$

$\qquad\qquad x_1 = +x_2 \text{ or } - x_2$ $\qquad\qquad$ (2)

Since from above, $f(x_1) = f(x_2)$, does **not** imply $x_1 = x_2$ (from equation (2)), $f(x)$ is **not** one-to-one and therefore does **not** have an inverse function.

Finding the inverse of a function specified by an equation

Example 3. Find the inverse function of $f(x) = 3x + 2$. $\qquad\qquad$ (1)

Solution

$\quad$ Step 1: Let $f(x) = y$. Then $y = 3x + 2$.

$\qquad\qquad$ Interchange x and y in the given equation.

$\qquad\qquad$ Then, we obtain $x = 3y + 2$. $\qquad\qquad\qquad$ (2)

$\qquad\qquad$ By tradition, we want to keep the x-axis horizontal and express y as a function of x.

$\quad$ Step 2: Solve equation (2) for y.

$\qquad\qquad$ Then from $x = 3y + 2$.

$$\frac{x-2}{3} = y$$

$$\text{or } y = \frac{x-2}{3}$$

The inverse $f^{-1}(x) = \frac{x-2}{3}$

$\quad$ Alternatively,

$\quad$ Step 1: Solve $y = 3x + 2$. for x.

$\qquad\qquad$ Then $x = \frac{y-2}{3}$

$\quad$ Step 2: Interchange x and y $\qquad$ (by definition of the inverse)

$$y = \frac{x-2}{3}$$

The inverse $f^{-1}(x) = \frac{x-2}{3}$

We can observe from above that Steps 1 and 2 are interchangeable.

Since the inverse $y = \frac{x-2}{3}$ is also a function, we can say that $y = 3x + 2$. and $y = \frac{x-2}{3}$ are inverse functions (of each other). We may also add that $f(x) = 3x + 2$. is invertible.

Example 4 Find the inverse function of $f(x) = x^2 + 6$.

Solution

Step 1:   Let $f(x) = y$

Step 2: Solve for x.

$$y = x^2 + 6.$$
$$x = \pm\sqrt{y - 6} \qquad\qquad (2)$$

Step 3: Interchange x and y in equation (2).

Then $y = \pm\sqrt{x - 6}$ (same as $y = +\sqrt{x - 6}$ or $y = -\sqrt{x - 6}$)   (3)

Clearly, equation (3) does not represent a function, since for a given value of x, there are two y-values. We can test this by graphing and using the so called **vertical line tes**t (see page 141).

We can say that the given function has an inverse relation specified by $y = \pm\sqrt{x - 6}$. However, the given function does not have an inverse function.

In this example, if we were asked to find $f^{-1}(x)$, we would say that there is no $f^{-1}(x)$ (no inverse function) for the given function.

Example 5 Find the inverse function of $f(x) = x^2 + 6$ $x \geq 0$

Solution

Step 1:   Let $f(x) = y$

Then $y = x^2 + 6$ $x \geq 0$

Step 2: Solve for x.

Then $x = \pm\sqrt{y - 6}$ (That is, $x = +\sqrt{y - 6}$ or $x = -\sqrt{y - 6}$)   (1)

Since we are only interested in the domain $x \geq 0$, we reject the negative part of equation (1).

Then, we obtain $x = +\sqrt{y - 6}$ (2)

Step 3: Interchange (to obtain the inverse) x and y in equation (2)

Then $y = +\sqrt{x - 6}$ (3)

Equation (3) is the inverse of $y = x^2 + 6$ $x \geq 0$

Therefore, $f^{-1}(x) = +\sqrt{x - 6}$ $x \geq 6$

The graph of this inverse (Figure **2**) indicates that the inverse is a function.

Similarly, if the given function were $f(x) = x^2 + 6$ $x \leq 0$,

the inverse relation would be $f(x) = -\sqrt{x - 6}$ and this also is a function.

We may note from Examples 2 and 3 that, sometimes, by redefining (or restricting) the domain of a given function, a function which does not have an inverse function can be made to hav e an inverse function. See Figures 1, 2 and 3.

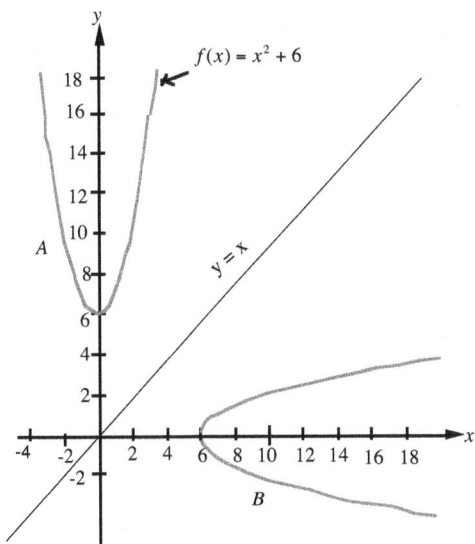

Figure 1: B, the inverse relation of A, is **not** a function .

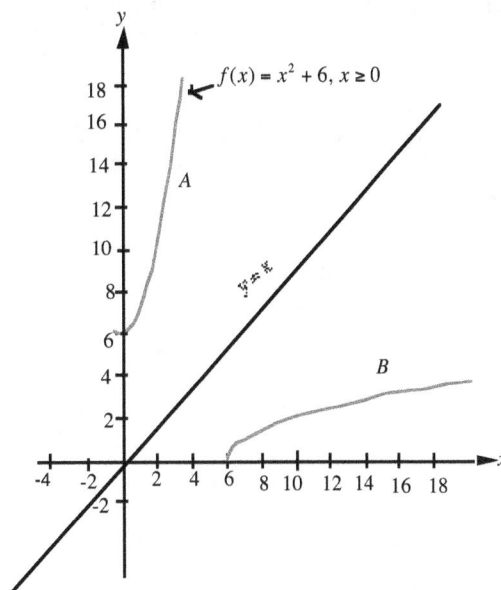

Figure 2: B, the inverse relation of A, is a function (by the vertical line test.)

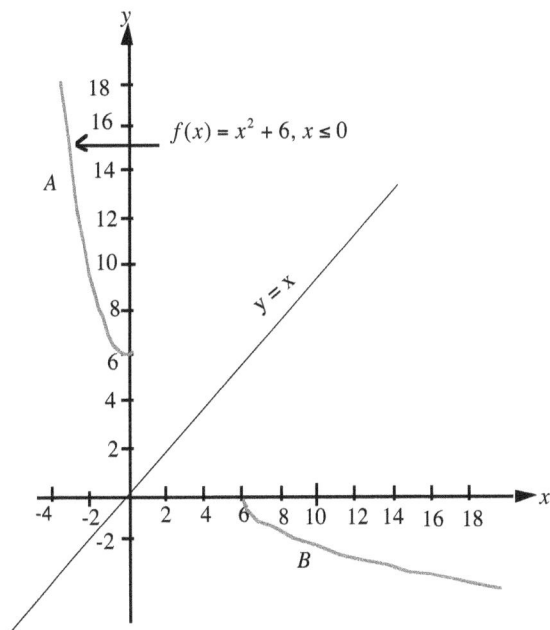

Figure 3: B, the inverse relation of A, is a function (by the vertical line test)

Given the graph of a function, how to sketch the inverse of the graph

Example 5 Given the graph of $y = 3x + 2$, sketch the graph of the inverse of this function.

Geometrically, a function (or relation) and its inverse are symmetric with respect to the line $y = x$. To obtain the inverse graph, we will reflect the given graph about the line $y = x$. The given function is a straight line and so, we will reflect two points about the line $y = x$, and then connect these points by a straight line.

Step 1: Interchange the x- and y-coordinates of any two points on the given line, say, $(1, 5)$ becomes $(5, 1)$; and $(-2, -4)$ becomes $(-4, -2)$.

Step 2: Plot the points $(5, 1)$ and $(-4, -2)$ on the same coordinate system of axes (as the given graph

Step 3: Connect the points from Step 2 by a straight line to obtain the inverse graph (Fig.)

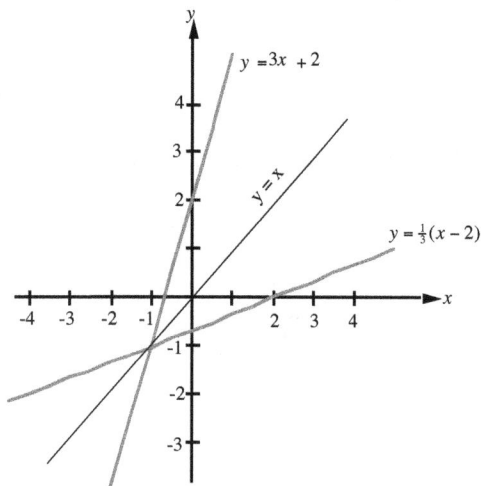

Figure 4: The graphs of $y = 3x + 2$ and $y = \frac{1}{3}(x - 2)$

Note above that, we could also find the equation of the graph, find its inverse equation and then use this equation to sketch the inverse graph.

Determining if two functions are inverses of each other
Case 1: Given the equations of the functions

Example Are the functions $y = 4x + 2$ and $y = \dfrac{x-2}{4}$ inverses of each other?

Solution

Method 1   We will use the principle of composition of functions which states that:

Two functions f_1 and f_2 are inverses of each other if

1. $f_1[f_2(x)] = x$ and   **2.** $f_2[f_1(x)] = x$ (Note that inverse functions reverse the action of
each other.)

Let $f_1 = 4x + 2$ and $f_2 = \dfrac{x-2}{4}$

$$f_1[f_2(x)] = 4[\dfrac{x-2}{4}] + 2$$

$$= 4[\dfrac{x-2}{4}] + 2$$

$$= x - 2 + 2$$

$$= x$$

$$f_2[f_1(x)] = [\dfrac{(4x+2)-2}{4}]$$

$$= \dfrac{4x+2-2}{4}$$

$$= \dfrac{4x}{4}$$

$$= x$$

Since $f_1[f_2(x)] = f_2[f_1(x)] = x$,

$y = 4x + 2$ and $y = \dfrac{x-2}{4}$ are inverses of each other.

Method 2 Find the inverse of one of the functions and compare this inverse with the other
function.

Solution

Step 1: We find the inverse of $y = 4x + 2$ (by interchanging x and y and solving for y)

$$x = 4y + 2$$

$$y = \dfrac{x-2}{4}$$ (solving for y)

Step 2: Clearly, this inverse is identical with the other function. .

Therefore, $y = 4x + 2$ and $y = \dfrac{x-2}{4}$ are inverses of each other.

Case 2: Given the graphs of the functions

Procedure: Fold the page along the line $y = x$ and if the two graphs coincide, then the
functions are inverses of each other. See Figures 2 and 3   (page 317, 318,)

Lesson 72 Exercises

A Find the inverse relation in each of the following and state if the inverse relation is a function:

1. $A = \{(2,3), (4,5), (6,7)\}$ 2. $B = \{(0,1), (2,5), (4,10)\}$ 3. $C = \{(b,a), (c,d), (a,b)\}$

4. $\{(1,3),(2,3),(4,5)\}$

5. Given the x- and y-values of a function, determine if the inverse relation in the following table represents an inverse function.

x	2	4	8	9
y	5	6	10	11

Determine if the inverse relation in each of the following graphs represents an inverse function:

6.

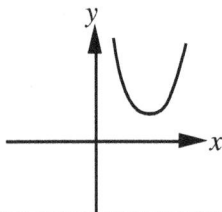

7.

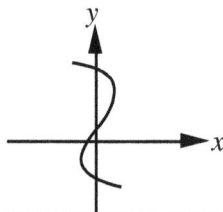

Answers: 1. $\{(3,2),(5,4),(7,6)\}$. It is a function; 2. $\{(1,0),(5,2),(10,4)\}$ .It is a function.
3. $\{(a,b),(d,c),(b,a)\}$ .It is a function; 4. $\{(3, 1),(3, 2),(5, 4)\}$ .It is not a function.
5. The inverse relation is a function; 6. The inverse relation is **not** a function;
7. The inverse relation is a function; even though the given relation is not a function.

B Find the inverse relation of the following and indicate which of the inverse relations are functions.

1. $y = -4x + 1$;   2. $y = x^2 - 3$ 3. $y - 5 = x^2$;   4. $y = x^3$;   5. $x - 2y = 6$

Determine which of the following pairs are inverses of each other.

6. $y = \dfrac{1}{x+2} - 3$ and $y = \dfrac{1}{x+1} - 4$;   7. $y = 5x + 2$ and $y = \dfrac{x-2}{5}$;   3. $y = x^3$ and $y = x^{\frac{1}{3}}$

Answers: 1. $y = -\dfrac{x}{4} + \dfrac{1}{4}$ (a function); 2. $y = \pm\sqrt{x+3}$ (**not** a function);

3. $y = \pm\sqrt{x-5}$ (**not** a function); 4. $y = x^{\frac{1}{3}}$ (a function);   5. $y = 2x + 6$ (a function); 6. No;
7. Yes;   8. Yes.

C

In each of the following graphs, sketch its inverse by reflecting the graph in the line $y = x$:

1.

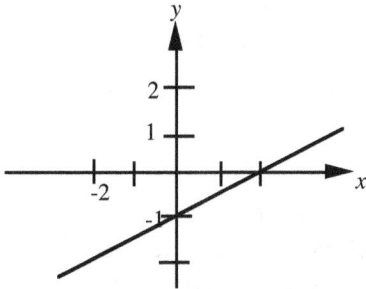

2.

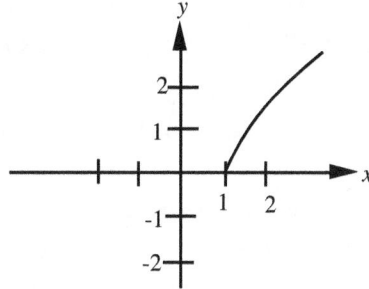

Answers:

1.

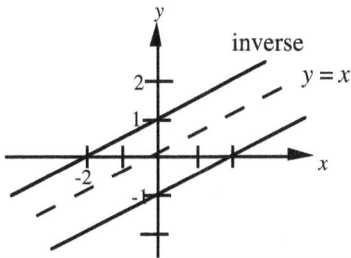

2.

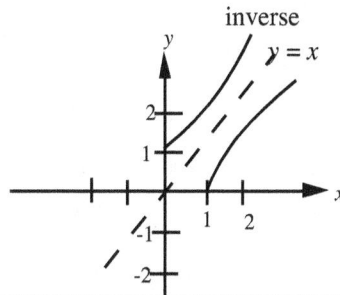

APPENDIX D

Lesson 73: **Graphs of Exponential Functions and Applications of Exponential Functions**

Lesson 74: **Graphs of Logarithmic Functions and Applications of Logarithmic Functions**

Introduction to Exponential and Logarithmic Functions

We use exponential functions in numerical calculations.

1. In physical chemistry, the exponential function is an important function. For example, the exponential equation $C = C_0 e^{-kt}$ relates the decrease in concentration, C, of a reacting material, per unit time, t, where C_0 is the concentration at time $t = 0$. There are similar equations for the decay of a radioactive material.

2. In optics, $I = I_0 e^{-kl}$, where I is the decrease in the intensity of light, l is the depth of the absorbing material.

3. In barometry, $P = P_0 e^{-kh}$, where P_0 is the pressure when h is zero (say at sea level).

4. In physics, the equation of the vibration of a material could be given by $y = Ae^{-kt}\cos 2\pi bt$

5. In chemistry, $\log_e k_p = -\dfrac{\Delta G^0}{RT}$ where k_p is the equilibrium constant, ΔG^0 is the free energy, R is the gas constant, and T is the temperature in degrees Kelvin.

5. In business, we use logarithms, to evaluate the compound interest formula $A = P(1 + \frac{r}{n})^{nt}$ where A is the amount accumulated after t years; P is the principal invested at r % annual interest. rate; and n is the number of times a year the interest is compounded.

Lesson 73

Graphs of Exponential Functions and Applications of Exponential Functions

Let b be a positive constant ($b > 0$), such that b is not equal to 1 ($b \neq 1$). Then the exponential function with a base b is defined by $f(x) = b^x$ (1)

The domain of the exponential function consists of all real numbers and the range consists of all **positive** real numbers.

If we let $f(x) = y$,, then equation (1) above becomes

$$y = b^x \qquad (2)$$

We shall cover two cases of the values of b.

Case 1: $b > 1$ and $y = b^x$. Example: $y = 2^x$

Case 2: $0 < b < 1$ and $y = b^x$. Example: $y = \left(\frac{1}{2}\right)^x$

Case 1: Properties of $y = b^x$, $b > 1$

1. The curve crosses the y-axis at $(0, 1)$. That is, the y-intercept $= 1$. (Note: $y = b^0 = 1$).

2. For all values of x, the function is positive (the curve is above the x-axis or all y-values are positive). The function is increasing

3. As x decreases without bound ($x \rightarrow -\infty$), y approaches 0 ($y \rightarrow 0$.). Thus the negative x-axis is a horizontal asymptote to the curve. Graphically, the curve is concave up to the left.

4. As x increases without bound ($x \rightarrow \infty$), the function increases without bound ($y \rightarrow \infty$)

5. When $x = 1$, $y = b$.

How to Sketch the Graph of an Exponential Function

Example 1 Sketch the graph of $f(x) = 2^x$.

Note: Do not be frightened by the type of equation involved in this example. Sketching the graph of an exponential equation is **not** more difficult than sketching the graphs of, say, the line $y = 3x - 5$ or the parabola $y = x^2$. What we need are some points (ordered pairs) on both sides of the y-axis to plot and connect. In this graph, always include $x = 0$ and $x = 1$.

Step 1: Let $f(x) = y$, Then $y = 2^x$.

Calculations   Ordered pairs (points)

Step 2: Find the y-intercept by letting $x = 0$.

$$\left.\begin{array}{l} x = 0, y = 2^0 \\ \quad = 1 \end{array}\right\} \Rightarrow (0, 1)$$

Step 3: Choose other convenient x-values (some positive and some negative values) and calculate the corresponding y-values.

$$\left.\begin{array}{l} x = 1, y = 2^1 \\ \quad = 2 \end{array}\right\} \Rightarrow (1, 2)$$

$$\left.\begin{array}{l} x = 2, y = 2^2 \\ \quad = 4 \end{array}\right\} \Rightarrow (2, 4)$$

$$\left.\begin{array}{l} x = 3, y = 2^3 \\ \quad = 8 \end{array}\right\} \Rightarrow (3, 8)$$

$$\left.\begin{array}{l} x = -1, y = 2^{-1} \\ \quad = \dfrac{1}{2} \end{array}\right\} \Rightarrow (-1, \tfrac{1}{2})$$

$$\left.\begin{array}{l} x = -2, y = 2^{-2} \\ \quad = \dfrac{1}{2^2} \\ \quad = \dfrac{1}{4} \end{array}\right\} \Rightarrow (-2, \tfrac{1}{4})$$

Step 4: Plot the points from Steps 2, and 3 (Figure...): That is plot $(0, 1)$, $(1, 2)$, $(2, 4)$, $(3, 8)$, $(-1, \frac{1}{2})$ 3 2 4
and $(-2, \frac{1}{4})$, noting that as x decreases without bound ($x \to -\infty$), y approaches 0 ($y \to 0$).
Thus the negative x-axis is a horizontal asymptote to the curve. Graphically, the curve is
concave up to the left. Also, note that as x increases without bound, the function increases without bound.

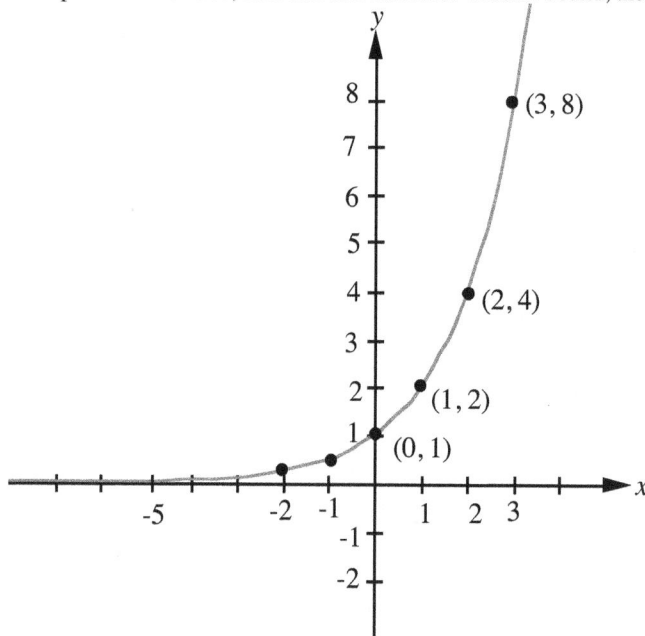

Figure 1: The graph of $y = 2^x$

Case 2: $y = b^x$, $0 < b < 1$

Examples:

1. $y = \left(\frac{1}{2}\right)^x$ or $y = 2^{-x}$ Note : $\left(\frac{1}{2}\right)^x = \left(2^{-1}\right)^x = 2^{-x}$

2. $y = \left(\frac{1}{3}\right)^x$ or $y = 3^{-x}$

Properties of $y = b^x$, $0 < b < 1$

1. The y-intercept is 1. That is, the curve crosses the y-axis at $(0, 1)$.

2. For all values of x the function is positive (All the y-values are positive).

3. The function is decreasing.

4. As x increases without bound ($x \to \infty$), the function approaches zero ($y \to 0$).
Thus the positive x-axis is a horizontal asymptote to the curve .

5. As x decreases without bound, ($x \to -\infty$), the function   increases without bound ($y \to \infty$).

6. Graphically, the graph is concave up to the right (Figure)

Example 2 Sketch the graph of $f(x) = \left(\frac{1}{2}\right)^x$ or $f(x) = 2^{-x}$

Step 1: Let $f(x) = y$, Then $y = \left(\frac{1}{2}\right)^x$ or $y = 2^{-x}$

Step 2: Find the y-intercept by letting $x = 0$. $\left. x = 0, \ y = \left(\frac{1}{2}\right)^0 = 1 \right\} \Rightarrow (0,1)$

Step 3: Choose some other convenient x-values (positive and negative values) ; calculate the corresponding y-values

$\left. x = 1, y = \left(\frac{1}{2}\right)^1 = \frac{1}{2} \right\} \Rightarrow (1,\frac{1}{2})$

$\left. x = -2, y = \left(\frac{1}{2}\right)^{-2} \right.$
$= \frac{1}{2^{-2}}$
$= 2^2$
$\left. = 4 \right\} \Rightarrow (-2,4)$

$\left. x = -3, y = \left(\frac{1}{2}\right)^{-3} \right.$
$= \frac{1}{2^{-3}}$
$= 2^3$
$\left. = 8 \right\} \Rightarrow (-3,8)$

$\left. x = 2, y = \left(\frac{1}{2}\right)^2 = \frac{1}{4} \right\} \Rightarrow (2,\frac{1}{4})$

$\left. x = 3, y = \left(\frac{1}{2}\right)^3 = \frac{1}{8} \right\} \Rightarrow (3,\frac{1}{8})$

$\left. x = -1, y = \left(\frac{1}{2}\right)^{-1} = \frac{1}{2^{-1}} \right.$
$\left. = 2^1 \right\} \Rightarrow (-1,2)$

Step 4: Plot the points from Steps 2, and 3 (Fig.2). That is plot $(0,1)$ $(1,\frac{1}{2})$ $(2,\frac{1}{4})(3,\frac{1}{8})$, $(-1,2)$, $(-2,4))$ and $(-3,8))$, noting that as x increases without bound ($x \to \infty$), y approaches 0 ($y \to 0$). Thus the positive x-axis is a horizontal asymptote to the curve. Graphically, the curve is concave up to the right. Also, note that as x decreases without bound ($x \to -\infty$), the function increases without bound ($y \to \infty$).

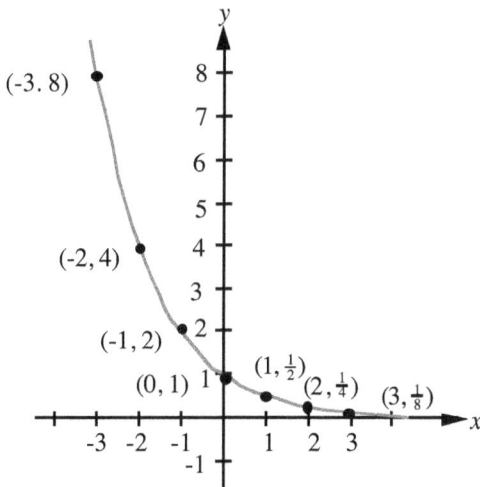

Fig. 2: The graph of $y = \left(\frac{1}{2}\right)^x$ or $y = 2^{-x}$

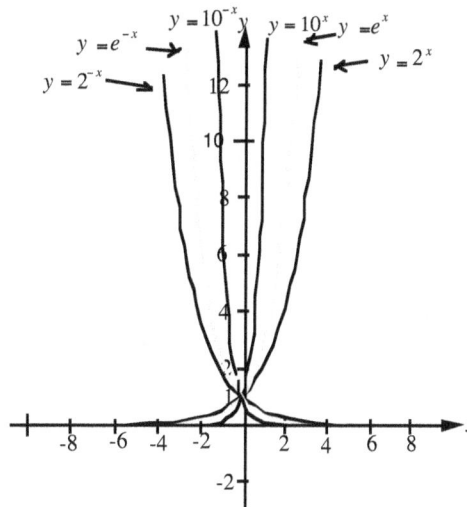

Fig. 3: The graphs of $y = e^x$, $y = 10^x$, $y = 2^x$, $y = e^{-x}$, $y = 10^{-x}$ and $y = 2^{-x}$

Other Exponential Graphs; **The graphs of** $y = e^x$, $y = 10^x$ $y = 2^x$, $y = 2^{-x}$,

The graph of $y = e^x$ is a special case of $y = a^x$ in which $a = e$ ($e = 2.72$, and e is the base of the natural logarithms. (see page 210). Similarly, when $a = 10$, we obtain $y = 10^x$. The graphs of $y = 10^x$, $y = 2^x$, $y = e^{-x}$, $y = 10^{-x}$ and $y = 2^{-x}$ on the same set of coordinate axes, are presented in Fig 3, above.

Applications of Exponential Functions

Exponential Growth and Decay

The exponential growth or decay model is given by the equation $y(t) = y_0 e^{kt}$,where y_0 is the initial value, $y(t)$ is the value at time t. If $k > 0$, y is said to grow exponentially , and k is then called the growth constant. If $k < 0$, y is said to decay exponentially , and k is then called the decay constant. The applications of this model include population growth, bacterial growth, radioactive decay and compound interest.

Compound Interest

The formula relating the principal P, the time, t, taken by P to grow to the amount A is given by
$$A(t) = P(1 + i)^t$$
Example If \$30,000 is invested at 5% interest and compounded annually, in t years, it will grow to an amount , A , given by $A(t) = 30,000(1.05)^t$

(a) How long will it take to accumulate \$65,000 in the account?
(b) Find the time required for \$30,000 to double itself.

Solution

a) $A = 30,000$

$$65,000 = 30,000(1.05)^t$$

$$\frac{65,000}{30,000} = (1.05)^t$$

$$2.17 = (1.05)^t$$

$$\log 2.17 = \log 1.05^t$$

$$\log 2.17 = t \log 1.05$$

$$\frac{\log 2.17}{\log 1.05} = t$$

$$15.9 = t$$

Scrapwork

$$\log 2.17 = 0.3365$$

$$\log 1.05 = 0.0212$$

$$\frac{\log 2.17}{\log 1.05} = \frac{0.3365}{0.0212} = 15.9$$

The time taken by \$30,000 to grow to \$65,000, at the interest rate of 5%, is 15.9 years.

(b) \$30,000 doubles to \$60,000 at the interest rate of 5% in t years.

Therefore, $P = 30,000$, $A = 60,000$; and substituting in $A(t) = P(1 + i)^t$, we obtain

$60,000 = 30,000(1.05)^t$, and next solve this exponential equation for t.

$$\frac{60,000}{30,000} = (1.05)^t$$

$$2 = (1.05)^t$$

$$\log 2 = \log (1.05)^t$$

$$\log 2 = t \log (1.05)$$

$$\frac{\log 2}{\log 1.05} = t$$

$$14.2 = t$$

Scrapwork

$$\log 2 = 0.3010$$

$$\log 1.05 = 0.0212$$

$$\frac{\log 2}{\log 1.05} = \frac{0.3010}{0.0212} = 14.2$$

Therefore, at the interest rate of 5%, \$30,000 doubles in 14.2 years.

Lesson 73 Exercises

Sketch the graphs of the following pairs on the same set of rectangular axes:

1. $y = 4^x$ and $y = 4^{-x}$;   **2.** $y = e^x$ and $y = e^{-x}$;   **3.** $y = 10^x$ and $y = 10^{-x}$

4. What can be said with respect to transformations about the relationship between each pair in Exercises 1-3?

5. If $30,000 is invested at 5% interest and compounded annually, in t years, it will grow to an amount , A, given by $A(t) = 30{,}000(1.05)^t$

  (a) How long will it take to accumulate $65,000 in the account?
  (b) Find the time required for $30,000 to double itself.

Answers 1-3

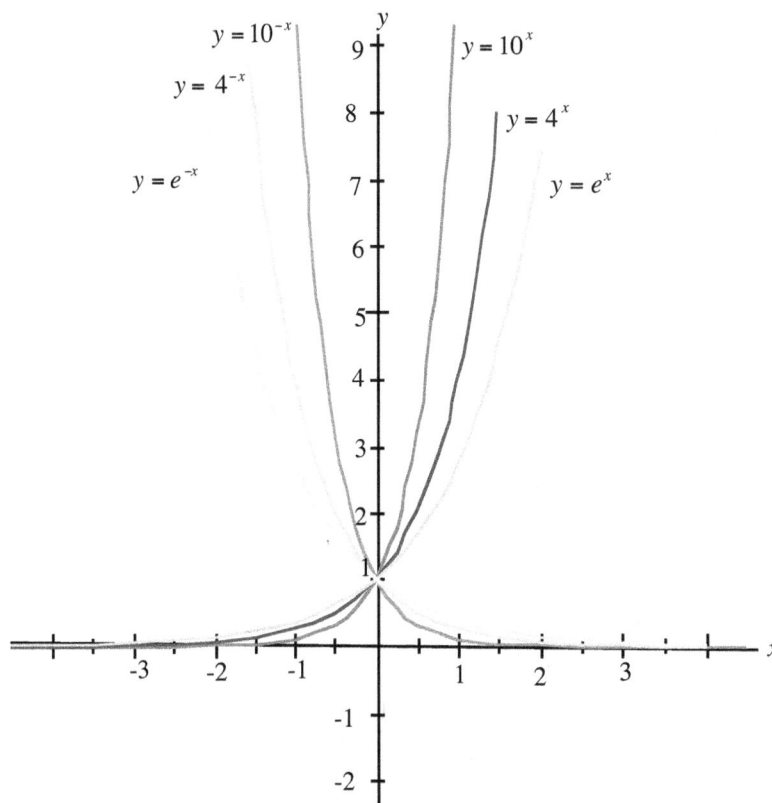

5. (a) The time taken by $30,000 to grow to $65,000, at the interest rate of 5%, is 15.9 years.
  (b At the interest rate of 5%, $30,000 doubles in 14.2 years.

Lesson 74
Graphs of Logarithmic Functions and Applications of Logarithmic Functions

The logarithmic function is the inverse of the exponential function.

The exponential function is given by $y = b^x$, $b > 0$, $b \neq 1$.. The inverse of this function is obtained by interchanging the x- and y-coordinates of the exponential function. If we interchange the coordinates, we obtain the logarithmic function given by

$$x = b^y \qquad \text{(A)}$$

Now, we solve equation (A) for y using the basic properties of logarithms,

$\log_b x = \log_b b^y$ (Taking logs of both sides of equation (A))

$\log_b x = y \log_b b$

$\log_b x = y$ ($\log_b b = 1$)

We call the function $\log_b x$ the logarithm function of x to the base b. We must note that

$x = b^y$ is equivalent to $\log_b x = y$ (Memorize this equivalent relationship)

Note that the logarithmic function is one-to-one.

Example 1 The graph of $y = \log_2 x$ or $x = 2^y$

The graph of $y = \log_2 x$ can be obtained from the graph of $y = 2^x$ by reflecting $y = 2^x$ about the line $y = x$.

Properties of $y = \log_2 x$ or $x = 2^y$

1. The x-intercept is at $(1, 0)$.

2. As $x \to 0$ (x approaches zero) $\log_2 x \to -\infty$. The curve decreases without bound. The negative y-axis is an asymptote to the curve (in the 4th quadrant).

3. The domain consists of the set of all positive numbers.

4. The range consists of the set of all real numbers.

5. The graph has the negative y-axis as a vertical asymptote (in the 4th quadrant).

6. The graph is concave down to the right (Figure 4)

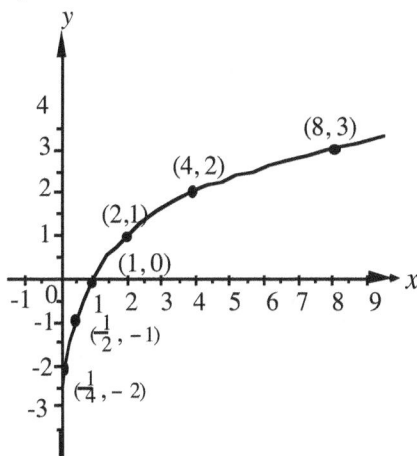

Figure 4: $y = \log_2 x$ or $x = 2^y$

Example Sketch the graph of $f(x) = \log_2 x$.

Step 1: Let $y = f(x)$.

Then $y = \log_2 x$.   (A)

Step 2: Write equation (A) in the equivalent exponential form.

Then $x = 2^y$ <--------- Exponential form.

Calculations   Ordered pairs (points)

$\Downarrow$ $\Downarrow$

Step 3: Choose y and calculate the corresponding values of x. It is more convenient to choose y and calculate x..(This may be the first time you are choosing y and calculating x.)
Find the x-intercept by letting $y = 0$. Note that there is no y-intercept for this function.

$$\left.\begin{array}{l} y = 0, x = 2^0 \\ \qquad = 1 \end{array}\right\} \Rightarrow (1,0)$$

Becareful of how you write the ordered pairs, since you are choosing y and calculating x. The first element of each ordered pair is always the x–value irrespective of the order in which the values are obtained.

Step 4: Choose other convenient y-values (positive and negative values) and calculate corresponding the x-values

$$\left.\begin{array}{l} y = 1, x = 2^1 \\ \qquad = 2 \end{array}\right\} \Rightarrow (2,1)$$

$$\left.\begin{array}{l} y = 2, x = 2^2 \\ \qquad = 4 \end{array}\right\} \Rightarrow (4,2)$$

$$\left.\begin{array}{l} y = 3, x = 2^3 \\ \qquad = 8 \end{array}\right\} \Rightarrow (8,3)$$

$$\left.\begin{array}{l} y = -1, x = 2^{-1} \\ \qquad = \dfrac{1}{2} \end{array}\right\} \Rightarrow (\tfrac{1}{2},-1)$$

$$\left.\begin{array}{l} y = -2, x = 2^{-2} \\ \qquad = \dfrac{1}{2^2} \\ \qquad = \dfrac{1}{4} \end{array}\right\} \Rightarrow (\tfrac{1}{4},-2)$$

Step 5: Plot the points from Steps 3, and 4 (Figure 4). (i.e., is plot $(1,0)$, $(2,1)$, $(4,2)$, $(8,3)$, $(\tfrac{1}{2},-1)$), $(\tfrac{1}{4},-2)$) and connect them, noting that as x approaches zero the function decreases without bound, and thus the negative y-axis is a vertical asymptote to this curve. Graphically, the curve is concave down to the right. Also, note that as x increases without bound, the function increases without bound.

Other Logarithmic Graphs

The graphs of $y = \log_a x$ and in particular the graphs of $y = \log_{10} x$ and $y = \log_e x$

The graph of $y = \log_{10} x$ or $x = 10^y$. can be obtained from the graph of $y = 10^x$ by reflecting $y = 10^x$ about the line $y = x$. Similarly, the graph of $y = \log_e x$ or $x = e^y$ can be obtained from the graph of $y = e^x$ by reflecting $y = e^x$ about the line $y = x$.

General Properties of $y = \log_a x$, $y = \log_{10} x$ and $y = \log_e x$

1. The x-intercept is at $(1, 0)$.

2. As $x \to 0$, $\log_{10} x \to -\infty$ and $y = \log_e x \to -\infty$

3. The domain consists of the set of all positive numbers.

4. The range consists of the set of all real numbers.

5. The graph has the negative y-axis as a vertical asymptote (in the 4th quadrant).

6. The graph is concave down to the right (Figure)

The graphs of $y = \log_{10} x$ and $y = \log_e x$ are presented below.

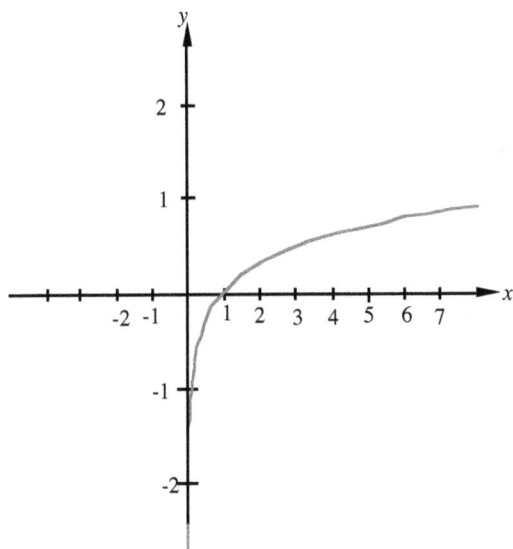

Figure 5: The graph of $y = \log_{10} x$ or $x = 10^y$.

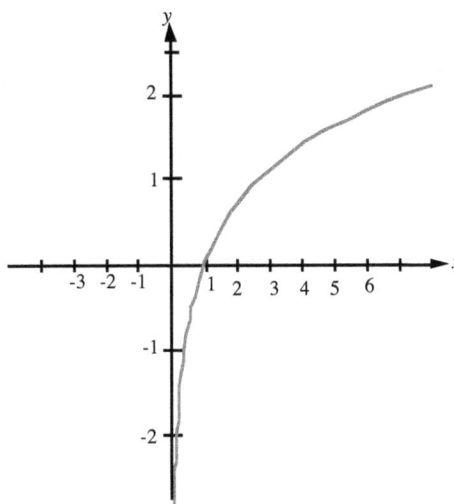

Figure 6: The graph of $y = \log_e x$ or $x = e^y$

We must note the following:

1. For an **exponential equation**, we must be careful not to confuse its equivalent equation in the logarithmic form with its inverse.

   Thus $y = e^x$ is equivalent to $x = \log_e y$ (but **not** equivalent to $y = \log_e x$, its inverse).

2. For a **logarithmic equation**, we must be careful not to confuse its equivalent equation in the exponential form with its inverse.

Thus $y = \log_e x$ is equivalent to $x = e^y$, (but **not** equivalent to $y = e^x$, its inverse).
See Figures below.

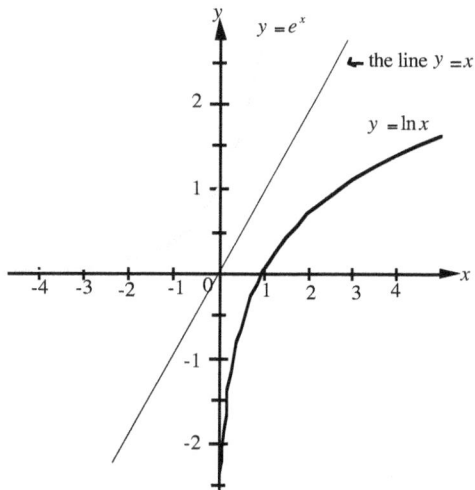

Figure 7: The graph of $y = e^x$ and $\ln x$

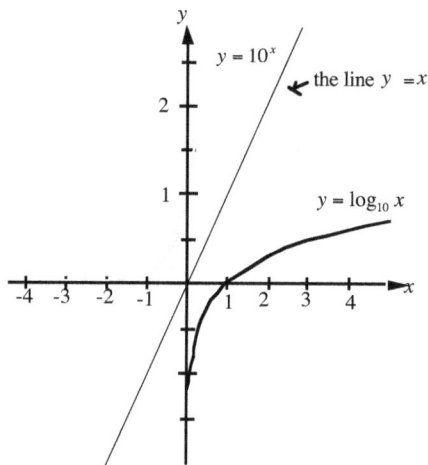

Figure 8 : Graph of $y = 10^x$ and $\log x$

Applications of Logarithmic Functions

Sound Level

Formula for sound level $L = 10 \log \dfrac{I}{I_0}$, where L (in decibels, dB) is the loudness of sound, I is the sound intensity in watts per square meter, and $I_0 = 10^{-12}\, W/m^2$

$\quad$ (a) If $I = 10^{-3}\, W/m^2$, find L. $\quad$ (b) If $L = 75$, find I.

Solution

(a) $L = 10 \log_{10} \dfrac{10^{-3}}{1} \dfrac{w}{m^2} \cdot \dfrac{m^2}{10^{-12} w}$

$\quad = 10 \log_{10} \dfrac{10^{-3}}{10^{-12}}$

$\quad = 10 \log_{10} 10^9$

$\quad = 9(10 \log_{10} 10)$

$\quad = 10(9) \quad$ (Note: $\log_{10} 10 = 1$)

$\quad = 90 \quad$ decibels (dB)

(b) $\quad L = 75$

Step 1; Substituting in

$\quad L = 10 \log \dfrac{I}{I_0}$

$\quad 75 = 10 \log \dfrac{I}{10^{-12}}$

$\quad 75 = 10 \log 10^{12} I$

$\quad \dfrac{75}{10} = \log_{10}(10^{12} I)$

Step 2: $7.5 = \log_{10} 10^{12} + \log_{10} I$

$\quad 7.5 = 12 + \log_{10} I$

$\quad (\log_{10} 10^{12} = 12 \log_{10} 10 = 12)$

$\quad 7.5 - 12 = \log_{10} I$

$\quad -4.5 = \log_{10} I$

$\quad 10^{-4.5} = I$

$\quad I = 10^{-4.5} \dfrac{w}{m^2}$

Hydrogen Ion Concentration

The *pH* of a solution is a measure of the acidity or alkalinity of the liquid. The *pH* of a solution is defined as follows: $pH = -\log_{10}\left[H^+\right]$, where $\left[H^+\right]$ is the hydrogen ion concentration (concentration per liter).

Example 1 If $\left[H^+\right]$ for a solution is 2.45×10^{-8} moles/liter, find the *pH* of this solution.

Solution $\quad pH = -\log_{10}\left[2.45 \times 10^{-6}\right]$

$= -\left(\log_{10} 2.45 + \log_{10} 10^{-6}\right)$

$= -\left(\log_{10} 2.45 + (-6)\log_{10} 10\right)$

$= -\left(\log_{10} 2.45 + (-6)(1)\right) \quad$ Note: $\log_{10} 10 = 1$

$= -(0.389 - 6) \quad$ Note: From a calculator, $\log_{10} 2.45 = 0.389$

$= -(-5.61)$

$= 5.61$

The *pH* of the 2.45×10^{-8} moles/liter solution is 5.61

Example 2 If the *pH* of a solution is 7.5, find the $\left[H^+\right]$ of this solution.

Solution

Step 1: $pH = -\log_{10}\left[H^+\right]$

$7.5 = -\log_{10}\left[H^+\right] \quad$ (substituting for *pH* = 7.5)

Step 2: $-7.5 = \log_{10}\left[H^+\right]$

$\quad 10^{-7.5} = \left[H^+\right]$

The $\left[H^+\right]$ of the solution is $10^{-7.5}$.

Lesson 74 Exercises

Sketch the graphs of the following:

1. $y = \log x$.; **2.** $y = \log(x-1)$; **3.** $y = \log(x+2)$; **4.** $y = \log_3 x$; **5.** $y = \log_2(x+3)$.

6. If the hydrogen ion concentration, $\left[H^+\right]$ for a solution is 2.45×10^{-8} moles/liter, find the *pH* of this solution.

7. If the *pH* of a solution is 7.5, find the $\left[H^+\right]$ of this solution.

8. Formula for sound level $L = 10\log\dfrac{I}{I_0}$,where L (in decibels, dB) is the loudness of sound, I is the sound intensity in watts per square meter, and $I_0 = 10^{-12}\,W/m^2$

(a) If $I = 10^{-3}\,W/m^2$, find L. (b) If $L = 75$, find I .

Answers:

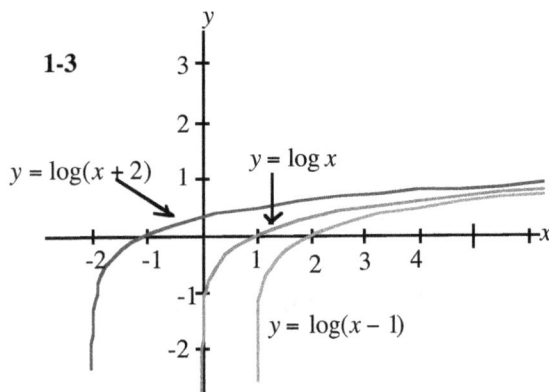

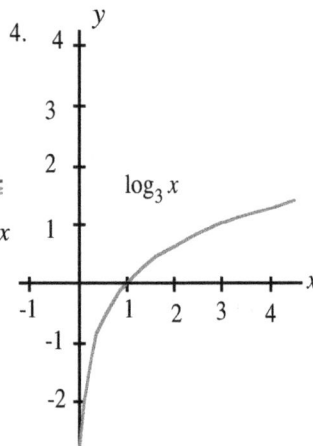

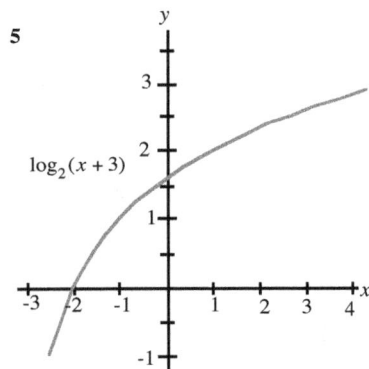

6.: The *pH* of the 2.45×10^{-8} moles/liter solution is 5.61

7. The hydrogen ion concentration, $\left[H^+\right]$ of the solution is $10^{-7.5}$.

8. (a) $= 90$ decibels (dB); (b) $I = 10^{-4.5}\dfrac{w}{m^2}$

APPENDIX E
About Measurements
Standard Unit, Error, Rounding-off Numbers, Significant Digits, Scientific Notation

To determine the size of a physical quantity, we compare its size with a standard quantity called a unit.
Example: To determine the length of the cover of a book in inches or in meters, we can use a ruler with its scale in inches or in meters.
A measurement is the ratio of the magnitude of a physical quantity to that of a standard unit

Standard unit
A standard unit is a measure with which other quantities are compared. A standard unit of measure is defined by a legal authority (such as the US Bureau of Weights and Measures) or by a conference of scientists.

Some universally accepted standards:
1. For mass, the standard (primary standard) is the kilogram (kg).
2. For length, the standard is the meter (m).
3. For time, the standard is the second (s)

Some devices for taking measurements
Examples: Rulers (for length), chemical balances (for mass), stop watches (for time), ammeters, (for electric current) voltmeters (for electric voltage) , thermometers (for temperature) a nd barometers (for pressure).

Experimental Errors (or Uncertainties)
There are two main types of errors, namely, systematic errors, and random errors.

Systematic Errors (constant errors):
These errors are due to faulty measuring devices. Systematic errors make the measurements either too small or too large:

Examples of faulty devices:
1. Instruments with needles off the zero mark; 2. Faulty clocks (stop clock)
3, Corroded weights; 4. Faulty thermometers. Heat leaking equipment

Random errors (accidental errors or indeterminate errors)
Random errors may be due to chance variations of the physical quantity being measured, or chance variations in the measuring device. Random errors may also be due to failure to take into account variables such as temperature fluctuations; and environmental effects. We can reduce random errors by making a large number of measurements and taking the average of the measurements.

Absolute error (or absolute uncertainty)

Absolute error Experimental value − accepted (true value)
Example: In an experiment to determine the acceleration due to gravity, g, the experimental value of g, was 986 cm/s^2. The accepted value of g is 980 cm/s^2. Find the absolute error.

Solution   Experimental value = 986 cm/s^2

Accepted value = 980 cm/s^2
Absolute error = experimental value - accepted value

= (986 - 980) cm/s^2

= 6 cm/s^2

The positive value indicates that the experimental value is greater than the accepted value.

Note: If the experimental value = 964 cm/s^2

The absolute error = (964 - 980) cm/s^2

$$= -16 \text{ cm}/s^2$$

The negative value means that the experimental value is less than the accepted value.

Relative error (or relative uncertainty)

$$\text{Relative error} = \frac{\text{absolute error}}{\text{accepted value}}$$

Example 1: If the absolute error = 6 cm/s^2, and

the accepted value = 980 cm/s^2,

$$\text{Relative error} = \frac{\text{absolute error}}{\text{accepted value}}$$

$$\text{Then relative error} = \frac{6 \ cm/s^2}{980 \ cm/s^2}$$

$$= 0.00612$$

Example 2: If the absolute error = -16 cm//s^2 and

the accepted value = 980 cm/s^2

$$\text{Then relative error} = \frac{-16 \ cm/s^2}{980 \ cm/s^2}$$

$$= -0.00612$$

$$= -0.0163$$

Note: If an accepted value is not known, and we have two or more experimental values, then the average of the experimental values would be used as the "accepted value" in calculations.

Rounding-Off Numbers

The rules for rounding-off a number may differ slightly depending upon the field. For instance, in accounting the rule may be slightly different from the rule in chemistry.

1. If the digit or group of digits to be dropped is more than 500...,hen drop that digit or group and add 1 to the last digit retained.

2. If the digit or group of digits to be dropped is less than 500...,then drop that digit or group and leave the last digit retained unchanged.

3. If the digit or group of digits to be dropped is exactly 500..., then drop that portion and add 1 to the last digit retained if this digit is odd but if this digit is even, then this digit remains unchanged.

Rounding off Whole numbers
Procedure:
Step 1: Locate the digit in the round-off place.
(The round-off place is the place to which we want to round-off the number)

Step 2: Drop all digits to the right of the round-off place, and if the digit immediately to the right of the round-off place is more than 5 or is 5 followed by non-zero digits, , add 1 to the round-off place digit (i.e. we round-up); but if the digit immediately to the right of the round-off place is less than 5, the round-off place digit remains unchanged. However, if the digit immediately to the right of the round-off place digit is 5 or 5 followed by zeros, we add 1 to the round-off place digit if it is odd, but if it is even, it remains unchanged.(i.e. we round-down). Also, replace each digit dropped by a zero.

Rounding-off Decimals

The procedure is the same as that for rounding-off whole numbers, except that after the decimal point, we do not replace any digits dropped by zeros.

Procedure:
Step 1: Locate the digit in the round-off place.
(The round-off place is the place to which we want to round-off the number)

Step 2: Drop all digits to the right of the round-off place, and if the digit immediately to the right of the round-off place is more than 5 or is 5 followed by non-zero digits, add 1 to the round-off place digit (i.e., we round-up); but if the digit immediately to the right of the round-off place is less than 5, the round-off place digit remains unchanged. However, if the digit immediately to the right of the round-off place digit is 5 or 5 followed by zeros, we add 1 to the round-off place digit if it is odd, but if it is even, it remains unchanged.(i.e., we round-down).

Rounding-off (Alternatively)

When we round-off a number, we drop some of the digits explicitly or implicitly specified.
We must distinguish between rounding-off to a specified number of decimal places
(or significant digits) and the implicit rounding-off which we must determine from the
numbers involved in the calculation.

The rules for rounding-off a number may differ slightly depending upon the field.
For instance, in accounting the rule may be slightly different from the rule in chemistry.

   1. If the digit or group of digits to be dropped is more than 500...then drop that
      digit or group and add 1 to the last digit retained.

   2. If the digit or group of digits to be dropped is less than 500...then drop that digit or
      group and leave the last digit retained unchanged.

   3. If the digit or group of digits to be dropped is exactly 500..., then drop that portion
      and add 1 to the last digit retained if this digit is odd but if this digit is even, then this
      digit remains unchanged.

Example: The following have been rounded-off to three decimal places.

   (1) .4398. ≈ **.440**

   (2) .43652 ≈ **.437**

   (3) .43637 ≈ **.436**

   (4) .43750 ≈ **.438**

   (5) .43650 ≈ **.436**

   (6) .43946 ≈ **.439**

   (7) .43650001 ≈ **.437**

   (8) .4365001 ≈ **.437**

Example We round off **85376.7463** to the following places, using the simple "5 or greater or
        less than 5 rule"

1. 85376.7463 to the nearest **thousandth** becomes **85376.746** (We do **not** replace the 3 dropped by a zero)

2. 85376.7463 to nearest **hundredth** becomes **85376.75** (We added 1 to the digit in the round-off place)

3. 85376.7463 to the nearest **tenth** becomes **85376.7** (The 7 is unchanged since the 4 dropped is less than 5)

4. 85376.7463 the nearest **unit** becomes **85377.**   (Adding 1 to the 6)

5. 85376.7463 to the nearest **ten** becomes **85380.** (Replacing the 6 dropped by a zero)

6. 85376.7463 to the nearest **hundred** becomes **85400.** (Replacing the digits (6 and 7) dropped by zeros)

7. 85376.7463 to the nearest **whole number** becomes **85377.** (same as to the nearest unit)

Estimation

In estimation, we round-off the numbers before carrying out the operations of addition, subtraction, multiplication , division etc. For convenience, we will round-off each number to the first non-zero digit, unless specified.

Approximate Numbers, Significant Digits, Scientific Notation,

A measurement consists of a numerical value and a unit of that measurement.
Example: 4 kilograms, where the 4 is the numerical value and the kilograms is the unit of the measurement.

Numbers obtained from a measurement are never exact (i.e., are approximate) due to the limitations of the measuring instrument as well as the skill of the person making the measurement. As such, when one records a measurement, one should indicate the reliability of the measurement. All measurements may be assumed to have an uncertainty in at least one unit in the last digit of the measurement, since in making a measurement, we usually estimate the last digit.

Results obtained from calculations using measurements are also as uncertain as the measurements themselves.
In summary, the numbers that we deal with in calculations are obtained from observations. Some of the numbers are exact and some are approximate, The approximate numbers are t hose numbers obtained from making measurements.

Exact Numbers: An exact number is a number that contains no uncertainties. It is assumed to be infinitely accurate. We can obtain exact numbers from definitions and from direct count. For example, the number of students in a math class by count is 25. In this case, there is no uncertainty, since we know that there are exactly 25 students. Similarly, when one counts 200 dollars, one knows that one has exactly 200 dollars, and there is therefore, no uncertainty. Also by definition, 60 minutes = 1 hour; 2.54 centimeters = 1 inch. Since these numbers are d efined, this 60 and 2,54 are exact and contain no uncertainties. We can also add that this 60 has an infinite number of significant digits, (we can write 60 as 60.000...) and therefore the zero in the 60 is significant. However if you make your own measurement and by coincidence obtain 2.54 centimeters, then this 2.54 would not be exact. We can generalize that all the conversion factors (from tables) are exact.

Significant Digits (Significant Figures), Digits obtained in a measurement

A significant digit (or figure) is one which is known to be reasonably reliable (or correct). When we make measurements, the digits we read and estimate on a scale are also called significant digits (or significant figures). These digits include digits that we are certain of, and one additional digit that we are uncertain of. This uncertain digit is obtained by the estimation of the fractional part of the smallest subdivision on the scale being used. As such, the rightmost digit is assumed to be uncertain.

Significant figure notation is an approximate method of indicating the uncertainty of a measurement. when recording a measurement.
We agree to the following:

1. The digits 1, 2, 3, 4, 5, 6, 7, 8, 9 are always significant.

2. The digit zero, 0. may or may not be significant according to its position in the number as follows:

  (a) Zeros before the first non-zero digits are **not** significant.

For example: (i) .0450 has **three** significant digits: The first zero is not significant; but the last zero is significant since if it were not we would not write it.

(ii) .0012 has **two** significant digits. The first two zeros are not significant

. (b) Zeros between non-zero digits **are** significant.

Example: 3.045 has **four** significant digits. The zero between 3 and 4 is significant.
40.240 has **five** significant digits. The last zero is significant because if it were not we would not write it.

More examples: The numbers referred to below are assumed to have been obtained from measurements. 23.00 has four significant digits: he zeros in this case are significant since we do not have to write the zeros if they were not significant. If the zeros were not significant we would have written 23. 2300 has two significant digits, and 600 has one significant digit; however, in each of these two examples, the number of significant digits is sometimes ambiguous.
It is suggested that when the number of significant digits is in doubt , the maximum number of significant digits is to be assumed. Also, in recording data, if we know the number of significant digits, we will use the scientific notation.
Note above that if 600 and 2300 had been obtained by counting, or by definition, all the zeros would have been significant. As a reminder, significant notation generally pertains to numbers obtained from measurements.
Using **scientific notation** avoids all ambiguities with respect to the number of significant digits.. For instance, if we know that the above number, 600 were measured to two significant digits, t hen we would write 6.0×10^2. If the measurement were to three significant digits, we would write 6.00×10^2
When we deal with very large or very small numbers we prefer to write the numbers in scientific notation form. In this form, the significant digits (digits) including the zeros can be unambiguously indicated. In scientific notation:

(1)   125000 would be written 1.25×10^5

(2)   1467 would be written 1.467×10^3

(3)   .032500 would be written 3.2500×10^{-2}

(4)   .0325 would be written 3.25×10^{-2}

(5)   If 125000 were known to four significant digits it would be written 1.250×10^5.

Note: Some authors indicate which zeros are significant by underlining the last significant zero. For example, in 23000 the first two zeros are significant but the last zero is not significant. Note also that in some books, a decimal point placed after the last zero makes all the zeros significant. For example, 23000. has five significant digits, but 23000 has two significant digits. However, but there may still be ambiguity if the number is at the end of a sentence.

Accuracy and Precision in Measurements
Two contributions to uncertainty in measurement are limitations of precision, and limitations of accuracy. **Accuracy** indicates how close a measured value is to the true value but **precision** indicates how close two measurements of the same quantity are close to each other. Generally, more precision implies more accuracy. However, there are instances in which numbers may be more precise, but may not be more accurate.. For example, if a measuring device is incorrectly calibrated (having incorrect scale).

Accuracy and Precision in Calculations

With respect to significant digits, **accuracy** refers to the number of significant digits but **precision** refers to the number of decimal places. The larger the number of significant digits in a number, the more accurate the number. The larger the number of decimal places, the more precise the number.

Note: In calculations, the approximate numbers determine how the rounding-off is done

Rounding-off to Significant Digits or Figures in Arithmetic Operations
(Implicit Specification)

1. In **multiplication and division** involving significant digits, the product or quotient (answer) should be rounded-off so that the number of significant digits in the answer is equal to the number of significant digits of the number with the least number of significant digits. (In other words, the answer should not be more accurate than the number with the least accuracy)

Example 1: Multiply 2.34 cm by 5.6 cm

Step 1: $2.34 \times 5.6 = 13.104$

Step 2: The number with the fewest number of significant digits is 5.6 and it has two significant digits. Therefore, the product (answer) should contain only two significant digits.
Thus, 13.104 becomes 13.

Answer: 13 cm^2

In the above case the "4" in 2.34 is considered to be reasonably reliable. The " 6" in 5.6 is considered to be reasonably reliable.

Example 2: Maria determined the length of a piece of wood to be 6.47 yards. What is the length of this wood in feet?

Solution
By definition, 3 feet = 1 yard
6.47 yards is used to determine the number of significant digits in the answer,
(The 3 feet is exact and has an infinite number of significant digits)

$$\frac{6.47 \text{ yards}}{1} \times \frac{3 \text{ feet}}{1 \text{ yard}}$$

= 19.41 feet

= 19.4 feet (6.47 has three significant figures)

2. In addition and subtraction, we shall round-off so that the number of decimal place in the answer equals the number of decimal places in the number with the least number of d ecimal places. (In other words, the answer should not be more precise than the number with the least precision)

Example 1: Add: 143.54, 172.3, and 64.62
Solution
143.54 < - - - - has two decimal places
172.3 < - - - - -has one decimal place (determines the number of decimal places in answer)
  64.62 < - - - - - has two decimal places
380.46
Answer: 380.5 (has one decimal place)

Example 2 Subtract 12.4 from 143.63
Solution
143.63 < - - - - has two decimal places

  12.4 < - - - - -has one decimal place (determines the number of decimal places in answer)

131.23
Answer: 131.2 (has one decimal place)

3. In finding powers and roots, the root or power should be rounded-off so that the number of significant digits in the answer is equal to the number of significant digits in the number.

Example 1: Find $\sqrt{26.9}$
Radicand has three significant digits , and therefore the root should have three significant digits.
From a calculator, $\sqrt{26.9} = 5.1865$
     $\sqrt{26.9} = 5.19$ (has three significant digits. Same as in the radicand)

Note: When two or more different operations are involved, the final operation determines how the final result is rounded-off.
Example Simplify: 38.3 + 12.9(3.58)
Solution According to order of operations, we multiply 12.9 by 3.58 first and then add 38.3
    38.3 + 12.9(3.58)

= 38.3 + 46.182

= 84.485

= 84.5 (has one decimal place as in 38.3)

Since addition was the last step, we use precision (the number of decimal places) to round-off.

Addition and Subtraction Involving Scientific Notation

Before adding or subtracting the numbers must have the **same powers** of 10. We will rewrite the expression so that the power of 10 is that of the highest power in the expression

Example: **1.** $4 \times 10^2 + 3 \times 10^2 = (4 + 3) \times 10^2$
$$= 7 \times 10^2$$

Example: **2.** $4 \times 10^2 + 2 \times 10^3 = 0.4 \times 10^3 + 2 \times 10^3$
$$= (0.4 + 2) \times 10^3$$
$$= 2.4 \times 10^3$$

or

$$4 \times 10^2 + 2 \times 10^3 = 4 \times 10^2 + 20 \times 10^2$$
$$= (4 + 20) \times 10^2$$
$$= 24 \times 10^2$$
$$= 2.4 \times 10^3 \quad \text{(Again, we obtain the same result)}$$

Order of Magnitude (for comparing relative sizes using powers of 10).

The order of magnitude is the power of 10 closest to the given number.

(It is an approximation to the number. Note the sequence, ..., $10^{-2}, 10^{-1}, 10^0, 10^1, 10^2, ...$)

If a given quantity is 1000 times another quantity, the given quantity is larger by three orders of magnitude.

Examples:

1. The order of magnitude of 123 is 10^2, since 123 is closer to 100 than to 1000.

2. Find the order of magnitude of 0.00352.

Solution

Step 1: Write the number in scientific notation.

$\qquad 0.00352 = 3.52 \times 10^{-3}$.

Step 2: Since the integer before the decimal point, 3, is less than 5, we replace 3.52 by 1

$\qquad$ (since this is closer to 1 or 10^0, than it is to 10 or 10^1)

Step 3: 3.52×10^{-3}

$\qquad = 10^0 \times 10^{-3}$

$\qquad = 1 \times 10^{-3}$

$\qquad = 10^{-3}$

The order of magnitude of 0.00352 is 10^{-3}. (since by definition, the order of magnitude is the power of 10 closest to the given number.).

We can use the order of magnitude in estimation by rounding-off to the orders of magnitude.

More examples: Round-off to the nearest order of magnitude.

1. 1.32×10^2

2. 8.02×10^4

3. 0.0009

4. 0.0302

Solution:

1. Since 1.32 is closer to 1 than to 10,

1.32×10^2

$= 10^0 \times 10^2$

$= 1 \times 10^2$

$= 10^2$

The order of magnitude of 1.32×10^2 is 10^2

2. 8.02 is closer to 10 than to 1

8.02×10^4

$= 10^1 \times 10^4$

$= 10^5$

The order of magnitude is 10^5

3. $0.0009 = 9 \times 10^{-4}$

$= 10^1 \times 10^{-4}$

$= 10^{-3}$

The order of magnitude is 10^{-3}.

4. $0.0201 = 2.01 \times 10^{-2}$

$= 10^0 \times 10^{-2}$

$= 1 \times 10^{-2}$

$= 10^{-2}$

The order of magnitude is 10^{-2}

Summary for rounding-off a number to the order of magnitude.

Step !: Write the number in scientific notation

Step 2: Ignoring the power of 10, if the integer before the decimal point is 5 or greater, replace the non-power of 10 part by 10 ((i.e. 10^1); but if the integer is less than 5, replace the non-power of 10 part by 1 (10^0)

Step 3: Simplify

International System of Units

The International System of Units (SI) has adopted a set of seven base (or primary) units.

Quantity	Unit	Symbol
Length	meter	m
Mass	kilogram	kg
Time	seconds	s
Electric current	ampere	A
Temperature	Kelvin	K
Amount of substance	mole	mol
Luminous Intensity	candela	cd

Derived units

In addition to the seven base units, there are derived units which are combinations of the base units

Example:

From the SI base unit, m (meter). for length, the unit for area is $m \times m = m^2$,
Since area = length × width, and the unit of length is m and the unit of with is m.

For more practical or convenient units, we use prefixes and multiplication factors (in powers of 10) to express other units.

Example: 1 kilometer = $10^3 m$, where kilo = 10^3

1 km = 10^3m or I km = 1000m.

INDEX

A

B

C

D

E

M

N

O

P

Q

R

Square root
    of negative numbers 195
Square Root Method 127
square root method) 47
Square Roots 91
Straight angle 220
Subtraction
    of Polynomials 12
    of Signed Numbers 2
Supplementary angles 221

T

Term 11
Terminal Side of Angle 278
Theorems 224
Transversal 225
Trapezoid
    area and perimeter' 234
Trapezoids 232
Triangles 228
Triangles, 232
Trigonometric Equations 294
Trigonometric Functional Value 276
Trigonometric Functional Value of any Angle 274
trigonometric functional values 281
Trigonometric Functions 268
Trigonometric Identities 290
Trigonometric Identities: Proving 292
Trigonometric Ratios 268
Trinomial 10

U

Union of two sets 119
Units
    of Measuring angles 219
Unlike Fractions 110

V

Variable 10
Variables 5
Vertex
    parabola 162
Vertical angles 222
Vertical line
    equation 71
Vertical line test 140, 141

W

Word Problems
    inequalities 125

X

Y

Z

Some useful conversion factors

Units of length

1 cm = .3937 in = .0328ft = .01094 yd
1 m = 100 cm = 39.3701 in = 3.2808 ft.= 1.0936 yd
1 in. = 2.54 cm = .0833 ft = .0254 m
1 ft = 12 in. = 30.48 cm = .3048 m = .3333 yd
1 mile = 1760 yd = 5280 ft = 1.6093 km
1 yd = 3 ft = 36 in.= .9144 m = 91.44 cm
1 km = .62137 mile = 1000 m = 100,000 cm

Units of mass

1 kg = 1000 g = 2.2046 lb =35.274 oz
1 lb (avdp) = 453.592 g = 16 oz
1 metric ton = 1000 kg = 10^6 g=1.1023 ton
1 ton = 907.1847 kg = 2000 lb = .9072 metric ton
1 gm = .03527 oz = 1000 mg =.0022046 lb
1 oz = 28.3495 g = .0625 lb = 16 drams
1 long ton (British) = 2240 lb

Units of volume

1 liter = 1000 cm^3= 61.0237 $in.^3$ = .26417 gal = 1.0567 qt = .03531 ft^3 = 2.113 pt.
1 gal (U.S.) = 4 qt = 3.7854 liter = 8 pt. = 231 $in.^3$ = .13368 ft^3
1 qt = 2 pt.= .946353 liter = 946.353 cm^3 = 57.75 $in.^3$ = .25 gal = .034201 ft^3
1 cord = 128 ft^3
1 pt = .473 liter
1 ft^3 = 1728 $in.^3$
1 yd^3 = 27 ft^3 = 46656 $in.^3$

Units of area

1 ft^2 (sq. ft.) = 144 $in.^2$ (sq. in.)
1 yd^2 (sq. yd.) = 9 ft^2 (sq. ft.) = 1296 sq. in.
1 $mile^2$ (sq. mile) = 640 acres
1 m^2 = 10^4 cm^2
1 acre = 4840 yd^2

Symbols for units

cm = centimeter
m = meter
in. = inch
ft = foot
yd = yard
km = kilometer
mi = mile

gal = gallon
qt = quart
pt = pint
oz = ounce

g = gram
kg = kilogram
lb = pound
mg = milligram

Some prefixes (International System)

Prefix	Power
tera	10^{12}
giga	10^9
mega	10^6
kilo	10^3
hecto	10^2
deka	10^1
deci	10^{-1}
centi	10^{-2}
milli	10^{-3}
micro	10^{-6}
nano	10^{-9}
pico	10^{-12}

Conversion Factors for Measurements

American System (British System) Interconversion (Factors) Metric System

↓ ↓ ↓

Some "**bridge**s" for converting from one system to the other

↓

Length

| 12 inches (in) = 1 foot (ft.) |
| 3 feet (ft.)   = 1 yard (yd) |
| 5280 feet   = 1 mile (mi) |
| 1760 yards   = 1 mile |

| 1 in. = 2.54 cm |
| 1 yd = 0.9144 m |
| 1 mi = 1.61 km |
| 1 km = 0.62 mi |

| 1 kilometer (km) $= 10^3$ m = 1000 m |
| 1 hectometer (hm) $= 10^2$ m = 100 m |
| 1 dekameter (dam) $= 10^1$ m = 10 m |
| **1 meter (m)** $= 10^0$ m = 1m |
| 1 decimeter (dm) $= 10^{-1}$ m = 0.1m |
| 1 centimeter (cm) $= 10^{-2}$ m = 0.01m |
| 1 millimeter (mm) $= 10^{-3}$ m = 0.001m |

Some "**bridge**s" for converting from one system to the other

↓

Mass

| 1 lb = 16 oz |
| 1 ton = 2000 lb |
| 1 long ton = 2240 lb |

| 1 kg = 2.2 lb |
| 1 lb = 454 g |
| 1 oz = 28.4 g = 16 drams |
| 1 ton = 0.9072 metric ton |

| 1 kilogram (kg) $= 10^3$ g = 1000 g |
| 1 hectogram (hg) $= 10^2$ g = 100 g |
| 1 dekagram (dag) $= 10^1$ g = 10 g |
| **1 gram (g)** $= 10^0$ g = 1 g |
| 1 decigram (dg) $= 10^{-1}$ g = 0.1g |
| 1 centigram (cg)) $= 10^{-2}$ g = 0.01g |
| 1 milligram (mg) $= 10^{-3}$ g = 0.001g |

Some "**bridge**s" for converting from one system to the other

↓

Volume

| 16 fluid oz (fl-oz)= 1 pint (pt) |
| 2 pints (pt) = 1 quart (qt) |
| 4 quarts   = 1 gallon (gal) |

| 1 liter (l) = 1.057 qt |
| 1 gal = 3.785 l |
| 1 liter = 2.1 pt |
| 1 pt = .473 l |

| 1 kiloliter (kl) $= 10^3$ l = 1000 l |
| 1 hectoliter (hl) $= 10^2$ l = 100 l |
| 1 dekaliter (da$l$) $= 10^1$ l = 10 l |
| **1 liter (l)** $= 10^0$ l = 1 l |
| 1 deciliter (dl) $= 10^{-1}$ l = 0.1l |
| 1 centiliter (cl) $= 10^{-2}$ l = 0.01l |
| 1 milliliter (ml) $= 10^{-3}$ l = 0.001 l |

Must remember the following (metric system:)

| 100 cm = 1 m |
| 1000 m = 1 km |

| 1000 mg = 1 g |
| 1000 g = 1 kg |

| 1000 ml = 1 l |
| 1 ml = 1 cc = 1cm^3 |
| 1000 cc = 1 l |

Mnemonic device (metric system)

| k – ilo – 10^3 |
| h – ecto – 10^2 |
| d – eka – 10^1 |
| d – ec$i$ – 10^{-1} |
| c – enti – 10^{-2} |
| m – illi– 10^{-3} |

Say the following aloud:

Step 1: First go down vertically as kei-eitch-dii-dii-see-em, then Step 2

Step 2: Kilo-hecto-deka-deci-centi-milli, and then note how the powers decrease vertically downwards.

Examples: 1 Kilometer = 10^3 meter; 1 milligram = 10^{-3} gram;

1 centimeter = 10^{-2} meter = $\frac{1}{100}$ meter ---> 100 centimeters = 1 meter.

Mathematical Modeling
Some Reciprocal Relationships

1. Arithmetic If A working alone can do a piece of work in time t_A; B working alone can do the same work in time t_B; C working alone can do the same work in time t_C, and if A, B, and C working together, can do the same work in time t_{ABC}, then

$$\frac{1}{t_{ABC}} = \frac{1}{t_A} + \frac{1}{t_B} + \frac{1}{t_C}$$

That is, the reciprocal of the working-together time equals the sum of the reciprocals of working-alone times (individual times).

2. Geometry: For any triangle, the reciprocal of the inradius (R) equals the sum of the reciprocals of the exradii $(r_1, r_2,$ and $r_3)$.

Thus
$$\frac{1}{R} = \frac{1}{r_1} + \frac{1}{r_2} + \frac{1}{r_3}$$

3. Physics (Electricity) For electrical resistances in parallel (in an electric circuit), the reciprocal of the combined resistance, R, equals the sum of the reciprocals of the separate resistances, $r_1, r_2,$ and r_3.

Thus
$$\frac{1}{R} = \frac{1}{r_1} + \frac{1}{r_2} + \frac{1}{r_3}$$

4. Physics (Optics)

For two thin lenses in contact, the reciprocal of the combined focal length, F, equals the sum of the reciprocals of the separate focal lengths, f_1 and f_2, .

Thus
$$\frac{1}{F} = \frac{1}{f_1} + \frac{1}{f_2}$$

5. Physics (Optics) For spherical mirrors and thin lenses, the reciprocal of the focal length F equals the sum of the reciprocals of the object distance, d_o and the image distance d_i.

Thus
$$\frac{1}{F} = \frac{1}{d_o} + \frac{1}{d_i}$$

6. Physics (Mechanics). If two bubbles of radii $r_1, r_2,$ coalesce into a double bubble, the radius, R, of the partition is given by

$$\frac{1}{R} = \frac{1}{r_1} - \frac{1}{r_2}$$